Solutions Manual

for

Algebra 2
and Pre-Calculus
(Volume I)

Build Your Self-Confidence and Enjoyment of Math!

Problems Included

Aejeong Kang

MathRadar

Send all inquiries to:

MathRadar, LLC
5705 Spring Hill Dr.
Mckinney, Texas 75072

Visit www.mathradar.com for more information and a sneak preview of the MathRadar series of math books.

Send inquires via email: info@mathradar.com

Solutions Manual for Algebra 2 and Pre-Calculus (Volume I)

ISBN-13: 978-0-9893689-7-1

ISBN-10: 0989368971

Printed in the United States of America.

Preface

I wrote these books because I am a mother and I have a strong academic background in mathematics. I have a BS degree in Mathematics and Master's degree in Mathematics as well. I have completed Ph.D. program in Biostatistics.

After receiving the big blessing of our first child, a daughter, I decided to forgo my personal career goals to become a full-time mother. When our daughter entered 7th grade, that meant lots of help with her study of math-my passion. However, I struggled to find good math books that would help her understand difficult concepts both clearly and quickly. About two years ago, I talked with my husband and my kids (now I have 2 children 8th grader, Nichole and 1st grader, Richard) about an idea that it would be better to write math books myself at least for my kids because I really want my kids study math with best books. After the conversation, I decided that the best way to help my children was by writing math books for them myself. They wholeheartedly agreed.

That's why I've been able to pour all my knowledge, energy, and soul into these books. Because I'm a mom, I would do anything for my children. Thanks to my family's endless support, I wrote them eight books, designed for use in junior high and high-school mathematics.

And that would have been the end of my journey, but my husband and children insisted that I share my work outside of our family. They encouraged me to make my work available to other parents looking, as I was, for well-written, great mathematics books for their children.

So I finally decided to publish these books. I do so with the hope that they will help your children find success and confidence in learning and studying mathematics.

But I would never have begun or finished this project without the support of my family. Kyungwan, Nichole, and Richard, you are my world. Thank you.

Introduction

✔ *After reading several pages of explanation/description about a certain mathematical concept, you still don't get it.*

✔ *You have worked on many related problems to understand mathematical concepts, but you still feel completely lost in the mathematical jungle.*

✔ *You bought a math book with good reviews, but it only offers short answers without detailed solutions. You feel confused and frustrated.*

✔ *You've tried multiple learning math books, but you've still not getting good grades in math. It seems like math is just not for you.*

If any one of these situation sound familiar, the MathRadar series will help you escape!

Everyone has different learning abilities and academic skill. MathRadar series is written and organized with emphasis on helping each individual study mathematics at his/her own pace.

Each book consists of clean and concise summaries, callouts, additional supporting explanations, quick reminders and/or shortcuts to facilitate better understanding. Each concept is thoroughly explained with step-by-step instruction and detailed proofs.

With the numerous examples and exercises, students can check their comprehension levels with both basic and more advanced problems.

Carry the MathRadar series with you!

Work on them anytime and anywhere!

Finally, you can start to enjoy mathematics!

Whether you are struggling or advanced in your math skills, the MathRadar series books will build your self-confidence and enjoyment of math.

I hope Math Radar is what you need and will be a great tool for your hard work.

Your comments or suggestions are greatly appreciated.

Please visit my website at www. mathradar.com or email me at ae-jeong@mathradar.com

Thank you very much. And remember, math can be fun!

Aejeong Kang

Algebra 2 and Pre-Calculus ⋯⋯  Volume I

Algebra 2 and Pre-Calculus ⋯⋯  Volume II

Volumes I and II have 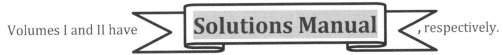 Solutions Manual , respectively.

The solutions manuals make it possible for students to study difficult concepts on their own. With the solutions manuals, students will be able to better understand how to solve problems through step-by-step for each problem.

Chapter 1. The Number System

#1 Simplify the following given expressions :

(1) When $-1 < a < 1$, $|a + 1| + |1 - a|$

$-1 < a < 1 \Rightarrow a + 1 > 0$ and $1 - a > 0$

$\therefore |a + 1| + |1 - a| = (a + 1) + (1 - a) = 2$

(2) When $-1 < a < 1$, $|a - 1| - |1 + a|$

$-1 < a < 1 \Rightarrow a - 1 < 0$ and $1 + a > 0$

$\therefore |a - 1| - |1 + a| = -(a - 1) - (1 + a) = -2a$

(3) When a is a real number, $\big|a + |a|\big| - \big|a - |a|\big|$

Case 1. $a \geq 0 \Rightarrow \big|a + |a|\big| = |a + a| = 2a$

$$\big|a - |a|\big| = |a - a| = |0| = 0$$

$$\therefore \big|a + |a|\big| - \big|a - |a|\big| = 2a - 0 = 2a$$

Case 1. $a < 0 \Rightarrow \big|a + |a|\big| = |a - a| = |0| = 0$

$$\big|a - |a|\big| = |a + a| = |2a| = -2a$$

$$\therefore \big|a + |a|\big| - \big|a - |a|\big| = 0 - (-2a) = 2a$$

Therefore, $\big|a + |a|\big| - \big|a - |a|\big| = 2a$.

(4) When $-2 < a < 2$, $\sqrt{(a - 2)^2} + |a + 2|$

$-2 < a < 2 \Rightarrow a - 2 < 0$ and $a + 2 > 0$

Since $a - 2 < 0$, $\sqrt{(a - 2)^2} = -(a - 2)$

Since $a + 2 > 0$, $|a + 2| = a + 2$

Therefore, $\sqrt{(a - 2)^2} + |a + 2| = -(a - 2) + a + 2 = 4$

(5) When $a < b < c$, $\sqrt{(a - b)^2} - \sqrt{(b - c)^2} - \sqrt{(c - a)^2}$

Since $a - b < 0$, $b - c < 0$, and $c - a > 0$,

$$\sqrt{(a - b)^2} - \sqrt{(b - c)^2} - \sqrt{(c - a)^2} = -(a - b) + (b - c) - (c - a) = 2b - 2c$$

(6) When $a + b < 0$ and $ab > 0$, $\sqrt{(-a)^2} - \sqrt{(-b)^2} - \sqrt{(-2a)^2} + \sqrt{(-2b)^2}$

Since $ab > 0$, $(a > 0$ and $b > 0)$ or $(a < 0$ and $b < 0)$

Since $a + b < 0$, $a < 0$ and $b < 0$

So, $\sqrt{(-a)^2} - \sqrt{(-b)^2} - \sqrt{(-2a)^2} + \sqrt{(-2b)^2} = \sqrt{a^2} - \sqrt{b^2} - \sqrt{4a^2} + \sqrt{4b^2}$

$= -a - (-b) - 2(-a) + 2(-b) = -a + b + 2a - 2b = a - b$

(7) When $a < 0$ and $b < 0$, $\sqrt{(-a)^2} + |b| - \sqrt{(-b)^2} - 2|a+b|$

$$\sqrt{(-a)^2} + |b| - \sqrt{(-b)^2} - 2|a+b|$$

$$= \sqrt{a^2} + |b| - \sqrt{b^2} - 2|a+b| = (-a) - b - (-b) + 2(a+b) = a + 2b$$

#2 Find the value of $a + b$ for two real numbers a and b which satisfy the following :

(1) $\sqrt{(a+5)^2} + |2a - 3b + 4| = 0$

Note that : For any real numbers a and b, $|a| + |b| = 0 \Rightarrow a = b = 0$

$$\sqrt{(a+5)^2} + |2a - 3b + 4| = 0 \Rightarrow |a+5| + |2a - 3b + 4| = 0$$

$$\Rightarrow a + 5 = 0,\ 2a - 3b + 4 = 0$$

$$\Rightarrow a = -5,\ b = -2$$

$$\therefore\ a + b = -7$$

(2) $|2a - b + 3| + |a - 2b - 3| = 0$

$$|2a - b + 3| + |a - 2b - 3| = 0 \Rightarrow 2a - b + 3 = 0 \text{ and } a - 2b - 3 = 0$$

$$\Rightarrow b = 2a + 3,\qquad a - 2(2a + 3) - 3 = 0$$

$$\Rightarrow a - 4a - 9 = 0 \Rightarrow a = -3 \text{ and } b = 2(-3) + 3 = -3$$

$$\therefore\ a + b = -6$$

#3 Find the square roots of each number.

(1) $2 = \pm\sqrt{2}$ **(2)** $0.01 = \pm 0.1$ **(3)** $\dfrac{16}{9} = \pm\dfrac{4}{3}$ **(4)** $0 = 0$

#4 Evaluate the expression.

(1) $\left(\sqrt{3}\right)^2 = 3$

(2) $\left(-\sqrt{5}\right)^2 = \left(\sqrt{5}\right)^2 = 5$

(3) $\sqrt{(-8)^2} = \sqrt{8^2} = 8$

(4) $-\sqrt{36} = -\sqrt{6^2} = -6$

(5) $-\sqrt{\left(-\dfrac{2}{3}\right)^2} = -\sqrt{\left(\dfrac{2}{3}\right)^2} = -\dfrac{2}{3}$

(6) $\sqrt[3]{(-2)^2} = \sqrt[3]{2^2} = (2^2)^{\frac{1}{3}} = 2^{\frac{2}{3}}$

(7) $\sqrt[3]{(-5)^3} = \sqrt[3]{-5^3} = -\sqrt[3]{5^3} = -5$

(8) $-\sqrt[3]{-27} = -\sqrt[3]{-3^3} = \sqrt[3]{3^3} = 3$

(9) $\sqrt{(-2)^2} + \sqrt[3]{(-2)^3} - 5\sqrt{(-2)^2} = 2 - \sqrt[3]{2^3} - 5 \cdot 2 = 2 - 2 - 10 = -10$

(10) $\sqrt[3]{(-5)^3} - \sqrt{(-5)^2} + \sqrt{5^3} = -\sqrt[3]{5^3} - \sqrt{5^2} + \sqrt{5^3} = -5^{\frac{3}{3}} - 5^{\frac{2}{2}} + 5^{\frac{3}{2}} = -5^1 - 5^1 + 5^{\frac{3}{2}}$

$$= -10 + 5^{\frac{3}{2}}$$

(11) $3\sqrt[4]{4^2} - \sqrt[4]{(-2)^4} = 3\sqrt[4]{(2^2)^2} - \sqrt[4]{2^4} = 3\sqrt[4]{2^4} - \sqrt[4]{2^4} = (3-1)\sqrt[4]{2^4} = 2 \cdot 2^{\frac{4}{4}} = 2 \cdot 2^1 = 4$

(12) $\dfrac{\sqrt[3]{20}}{\sqrt[3]{10}} = \dfrac{20^{\frac{1}{3}}}{10^{\frac{1}{3}}} = \dfrac{(4 \cdot 5)^{\frac{1}{3}}}{(2 \cdot 5)^{\frac{1}{3}}} = \dfrac{(2^2 \cdot 5)^{\frac{1}{3}}}{(2 \cdot 5)^{\frac{1}{3}}} = \dfrac{2^{\frac{2}{3}} \cdot 5^{\frac{1}{3}}}{2^{\frac{1}{3}} \cdot 5^{\frac{1}{3}}} = 2^{\frac{2}{3} - \frac{1}{3}} = 2^{\frac{1}{3}}$ or $\dfrac{\sqrt[3]{20}}{\sqrt[3]{10}} = \sqrt[3]{\dfrac{20}{10}} = \sqrt[3]{2} = 2^{\frac{1}{3}}$

(13) $\dfrac{\sqrt[3]{20}}{\sqrt{10}} = \dfrac{20^{\frac{1}{3}}}{10^{\frac{1}{2}}} = \dfrac{(4 \cdot 5)^{\frac{1}{3}}}{(2 \cdot 5)^{\frac{1}{2}}} = \dfrac{(2^2 \cdot 5)^{\frac{1}{3}}}{(2 \cdot 5)^{\frac{1}{2}}} = \dfrac{2^{\frac{2}{3}} \cdot 5^{\frac{1}{3}}}{2^{\frac{1}{2}} \cdot 5^{\frac{1}{2}}} = 2^{\frac{2}{3} - \frac{1}{2}} 5^{\frac{1}{3} - \frac{1}{2}} = 2^{\frac{1}{6}} 5^{-\frac{1}{6}} = \dfrac{2^{\frac{1}{6}}}{5^{\frac{1}{6}}} = \left(\dfrac{2}{5}\right)^{\frac{1}{6}}$

#5 Simplify each of the following :

(1) $\sqrt{12} \times \sqrt{\dfrac{3}{4}} = 2\sqrt{3} \times \dfrac{\sqrt{3}}{2} = 3$

(2) $2\sqrt{10} \div 8\sqrt{5} = \dfrac{2\sqrt{10}}{8\sqrt{5}} = \dfrac{1}{4}\sqrt{\dfrac{10}{5}} = \dfrac{\sqrt{2}}{4}$

(3) $\dfrac{3}{\sqrt{5}} \times \dfrac{\sqrt{15}}{9} \div \dfrac{\sqrt{18}}{12} = \dfrac{3}{\sqrt{5}} \times \dfrac{\sqrt{15}}{9} \times \dfrac{12}{\sqrt{18}} = 4\sqrt{\dfrac{15}{5 \cdot 18}} = 4\sqrt{\dfrac{1}{6}} = 4\dfrac{\sqrt{6}}{6} = \dfrac{2\sqrt{6}}{3}$

(4) $\sqrt[3]{20} \cdot \sqrt{15} = 20^{\frac{1}{3}} \cdot 15^{\frac{1}{2}} = (2^2 \cdot 5)^{\frac{1}{3}} \cdot (3 \cdot 5)^{\frac{1}{2}} = 2^{\frac{2}{3}} \cdot 5^{\frac{1}{3}} \cdot 3^{\frac{1}{2}} \cdot 5^{\frac{1}{2}}$

$$= 2^{\frac{2}{3}} \cdot 5^{\frac{1}{3} + \frac{1}{2}} \cdot 3^{\frac{1}{2}} = 2^{\frac{2}{3}} \cdot 3^{\frac{1}{2}} \cdot 5^{\frac{5}{6}}$$

(5) $\dfrac{\sqrt{15}}{\sqrt{5}} \div \dfrac{5}{3\sqrt{2}} = \dfrac{\sqrt{15}}{\sqrt{5}} \cdot \dfrac{3\sqrt{2}}{5} = \sqrt{\dfrac{15}{5}} \cdot \dfrac{3\sqrt{2}}{5} = \sqrt{3} \cdot \dfrac{3\sqrt{2}}{5} = \dfrac{3\sqrt{6}}{5}$

(6) $\sqrt{\dfrac{3}{4}} \div \sqrt{\dfrac{15}{10}} \div \dfrac{1}{\sqrt{6}} = \sqrt{\dfrac{3}{4}} \cdot \sqrt{\dfrac{10}{15}} \cdot \dfrac{\sqrt{6}}{1} = \sqrt{\dfrac{3}{4} \cdot \dfrac{10}{15} \cdot \dfrac{6}{1}} = \sqrt{3}$

(7) $\sqrt[3]{6} \cdot \sqrt[3]{9} = \sqrt[3]{6 \cdot 9} = \sqrt[3]{2 \cdot 3^3} = 3\sqrt[3]{2}$

(8) $\dfrac{|2-4|}{|-3\sqrt{2}|-2} = \dfrac{|-2|}{|-3\sqrt{2}|-2} = \dfrac{2}{3\sqrt{2}-2} = \dfrac{2(3\sqrt{2}+2)}{(3\sqrt{2}-2)(3\sqrt{2}+2)} = \dfrac{6\sqrt{2}+4}{18-4} = \dfrac{6\sqrt{2}+4}{14} = \dfrac{3\sqrt{2}+2}{7}$

(9) $\left|\dfrac{-|\sqrt{2}-\sqrt{5}|}{\sqrt{5}-\sqrt{2}}\right| = \left|\dfrac{(\sqrt{2}-\sqrt{5})}{\sqrt{5}-\sqrt{2}}\right|$, because $\sqrt{2} < \sqrt{5}$; $\sqrt{2} - \sqrt{5} < 0$

$$= \left|\dfrac{-(\sqrt{5}-\sqrt{2})}{\sqrt{5}-\sqrt{2}}\right| = |-1| = 1$$

#6 Rationalize the denominator. Then, simplify the result.

(1) $\dfrac{3}{\sqrt{2}} = \dfrac{3 \cdot \sqrt{2}}{\sqrt{2} \cdot \sqrt{2}} = \dfrac{3\sqrt{2}}{2}$

(2) $\dfrac{3}{2\sqrt{5}} = \dfrac{3 \cdot \sqrt{5}}{2\sqrt{5} \cdot \sqrt{5}} = \dfrac{3\sqrt{5}}{2 \cdot 5} = \dfrac{3\sqrt{5}}{10}$

(3) $\dfrac{\sqrt{18}}{\sqrt{12}} = \dfrac{\sqrt{18} \cdot \sqrt{12}}{\sqrt{12} \cdot \sqrt{12}} = \dfrac{3\sqrt{2} \cdot 2\sqrt{3}}{12} = \dfrac{6\sqrt{6}}{12} = \dfrac{\sqrt{6}}{2}$

(4) $\dfrac{3}{2\sqrt{3}} - 4\sqrt{3} = \dfrac{3 \cdot \sqrt{3}}{2\sqrt{3} \cdot \sqrt{3}} - 4\sqrt{3} = \dfrac{3 \cdot \sqrt{3}}{2 \cdot 3} - 4\sqrt{3} = \dfrac{\sqrt{3}}{2} - 4\sqrt{3} = \left(\dfrac{1}{2} - 4\right)\sqrt{3} = -\dfrac{7}{2}\sqrt{3}$

(5) $\dfrac{3\sqrt{5}-\sqrt{3}}{\sqrt{3}} = \dfrac{(3\sqrt{5}-\sqrt{3})\sqrt{3}}{\sqrt{3}\cdot\sqrt{3}} = \dfrac{3\sqrt{5}\cdot\sqrt{3}-\sqrt{3}\cdot\sqrt{3}}{\sqrt{3}\cdot\sqrt{3}} = \dfrac{3\sqrt{15}-3}{3} = \sqrt{15}-1$

(6) $\dfrac{5}{\sqrt[3]{2}} = \dfrac{5}{\sqrt[3]{2}} \cdot \dfrac{\sqrt[3]{2^2}}{\sqrt[3]{2^2}}$ (to produce a perfect cube radicand) $= \dfrac{5\sqrt[3]{4}}{\sqrt[3]{2^3}} = \dfrac{5\sqrt[3]{4}}{2}$

(7) $\sqrt{\dfrac{2}{3}} - \sqrt{5} + \sqrt{\dfrac{3}{2}} = \dfrac{\sqrt{2}\cdot\sqrt{3}}{\sqrt{3}\cdot\sqrt{3}} - \sqrt{5} + \dfrac{\sqrt{3}\cdot\sqrt{2}}{\sqrt{2}\cdot\sqrt{2}} = \dfrac{\sqrt{6}}{3} - \sqrt{5} + \dfrac{\sqrt{6}}{2} = \dfrac{2\sqrt{6}}{6} + \dfrac{3\sqrt{6}}{6} - \sqrt{5} = \dfrac{5\sqrt{6}}{6} - \sqrt{5}$

(8) $\dfrac{4}{2+\sqrt{3}} = \dfrac{4(2-\sqrt{3})}{(2+\sqrt{3})(2-\sqrt{3})} = \dfrac{4(2-\sqrt{3})}{2^2-(\sqrt{3})^2} = \dfrac{4(2-\sqrt{3})}{4-3} = 4(2-\sqrt{3})$

(9) $\dfrac{2\sqrt{3}}{\sqrt{3}-\sqrt{2}} + \dfrac{3\sqrt{2}}{\sqrt{3}+\sqrt{2}} = \dfrac{2\sqrt{3}(\sqrt{3}+\sqrt{2})+3\sqrt{2}(\sqrt{3}-\sqrt{2})}{(\sqrt{3}-\sqrt{2})(\sqrt{3}+\sqrt{2})} = \dfrac{6+2\sqrt{6}+3\sqrt{6}-6}{3-2} = \dfrac{5\sqrt{6}}{1} = 5\sqrt{6}$

#7 Find the distance d between the points and find the midpoint m of the line segment joining the points.

(1) $a(1,2)$, $b(3,4)$

$d = \sqrt{(3-1)^2 + (4-2)^2} = \sqrt{4+4} = \sqrt{8} = 2\sqrt{2}$

$m = \left(\dfrac{1+3}{2}, \dfrac{2+4}{2}\right) = (2,3)$

(2) $a(-2,3)$, $b(3,-5)$

$d = \sqrt{(3-(-2))^2 + (-5-3)^2} = \sqrt{25+64} = \sqrt{89}$

$m = \left(\dfrac{-2+3}{2}, \dfrac{3-5}{2}\right) = (\tfrac{1}{2}, -1)$

(3) $a(3,4)$, $b(-2,-3)$

$d = \sqrt{(-2-3)^2 + (-3-4)^2} = \sqrt{25+49} = \sqrt{74}$

$m = \left(\dfrac{3-2}{2}, \dfrac{4-3}{2}\right) = (\tfrac{1}{2}, \tfrac{1}{2})$

(4) $a(-\sqrt{2}, 3)$, $b(3, \sqrt{2})$

$d = \sqrt{(3-(-\sqrt{2}))^2 + (\sqrt{2}-3)^2} = \sqrt{9+6\sqrt{2}+2+2-6\sqrt{2}+9} = \sqrt{22}$

$m = \left(\dfrac{3-\sqrt{2}}{2}, \dfrac{3+\sqrt{2}}{2}\right)$

(5) $a(0,-1)$, $b\left(-4, \dfrac{1}{2}\right)$

$d = \sqrt{(-4-0)^2 + (\tfrac{1}{2}-(-1))^2} = \sqrt{16+\dfrac{9}{4}} = \sqrt{\dfrac{73}{4}}$

$m = \left(\dfrac{0-4}{2}, \dfrac{-1+\frac{1}{2}}{2}\right) = \left(-2, -\dfrac{1}{4}\right)$

#8 For rational numbers a and b, each expression is a rational number. Find the value of ab.

(1) $\dfrac{3+b\sqrt{2}}{\sqrt{2}-a} = \dfrac{(3+b\sqrt{2})(\sqrt{2}+a)}{(\sqrt{2}-a)(\sqrt{2}+a)} = \dfrac{3\sqrt{2}+2b+3a+ab\sqrt{2}}{2-a^2} = \dfrac{(3a+2b)+(3+ab)\sqrt{2}}{2-a^2}$

Since $\dfrac{(3a+2b)+(3+ab)\sqrt{2}}{2-a^2}$ is a rational number, irrational terms should not exist.

$\therefore \ (3+ab)\sqrt{2} = 0 \ ; \ 3+ab = 0$

Therefore, $ab = -3$

(2) $\dfrac{\sqrt{3}+b}{a\sqrt{3}+2} = \dfrac{(\sqrt{3}+b)(a\sqrt{3}-2)}{(a\sqrt{3}+2)(a\sqrt{3}-2)} = \dfrac{3a+ab\sqrt{3}-2\sqrt{3}-2b}{3a^2-4} = \dfrac{(3a-2b)+(ab-2)\sqrt{3}}{3a^2-4}$

Since $\dfrac{(3a-2b)+(ab-2)\sqrt{3}}{3a^2-4}$ is a rational number, irrational terms should not exist.

$\therefore \ (ab-2)\sqrt{3} = 0 \ ; \ ab-2 = 0$

Therefore, $ab = 2$

#9 Find the real numbers a and b so that the statement is identity.

(1) $(3a - 2b) + (a + 3b)i = -5 + 2i$

Since $3a - 2b = -5$ and $a + 3b = 2$, $\ 3(2 - 3b) - 2b = -5$

$\therefore \ -11b = -11 \quad \therefore b = 1, \ a = 2 - 3 \cdot 1 = -1$

(2) $a(1 + 2i) - b(2 - 5i) = 10 + 8i$

Since $a(1 + 2i) - b(2 - 5i) = (a - 2b) + (2a + 5b)i$, $\ a - 2b = 10$ and $2a + 5b = 8$

$\therefore \ 2(2b + 10) + 5b = 8 \quad \therefore 9b = -12 \quad \therefore b = -\dfrac{4}{3}, \ a = 2\left(-\dfrac{4}{3}\right) + 10 = \dfrac{22}{3}$

(3) $\dfrac{a}{1+i} - \dfrac{b}{1-i} = 2 + i$

Since $\dfrac{a}{1+i} - \dfrac{b}{1-i} = \dfrac{a(1-i)-b(1+i)}{(1+i)(1-i)} = \dfrac{(a-b)-(a+b)i}{(1+i)(1-i)} = \dfrac{(a-b)-(a+b)i}{1+1} = \dfrac{a-b}{2} - \dfrac{(a+b)i}{2}$,

$\dfrac{a-b}{2} = 2$ and $\dfrac{a+b}{2} = -1$

$\therefore \ a - b = 4$ and $a + b = -2 \quad \therefore a = 1, \ b = -3$

(4) $\dfrac{a}{1+i} + \dfrac{b}{1-i} = \dfrac{3}{2-i}$

Since $\dfrac{a}{1+i} + \dfrac{b}{1-i} = \dfrac{a(1-i)+b(1+i)}{(1+i)(1-i)} = \dfrac{(a+b)+(b-a)i}{(1+i)(1-i)} = \dfrac{(a+b)+(b-a)i}{1+1} = \dfrac{a+b}{2} + \dfrac{(b-a)i}{2}$ and

$\dfrac{3}{2-i} = \dfrac{3(2+i)}{(2-i)(2+i)} = \dfrac{3(2+i)}{4+1} = \dfrac{6}{5} + \dfrac{3}{5}i$, $\ \dfrac{a+b}{2} = \dfrac{6}{5}$ and $\dfrac{b-a}{2} = \dfrac{3}{5}$

$\therefore \ a + b = \dfrac{12}{5}$ and $b - a = \dfrac{6}{5} \quad \therefore 2b = \dfrac{18}{5} \quad \therefore b = \dfrac{9}{5}, \ a = \dfrac{12}{5} - \dfrac{9}{5} = \dfrac{3}{5}$

(5) $|a - b| + (a - 3)i = 3b - 2 - bi$

    Case 1. $a \geq b$

        Since $|a - b| + (a - 3)i = (a - b) + (a - 3)i$, $a - b = 3b - 2$ and $a - 3 = -b$

        $\therefore \;\; a = 4b - 2 = 4(3 - a) - 2 = 10 - 4a$; $5a = 10$ $\therefore a = 2, \; b = 1$

    Case 2. $a < b$

        Since $|a - b| + (a - 3)i = -(a - b) + (a - 3)i$, $-a + b = 3b - 2$ and $a - 3 = -b$

        $\therefore \;\; a = -2b + 2 = -2(3 - a) + 2 = -4 + 2a$ $\therefore a = 4, \; b = -1$

        Since $a < b$, ($a = 4, \; b = -1$) is not possible.

    Therefore, $a = 2, \; b = 1$

#10   Simplify each of the following in standard form.

(1) $2 + \sqrt{-9} = 2 + \sqrt{9}\,i = 2 + 3i$

(2) $\sqrt{-80} = \sqrt{80}i = \sqrt{16 \cdot 5}\,i = 4\sqrt{5}\,i$

(3) $-3i^2 + 4i = -3 \cdot (-1) + 4i = 3 + 4i$

(4) $(12 - 3i) + (-4 + 8i) = (12 - 4) + (-3 + 8)i = 8 + 5i$

(5) $\sqrt{-5} \cdot \sqrt{-4} = \sqrt{5}\,i \cdot \sqrt{4}\,i = \sqrt{5 \cdot 4}\,i^2 = 2\sqrt{5} \cdot (-1) = -2\sqrt{5}$

(6) $\dfrac{1}{2+\sqrt{-2}} + \dfrac{1}{2-\sqrt{-2}} = \dfrac{1}{2+\sqrt{2}\,i} + \dfrac{1}{2-\sqrt{2}\,i} = \dfrac{2-\sqrt{2}\,i+2+\sqrt{2}\,i}{(2+\sqrt{2}\,i)(2-\sqrt{2}\,i)} = \dfrac{4}{2^2-(\sqrt{2}\,i)^2} = \dfrac{4}{4+2} = \dfrac{2}{3}$

(7) $(\sqrt{-75})^2 = (\sqrt{75}\,i)^2 = (\sqrt{75}\,)^2(\,i)^2 = 75 \cdot (-1) = -75$

(8) $(3 - 2i)(5 - 3i) = 15 + 6\,i^2 - 10i - 9i = 15 - 6 - 19i = 9 - 19i$

(9) $\dfrac{2+3i}{1-2i} = \dfrac{(2+3i)(1+2i)}{(1-2i)(1+2i)} = \dfrac{(2-6)+(3+4)i}{1+4} = \dfrac{-4+7i}{5} = \dfrac{-4}{5} + \dfrac{7}{5}i$

(10) $\dfrac{\sqrt{2}}{\sqrt{-3}} = \dfrac{\sqrt{2}}{\sqrt{3}\,i} = \sqrt{\dfrac{2}{3}}\dfrac{1}{i} = \sqrt{\dfrac{2}{3}}\dfrac{i}{i^2} = -\sqrt{\dfrac{2}{3}}\,i = -\dfrac{\sqrt{6}}{3}\,i$

(11) $\dfrac{\sqrt{-3}}{\sqrt{-5}} = \dfrac{\sqrt{3}\,i}{\sqrt{5}\,i} = \sqrt{\dfrac{3}{5}} = \dfrac{\sqrt{15}}{5}$

(12) $\dfrac{\sqrt{2}-\sqrt{-3}}{\sqrt{2}+\sqrt{-3}} = \dfrac{\sqrt{2}-\sqrt{3}\,i}{\sqrt{2}+\sqrt{3}\,i} = \dfrac{(\sqrt{2}-\sqrt{3}\,i)^2}{(\sqrt{2}+\sqrt{3}\,i)(\sqrt{2}-\sqrt{3}\,i)} = \dfrac{2-2\sqrt{6}\,i+3i^2}{2+3} = \dfrac{-1-2\sqrt{6}\,i}{5} = \dfrac{-1}{5} - \dfrac{2\sqrt{6}}{5}\,i$

(13) $(1 + i)^2(1 - i)^2 - (2 + 3i)(2 - 3i)$

      $= (1 + 2i - 1)(1 - 2i - 1) - (4 + 9) = 2i\,(-2i) - 13 = 4 - 13 = -9$

(14) $\left(\dfrac{1+i}{1-i}\right)^9 - \left(\dfrac{1-i}{1+i}\right)^{10}$

    Since $\dfrac{1+i}{1-i} = \dfrac{(1+i)^2}{(1-i)(1+i)} = \dfrac{1+2i-1}{1+1} = \dfrac{2i}{2} = i$ and $\dfrac{1-i}{1+i} = \dfrac{(1-i)^2}{(1+i)(1-i)} = \dfrac{1-2i-1}{1+1} = \dfrac{-2i}{2} = -i$,

    $\left(\dfrac{1+i}{1-i}\right)^9 - \left(\dfrac{1-i}{1+i}\right)^{10} = i^9 - (-i)^{10} = (i^4)^2 i - i^{10} = (i^4)^2 i - (i^4)^2 i^2 = i - i^2$

        $= i - (-1) = i + 1$

#11 For a complex number $z = -1 + \sqrt{3}i$, find the value of each number.

(1) $z^2 = (-1 + \sqrt{3}i)^2 = 1 - 2\sqrt{3}i + (\sqrt{3}i)^2 = 1 - 2\sqrt{3}i - 3 = -2 - 2\sqrt{3}i$

(2) $z^3 = z^2 z = (-2 - 2\sqrt{3}i)(-1 + \sqrt{3}i)$

$\qquad = (-2)(-1) - (2\sqrt{3}i)(\sqrt{3}i) + 2\sqrt{3}i - 2\sqrt{3}i = 2 + 6 = 8$

(3) $\left(\dfrac{z}{2}\right)^{100}$

Since $z^3 = 8$, $\left(\dfrac{z}{2}\right)^3 = \dfrac{z^3}{8} = 1$

$\therefore \left(\dfrac{z}{2}\right)^{100} = \left(\dfrac{z}{2}\right)^{3 \cdot 33}\left(\dfrac{z}{2}\right) = 1 \cdot \dfrac{z}{2} = \dfrac{-1 + \sqrt{3}i}{2} = -\dfrac{1}{2} + \dfrac{\sqrt{3}}{2}i$

#12 Find the value of a $(a > 0)$ so that the complex number $z = (a + \sqrt{3}i)^2 i$ is a real number.

$z = \left(a + \sqrt{3}i\right)^2 i = \left(a^2 + 2a\sqrt{3}i - 3\right)i$

$\quad = a^2 i - 2a\sqrt{3} - 3i = -2a\sqrt{3} + (a^2 - 3)i$

To get a real number z, $a^2 - 3 = 0$

$\therefore a^2 = 3 \qquad \therefore a = \pm\sqrt{3}$

Since $a > 0$, $a = \sqrt{3}$

#13 For a complex number $z = (2 - 3i)a - 2i(4 + 2i)$, find the value of a so that z^2 is a real number.

$z = (2 - 3i)a - 2i(4 + 2i) = 2a - 3ai - 8i + 4 = (2a + 4) - (3a + 8)i$

To get a real number z^2, z is a real number or z is a pure imaginary number.

Case 1. z is a real number.

$\Rightarrow 3a + 8 = 0 \Rightarrow a = -\dfrac{8}{3}$

Case 2. z is a pure imaginary number.

$\Rightarrow 2a + 4 = 0 \Rightarrow a = -2$

Therefore, $a = -\dfrac{8}{3}$ or $a = -2$

#14 For a non-zero complex number $z = (1 - i)x^2 - (5 + i)x + 6 + 6i$, find the value of real number x so that $z + \bar{z} = 0$.

$z = (1 - i)x^2 - (5 + i)x + 6 + 6i = (x^2 - 5x + 6) - (x^2 + x - 6)i$

Since $z + \bar{z} = 0$ and $z \neq 0$, z is a pure imaginary number.

Thus, $x^2 - 5x + 6 = 0$ and $x^2 + x - 6 \neq 0$

① $x^2 - 5x + 6 = 0 \Rightarrow (x-2)(x-3) = 0 \quad \therefore x = 2$ or $x = 3$

② $x^2 + x - 6 \neq 0 \Rightarrow (x+3)(x-2) \neq 0 \quad \therefore x \neq -3$ and $x \neq 2$

By ① & ②, $x = 3$

#15 For any non-zero real numbers $a, b, c,$ and d, $\sqrt{a}\,\sqrt{b} = -\sqrt{ab}$, $\dfrac{\sqrt{c}}{\sqrt{d}} = -\sqrt{\dfrac{c}{d}}$.

Simplify the expression $\sqrt{a^2} - |a+b| + \sqrt{b^2} - |d| - \sqrt{(d-c)^2}$.

Since $\sqrt{a}\,\sqrt{b} = -\sqrt{ab}$, $\quad a < 0$ and $b < 0$

Since $\dfrac{\sqrt{c}}{\sqrt{d}} = -\sqrt{\dfrac{c}{d}}$, $\quad c > 0$ and $d < 0$

$\therefore \sqrt{a^2} - |a+b| + \sqrt{b^2} - |d| - \sqrt{(d-c)^2} = |a| - |a+b| + |b| - |d| - |d-c|$

$= -a + (a+b) - b + d + (d-c) = 2d - c$

#16 Find all possible integers for x which satisfies $\dfrac{\sqrt{x+2}}{\sqrt{x-6}} = -\sqrt{\dfrac{x+2}{x-6}}$.

Since $\dfrac{\sqrt{x+2}}{\sqrt{x-6}} = -\sqrt{\dfrac{x+2}{x-6}}$, $x + 2 \geq 0$ and $x - 6 < 0$. $\quad \therefore -2 \leq x < 6$

Thus, the possible integers are $-2, -1, 0, 1, 2, 3, 4, 5$.

#17 Find the absolute value of each complex number.

(1) $-2i$

$|-2i| = |0 - 2i| = \sqrt{0^2 + (-2)^2} = \sqrt{2^2} = 2$

(2) $-3 + 4i$

$|-3 + 4i| = \sqrt{(-3)^2 + (4)^2} = \sqrt{9 + 16} = \sqrt{25} = 5$

(3) $\sqrt{5} + \sqrt{3}i$

$|\sqrt{5} + \sqrt{3}i| = \sqrt{(\sqrt{5})^2 + (\sqrt{3})^2} = \sqrt{5 + 3} = \sqrt{8} = 2\sqrt{2}$

#18 Solve the following :

(1) When $z = 2 - i$, find the value of $\left| z - \dfrac{1}{z} \right|$.

Note that $z = a + bi \Rightarrow |z| = \sqrt{a^2 + b^2}$ *and* $|z|^2 = z\,\bar{z}$

Since $z - \dfrac{1}{z} = 2 - i - \dfrac{1}{2-i} = 2 - i - \dfrac{2+i}{(2-i)(2+i)} = 2 - i - \dfrac{2+i}{4+1}$

$$= 2 - i - \left(\dfrac{2}{5} + \dfrac{i}{5}\right) = \dfrac{8}{5} - \dfrac{6}{5}i \,,$$

$$\left| z - \dfrac{1}{z} \right| = \left| \dfrac{8}{5} - \dfrac{6}{5}i \right| = \sqrt{\left(\dfrac{8}{5}\right)^2 + \left(-\dfrac{6}{5}\right)^2} = \sqrt{\dfrac{64}{25} + \dfrac{36}{25}} = \sqrt{\dfrac{100}{25}} = \sqrt{4} = 2$$

(2) For any complex numbers z_1 and z_2, $|z_1| = 1$ and $z_1 \neq z_2$.

Find the value of $\left| \dfrac{z_1 - z_2}{1 - \overline{z_1}\, z_2} \right|$.

Note that $\overline{(\overline{z_1})} = z_1 \,,\quad \overline{z_1 \pm z_2} = \overline{z_1} \pm \overline{z_2} \,,\quad \overline{z_1 \cdot z_2} = \overline{z_1} \cdot \overline{z_2} \,,\quad \overline{\left(\dfrac{z_1}{z_2}\right)} = \dfrac{\overline{z_1}}{\overline{z_2}} \,,\ z_2 \neq 0$

$$\left| \dfrac{z_1 - z_2}{1 - \overline{z_1}\, z_2} \right|^2 = \dfrac{z_1 - z_2}{1 - \overline{z_1}\, z_2} \cdot \overline{\left(\dfrac{z_1 - z_2}{1 - \overline{z_1}\, z_2}\right)} = \dfrac{z_1 - z_2}{1 - \overline{z_1}\, z_2} \cdot \left(\dfrac{\overline{z_1} - \overline{z_2}}{1 - z_1\, \overline{z_2}}\right)$$

$$= \dfrac{|z_1|^2 - \overline{z_1}\, z_2 - z_1\, \overline{z_2} - |z_2|^2}{1 - \overline{z_1}\, z_2 - z_1\, \overline{z_2} + |z_1|^2 |z_2|^2}$$

$$= \dfrac{1 - \overline{z_1}\, z_2 - z_1\, \overline{z_2} - |z_2|^2}{1 - \overline{z_1}\, z_2 - z_1\, \overline{z_2} + |z_2|^2} \quad (\because |z_1| = 1)$$

$$= 1$$

$$\therefore \left| \dfrac{z_1 - z_2}{1 - \overline{z_1}\, z_2} \right| = 1 \quad (\because |z|^2 = 1 \ \Rightarrow \ |z| = 1)$$

#19 Find the values of a and b for the complex number $z = a + bi$ such that $\overline{z - zi} = 3 - 5i$.

Since $z = a + bi$, $z - zi = a + bi - (a + bi)i = a + bi - ai + b = (a + b) + (b - a)i$

$\therefore \overline{z - zi} = \overline{(a + b) + (b - a)i} = (a + b) - (b - a)i = 3 - 5i$

$\therefore a + b = 3, \ b - a = 5$

$\therefore a = -1, \ b = 4$

#20 For any real numbers a and b, $\sqrt{(a - b - 2)^2} + |a - b - 2ab| = 0$.

Find the value of $\dfrac{b}{a} + \dfrac{a}{b}$.

$$\sqrt{(a - b - 2)^2} + |a - b - 2ab| = 0 \ \Rightarrow \ |a - b - 2| + |a - b - 2ab| = 0$$

$$\Rightarrow \ a - b - 2 = 0, \ \ a - b - 2ab = 0$$

$$\Rightarrow \ a - b = 2, \ \ ab = 1$$

$$\therefore \dfrac{b}{a} + \dfrac{a}{b} = \dfrac{a^2 + b^2}{ab} = \dfrac{(a - b)^2 + 2ab}{ab} = \dfrac{4 + 2}{1} = 6$$

Chapter 2. Polynomials

#1 Perform the indicated operation.

(1) $(2x + y) + (3x - 4y) = (2x + 3x) + (y - 4y) = 5x - 3y$

(2) $3(x - 2y) - (5x + y) = 3x - 6y - 5x - y = (3x - 5x) - (6y + y) = -2x - 7y$

(3) $(5x^3 - 2x + 8) + (2x^2 - x + 3)$

$\quad = 5x^3 + 2x^2 + (-2x - x) + (8 + 3) = 5x^3 + 2x^2 - 3x + 11$

(4) $(x - 2)(3x^2 - x + 1)$

$\quad = x(3x^2 - x + 1) - 2(3x^2 - x + 1) = 3x^3 - x^2 + x - 6x^2 + 2x - 2$

$\quad = 3x^3 - 7x^2 + 3x - 2$

(5) $(x + 3)(x - 1)(2x - 5)$

$\quad = (x^2 + 2x - 3)(2x - 5) = 2x^3 + 4x^2 - 6x - 5x^2 - 10x + 15 = 2x^3 - x^2 - 16x + 15$

(6) $(x + 2y + 1)(x - 2y - 1) = (x + A)(x - A)$ *Letting* $A = 2y + 1$

$\quad = x^2 - A^2 = x^2 - (2y + 1)^2 = x^2 - (4y^2 + 4y + 1) = x^2 - 4y^2 - 4y - 1$

(7) $(x^2 + x + 3)(x^2 + x - 1) = (A + 3)(A - 1)$ *Letting* $A = x^2 + x$

$\quad = A^2 + 2A - 3 = (x^2 + x)^2 + 2(x^2 + x) - 3 = x^4 + 2x^3 + x^2 + 2x^2 + 2x - 3$

$\quad = x^4 + 2x^3 + 3x^2 + 2x - 3$

(8) $(x + 2)(x - 2)(x - 1)(x - 5) = \{(x + 2)(x - 5)\}\{(x - 2)(x - 1)\}$

$\qquad\qquad\qquad\qquad\qquad\qquad\qquad$ *Regrouping to get* $x^2 - 3x$

$\quad = (x^2 - 3x - 10)(x^2 - 3x + 2) = (A - 10)(A + 2)$ *Letting* $A = x^2 - 3x$

$\quad = A^2 - 8A - 20 = (x^2 - 3x)^2 - 8(x^2 - 3x) - 20 = x^4 - 6x^3 + 9x^2 - 8x^2 + 24x - 20$

$\quad = x^4 - 6x^3 + x^2 + 24x - 20$

#2 $a = 2^{x+1}, b = 3^{x-1}$. **Express** 6^x **using** a **and** b.

$\quad 6^x = (2 \cdot 3)^x = 2^x \cdot 3^x$

Since $a = 2^{x+1} = 2^x \cdot 2$, $2^x = \dfrac{a}{2}$

Since $b = 3^{x-1} = 3^x \cdot 3^{-1} = \dfrac{3^x}{3}$, $3^x = 3b$

Therefore, $6^x = \dfrac{a}{2} \cdot 3b = \dfrac{3}{2}ab$

#3 Simplify $6^{\frac{x-y}{x+y}}$ **where** $10^x = 2$, $10^y = 3$.

$\quad 6 = 2 \cdot 3 = 10^x \cdot 10^y = 10^{x+y}$

$6^{\frac{x-y}{x+y}} = (10^{x+y})^{\frac{x-y}{x+y}} = 10^{x-y} = 10^x \div 10^y = \dfrac{2}{3}$

#4 For a positive integer n, compute $(-1)^{2n+1} \cdot (-1)^{3n-1} \cdot (-1)^{2n-1} \div (-1)^{3n}$

$$(-1)^{2n+1} \cdot (-1)^{3n-1} \cdot (-1)^{2n-1} \div (-1)^{3n}$$

$$= \frac{(-1)^{2n+1} \cdot (-1)^{3n-1} \cdot (-1)^{2n-1}}{(-1)^{3n}} = \frac{(-1)^{2n+1+3n-1+2n-1}}{(-1)^{3n}} = \frac{(-1)^{7n-1}}{(-1)^{3n}} = (-1)^{4n-1} = \frac{(-1)^{4n}}{(-1)}$$

$$= \frac{((-1)^4)^n}{-1} = \frac{1^n}{-1} = -1^n = -1$$

#5 $(2x^2y^3)^a \div 4xy \cdot \frac{1}{2}x^2y = bx^3y^3$. Find the value of $a+b$, where a and b are constants.

$$(2x^2y^3)^a \div 4xy \cdot \frac{1}{2}x^2y = \frac{2^a x^{2a} y^{3a} \cdot x^2 y}{4xy \cdot 2} = \frac{2^{a-1} x^{2a+2-1} y^{3a+1-1}}{4}$$

$$= \frac{2^{a-1} x^{2a+2-1} y^{3a+1-1}}{2^2} = 2^{a-1-2} x^{2a+1} y^{3a} = bx^3y^3$$

$3a = 3 \quad \therefore \quad a = 1, \qquad 2^{a-1-2} = 2^{1-1-2} = 2^{-2} = \frac{1}{4} \quad \therefore \quad b = \frac{1}{4}$

Therefore, $a + b = 1 + \frac{1}{4} = \frac{5}{4}$

#6 Determine values :

(1) Two polynomials $f(x) = (k^2 + k)x - (k^2 - k)y + (k-1)z$ and $g(x) = -3k + 5$ are always equal, not depending on the values of k. Determine the values of real numbers x, y, and .

Rewrite $f(x)$ with descending power of k. Then we have

$(x - y)k^2 + (x + y + z)k - z = -3k + 5$

$\therefore \ x - y = 0, \ x + y + z = -3$, and $z = -5$

Since $x = y$ and $x + y = 2$, $y = 1$ and $x = 1$

Therefore, $x = 1, \ y = 1, \ z = -5$

(2) Two polynomials $f(x) = x^2 - 3x + 6$ and

$g(x) = a(x-1)(x-2) + b(x-2)(x-3) + c(x-3)(x-1)$ are equal. Determine the values of constants a, b, and c.

If $x = 1$, then $f(x) = 1 - 3 + 6 = 4$ and $g(x) = b(-1)(-2) = 2b$

Since $f(x) = g(x)$, $b = 2$

If $x = 2$, then $f(x) = 4 - 6 + 6 = 4$ and $g(x) = c(-1)(1) = -c$

Since $f(x) = g(x)$, $c = -4$

If $x = 3$, then $f(x) = 9 - 9 + 6 = 6$ and $g(x) = a(2)(1) = 2a$

Since $f(x) = g(x)$, $a = 3$ Therefore, $a = 3, \ b = 2, \ c = -4$

#7 Factor each polynomial.

(1) $x^4 - 6x^2 + 5 = A^2 - 6A + 5$ *Letting* $A = x^2$

$$= (A - 1)(A - 5) = (x^2 - 1)(x^2 - 5) = (x + 1)(x - 1)(x^2 - 5)$$

(2) $x^4 + 7x^2 + 16 = (x^4 + 8x^2 + 16) - x^2 = A^2 + 8A + 16 - x^2$ *Letting* $A = x^2$

$$= (A + 4)^2 - x^2 = (A + 4 + x)(A + 4 - x) = (x^2 + x + 4)(x^2 - x + 4)$$

(3) $x^4 + 4x^2 + 16 = (x^4 + 8x^2 + 16) - 4x^2 = A^2 + 8A + 16 - (2x)^2$ *Letting* $A = x^2$

$$= (A + 4)^2 - (2x)^2 = (A + 4 + 2x)(A + 4 - 2x) = (x^2 + 2x + 4)(x^2 - 2x + 4)$$

(4) $x^4 + x^2y^2 + y^4 = (x^4 + 2x^2y^2 + y^4) - x^2y^2 = (x^2 + y^2)^2 - (xy)^2$

$$= (x^2 + y^2 + xy)(x^2 + y^2 - xy) = (x^2 + xy + y^2)(x^2 - xy + y^2)$$

(5) $\left(x^2 - 2x\right)^2 - 2\left(x^2 - 2x\right) - 8 = A^2 - 2A - 8$ *Letting* $A = x^2 - 2x$

$$= (A - 4)(A + 2) = (x^2 - 2x - 4)(x^2 - 2x + 2)$$

(6) $x(x - 1)(x^2 - x + 1) - 30 = (x^2 - x)(x^2 - x + 1) - 30$

$$= A(A + 1) - 30 \quad \textit{Letting } A = x^2 - x$$

$$= A^2 + A - 30 = (A + 6)(A - 5) = (x^2 - x + 6)(x^2 - x - 5)$$

(7) $(x - 1)(x - 2)(x + 3)(x + 4) - 84 = \{(x - 1)(x + 3)\}\{(x - 2)(x + 4)\} - 84$

$$= (x^2 + 2x - 3)(x^2 + 2x - 8) - 84$$

$$= (A - 3)(A - 8) - 84 \quad \textit{Letting } A = x^2 + 2x$$

$$= A^2 - 11A + 24 - 84 = A^2 - 11A - 60$$

$$= (A - 15)(A + 4) = (x^2 + 2x - 15)(x^2 + 2x + 4)$$

$$= (x + 5)(x - 3)(x^2 + 2x + 4)$$

(8) $x^2 + xy - 4x - 2y + 4 = (x^2 - 4x + 4) + xy - 2y = (x - 2)^2 + y(x - 2)$

$$= (x - 2)(x - 2 + y) = (x - 2)(x + y - 2)$$

(9) $x^3 - x^2y - xz^2 + yz^2 = x(x^2 - z^2) + y(z^2 - x^2) = (x^2 - z^2)(x - y)$

$$= (x + z)(x - z)(x - y)$$

(10) $x^2(y - z) + y^2(z - x) + z^2(x - y) = (y - z)x^2 - (y^2 - z^2)x + y^2z - yz^2$

$$= (y - z)x^2 - (y + z)(y - z)x + yz(y - z)$$

$$= (y - z)(x^2 - (y + z) + yz)$$

$$= (y - z)(x - y)(x - z)$$

(11) $x^2y^2 - 3xyz - xy + 3z = xy(xy - 1) - 3z(xy - 1) = (xy - 1)(xy - 3z)$

(12) $9x^2 + y^2 + 4 - 6xy - 12x + 4y$

$$= (3x)^2 + (-y)^2 + (-2)^2 + 2(3x)(-y) + 2(3x)(-2) + 2(-y)(-2)$$

$$= (3x - y - 2)^2 \quad \textit{by factoring formula}$$

(13) $-8x^3 + 12x^2 - 6x + 1 = (-2x)^3 + 3(-2x)^2(1) + 3(-2x)(1)^2 + (1)^3$

$$= (-2x + 1)^3 \quad by \ factoring \ formula$$

#8 Find the value of each expression :

(1) When $a - b = 3$ and $ab = -2$,

① $a^2 + b^2 = (a - b)^2 + 2ab = (3)^2 + 2(-2) = 9 - 4 = 5$

② $\dfrac{b}{a} + \dfrac{a}{b} = \dfrac{a^2 + b^2}{ab} = \dfrac{5}{-2}$

③ $(a + b)^2 = (a - b)^2 + 4ab = (3)^2 + 4(-2) = 9 - 8 = 1$

④ $a^3 - b^3 = (a - b)\left(a^2 + ab + b^2 \right) = 3(5 - 2) = 9$

(2) When $a + b = 2$ and $a^3 + b^3 = 32$, find the value of $a^2 + b^2$.

Since $(a + b)^3 = a^3 + b^3 + 3ab(a + b)$, $\ 8 = 32 + 6ab \ \therefore \ ab = -4$

So $a^2 + b^2 = (a + b)^2 - 2ab = 4 + 8 = 12$

(3) For any real number $a \ (a > 3)$, $a^2 + \dfrac{9}{a^2} = 15$. Find the value of $a^3 - \dfrac{27}{a^3}$.

$a^2 + \dfrac{9}{a^2} = 15 \Rightarrow \left(a - \dfrac{3}{a}\right)^2 + 6 = 15 \quad \therefore \left(a - \dfrac{3}{a}\right)^2 = 9$

$\therefore \ a - \dfrac{3}{a} = 3 \ \left(\because a > 3 \ ; \ \dfrac{3}{a} < 1\right)$

$a^3 - \dfrac{27}{a^3} = \left(a - \dfrac{3}{a}\right)^3 + 3a \cdot \dfrac{3}{a}\left(a - \dfrac{3}{a}\right) = 27 + 9 \cdot 3 = 54$

(4) When $A = x^3 + x - 2$ and $B = x - 2$, find the coefficient of x^3 in $A^3 - B^3$.

Since $A = x^3 + x - 2 = x^3 + B$,

$A^3 - B^3 = (x^3 + B)^3 - B^3 = x^9 + 3x^6 B + 3x^3 B^2 + B^3 - B^3 = x^9 + 3x^6 B + 3x^3 B^2$

Consider the term $3x^3 B^2$.

$3x^3 B^2 = 3x^3(x - 2)^2 = 3x^3(x^2 - 4x + 4) = 3x^5 - 12x^4 + 12x^3$

Therefore, the coefficient of x^3 is 12.

(5) When $a + b + c = 0$ and $a^2 + b^2 + c^2 = 1$, find the value of $a^2 b^2 + b^2 c^2 + c^2 a^2$.

Since $a^2 + b^2 + c^2 = (a + b + c)^2 - 2(ab + bc + ca)$, $\ 1 = 0 - 2(ab + bc + ca)$.

$\therefore ab + bc + ca = -\dfrac{1}{2}$

$\therefore \ a^2 b^2 + b^2 c^2 + c^2 a^2 = (ab + bc + ca)^2 - 2abc(a + b + c)$

$$= \left(-\dfrac{1}{2}\right)^2 - 2abc \cdot 0 = \dfrac{1}{4}$$

(6) For non-zero numbers $a, b,$ and c,

$a + b + c = 0$, $\dfrac{1}{a} + \dfrac{1}{b} + \dfrac{1}{c} = 3$, **and** $a^2 + b^2 + c^2 = 12$. **Find the value of** $a^3 + b^3 + c^3$.

Note that $a^3 + b^3 + c^3 = (a + b + c)(a^2 + b^2 + c^2 - ab - bc - ca) + 3abc$

Since $(a + b + c)^2 = a^2 + b^2 + c^2 + 2ab + 2bc + 2ca$, $0 = 12 + 2(ab + bc + ca)$

$\therefore\ ab + bc + ca = -6$

Since $\dfrac{1}{a} + \dfrac{1}{b} + \dfrac{1}{c} = \dfrac{bc + ac + ab}{abc} = \dfrac{ab + bc + ca}{abc} = \dfrac{-6}{abc}$, $3 = \dfrac{-6}{abc}$ $\therefore\ abc = -2$

Therefore, $a^3 + b^3 + c^3 = 0 \cdot (12 + 6) + 3(-2) = -6$

(7) For positive real numbers a and b, $a - b = 2$, $a^2 + ab + b^2 = 10$.

Find the value of $a^3 + b^3$.

$a^2 + ab + b^2 = (a - b)^2 + 3ab = 4 + 3ab = 10$ $\therefore\ ab = 2$

Since $(a + b)^2 = (a - b)^2 + 4ab = 4 + 8 = 12$, $a + b = 2\sqrt{3}$ $(\because a > 0,\ b > 0)$

$\therefore a^3 + b^3 = (a + b)^3 - 3ab(a + b) = \left(2\sqrt{3}\right)^3 - 3 \cdot 2\left(2\sqrt{3}\right) = 24\sqrt{3} - 12\sqrt{3} = 12\sqrt{3}$

(8) When $a^2 - 3a + 1 = 0$, find the value of $3a^2 + 4a^3 + a^4 + \dfrac{3}{a^2} + \dfrac{4}{a^3} + \dfrac{1}{a^4}$.

Since $a \neq 0$, divide both sides of $a^2 - 3a + 1 = 0$ by a.

Then $a - 3 + \dfrac{1}{a} = 0$ $\therefore\ a + \dfrac{1}{a} = 3$

$a^2 + \dfrac{1}{a^2} = \left(a + \dfrac{1}{a}\right)^2 - 2 = 9 - 2 = 7$

$a^3 + \dfrac{1}{a^3} = \left(a + \dfrac{1}{a}\right)^3 - 3a \cdot \dfrac{1}{a}\left(a + \dfrac{1}{a}\right) = 27 - 3 \cdot 3 = 18$

$a^4 + \dfrac{1}{a^4} = \left(a^2 + \dfrac{1}{a^2}\right)^2 - 2 = 49 - 2 = 47$

Therefore, $3a^2 + 4a^3 + a^4 + \dfrac{3}{a^2} + \dfrac{4}{a^3} + \dfrac{1}{a^4} = 3\left(a^2 + \dfrac{1}{a^2}\right) + 4\left(a^3 + \dfrac{1}{a^3}\right) + \left(a^4 + \dfrac{1}{a^4}\right)$

$$= 3 \cdot 7 + 4 \cdot 18 + 47 = 140$$

(9) For real numbers $a, b,$ and c, $a + b + c = \sqrt{3}$ and $a^2 + b^2 + c^2 = 1$.

Find the value of abc .

Note that $a^2 + b^2 + c^2 \pm ab \pm bc \pm ca = \dfrac{1}{2}\{(a \pm b)^2 + (b \pm c)^2 + (c \pm a)^2\}$

Since $(a + b + c)^2 = a^2 + b^2 + c^2 + 2(ab + bc + ca)$, $(\sqrt{3})^2 = 1 + 2(ab + bc + ca)$.

$\therefore ab + bc + ca = 1$

Since $a^2 + b^2 + c^2 - (ab + bc + ca) = 1 - 1 = 0$, $a^2 + b^2 + c^2 - ab - bc - ca = 0$.

$\therefore\ \dfrac{1}{2}\{(a - b)^2 + (b - c)^2 + (c - a)^2\} = 0$

Since $a, b,$ and c are real numbers, $a - b = 0$, $b - c = 0$, and $c - a = 0$

$\therefore\ a = b = c$

Since $a + b + c = 3a = \sqrt{3}$, $a = \dfrac{\sqrt{3}}{3}$

Therefore, $abc = a^3 = \left(\dfrac{\sqrt{3}}{3}\right)^3 = \dfrac{3\sqrt{3}}{27} = \dfrac{\sqrt{3}}{9}$

(10) For real numbers $a, b,$ and c, $a + b + c = 1$, $a^2 + b^2 + c^2 = 9$, and $a^3 + b^3 + c^3 = 28$.

Find the value of $(a + b)(b + c)(c + a)$.

Since $(a + b + c)^2 = a^2 + b^2 + c^2 + 2(ab + bc + ca)$, $\ 1 = 9 + 2(ab + bc + ca)$

$\therefore \ ab + bc + ca = -4$

Since $a^3 + b^3 + c^3 - 3abc = (a + b + c)(a^2 + b^2 + c^2 - ab - bc - ca)$,

$28 - 3abc = 1 \cdot (9 + 4) \quad \therefore abc = 5$

Since $a + b + c = 1$,

$$(a + b)(b + c)(c + a) = (1 - c)(1 - a)(1 - b)$$
$$= 1 - (a + b + c) + (ab + bc + ca) - abc$$
$$= 1 - 1 + (-4) - 5 = -9$$

(11) For real numbers $a, b,$ and c, $a + b + c = -2$, $abc = 6$, and $\dfrac{1}{a} + \dfrac{1}{b} + \dfrac{1}{c} = -\dfrac{3}{2}$.

Find the value of $(1 - a)(1 - b)(1 - c)$.

Since $\dfrac{1}{a} + \dfrac{1}{b} + \dfrac{1}{c} = \dfrac{ab + bc + ca}{abc} = -\dfrac{3}{2}$, $\ ab + bc + ca = -\dfrac{3}{2} abc = -\dfrac{3}{2} \cdot 6 = -9$

$\therefore (1 - a)(1 - b)(1 - c) = 1^3 - (a + b + c) \cdot 1^2 + (ab + bc + ca) \cdot 1 - abc$

$$= 1 + 2 - 9 - 6 = -12$$

#9 Evaluate each expression.

(1) When $x - \dfrac{1}{x} = 3$, find the values of $x^2 + \dfrac{1}{x^2}$ and $x^3 - \dfrac{1}{x^3}$.

$x^2 + \dfrac{1}{x^2} = \left(x - \dfrac{1}{x}\right)^2 + 2 = 9 + 2 = 11$

$x^3 - \dfrac{1}{x^3} = \left(x - \dfrac{1}{x}\right)\left(x^2 + 1 + \dfrac{1}{x^2}\right) = 3(11 + 1) = 36$

(2) When $\dfrac{x}{2} = \dfrac{y}{3}$, $xy \neq 0$, find the values of $\dfrac{xy}{x^2 + y^2}$ and $\dfrac{2x^2 + 5xy}{x^2 + y^2}$.

Let $\dfrac{x}{2} = \dfrac{y}{3} = k \ (k \neq 0)$. Then $x = 2k$, $y = 3k$

$\therefore \quad \dfrac{xy}{x^2 + y^2} = \dfrac{6k^2}{4k^2 + 9k^2} = \dfrac{6k^2}{13k^2} = \dfrac{6}{13}$

$\dfrac{2x^2 + 5xy}{x^2 + y^2} = \dfrac{2 \cdot 4k^2 + 5 \cdot 6k^2}{13k^2} = \dfrac{38k^2}{13k^2} = \dfrac{38}{13}$

(3) When $x^2 - 3x - 1 = 0$, find the value of $x + \frac{1}{x}$.

Divide both sides of $x^2 - 3x - 1 = 0$ by x. Then $x - 3 - \frac{1}{x} = 0$. $\therefore x - \frac{1}{x} = 3$

Since $\left(x - \frac{1}{x}\right)^2 = \left(x + \frac{1}{x}\right)^2 - 4$, $\left(x + \frac{1}{x}\right)^2 = 9 + 4 = 13$ $\therefore x + \frac{1}{x} = \pm\sqrt{13}$

(4) When $x^2 - 5x + 1 = 0$, find the value of $x^3 - 3x + 1 - \frac{3}{x} + \frac{1}{x^3}$.

Divide both sides of $x^2 - 5x + 1 = 0$ by x. Then $x - 5 + \frac{1}{x} = 0$. $\therefore x + \frac{1}{x} = 5$

$$x^3 - 3x + 1 - \frac{3}{x} + \frac{1}{x^3} = x^3 + \frac{1}{x^3} - 3\left(x + \frac{1}{x}\right) + 1$$

$$= \left(x + \frac{1}{x}\right)^3 - 3\left(x \cdot \frac{1}{x}\right)\left(x + \frac{1}{x}\right) - 3\left(x + \frac{1}{x}\right) + 1$$

(Note that $a^3 + b^3 = (a + b)^3 - 3ab(a + b)$)

$$= \left(x + \frac{1}{x}\right)^3 - 6\left(x + \frac{1}{x}\right) + 1$$

$$= (5)^3 - 6(5) + 1 = 125 - 30 + 1 = 96$$

(5) When $x^4 - 3x^2 + 1 = 0$ $(0 < x < 1)$, find the value of $x - \frac{1}{x}$.

Since $x \neq 0$, divide $x^4 - 3x^2 + 1 = 0$ by x^2. Then $x^2 - 3 + \frac{1}{x^2} = 0$ $\therefore x^2 + \frac{1}{x^2} = 3$

Since $\left(x - \frac{1}{x}\right)^2 = x^2 + \frac{1}{x^2} - 2 = 3 - 2 = 1$, $x - \frac{1}{x} = \pm 1$.

Since $0 < x < 1$, $\frac{1}{x} > 1$ $\therefore x - \frac{1}{x} < 0$ $\therefore x - \frac{1}{x} = -1$

(6) When $x = \frac{3 + \sqrt{5}}{2}$, find the value of $x^4 - x^3 - 6x^2 + 9x - 4$.

$x = \frac{3 + \sqrt{5}}{2}$ $\Rightarrow$ $2x - 3 = \sqrt{5}$

Squaring both sides, we have $(2x - 3)^2 = (\sqrt{5})^2$

$\therefore 4x^2 - 12x + 9 = 5$ $\therefore x^2 - 3x + 1 = 0$

$$
\Rightarrow \quad
\begin{array}{r}
x^2 + 2x - 1 \\
x^2 - 3x + 1 \overline{)\; x^4 - x^3 - 6x^2 + 9x - 4} \\
\underline{x^4 - 3x^3 + x^2} \\
2x^3 - 7x^2 + 9x - 4 \\
\underline{2x^3 - 6x^2 + 2x} \\
-x^2 + 7x - 4 \\
\underline{-x^2 + 3x - 1} \\
4x - 3
\end{array}
$$

$\therefore x^4 - x^3 - 6x^2 + 9x - 4 = (x^2 - 3x + 1)(x^2 + 2x - 1) + 4x - 3$

Since $x^2 - 3x + 1 = 0$,

$$x^4 - x^3 - 6x^2 + 9x - 4 = 4x + 3 = 4\left(\frac{3+\sqrt{5}}{2}\right) - 3 = 6 + 2\sqrt{5} - 3 = 3 + 2\sqrt{5}$$

(7) When $x = \dfrac{1+\sqrt{3}i}{2}$, find the value of $x^4 + 2x^2 - 3x - 1$.

$x = \dfrac{1+\sqrt{3}i}{2} \Rightarrow 2x - 1 = \sqrt{3}i$

Squaring both sides, we have $(2x - 1)^2 = (\sqrt{3}i)^2$

$\therefore 4x^2 - 4x + 1 = -3 \quad \therefore x^2 - x + 1 = 0$

$$
\begin{array}{r}
x^2 + x + 2 \\
x^2 - x + 1 \overline{\smash{)}\ x^4 + 2x^2 - 3x - 1} \\
\underline{x^4 - x^3 + x^2} \\
x^3 + x^2 - 3x - 1 \\
\underline{x^3 - x^2 + x} \\
2x^2 - 4x - 1 \\
\underline{2x^2 - 2x + 2} \\
-2x - 3
\end{array}
$$

$\therefore x^4 + 2x^2 - 3x - 1 = (x^2 - x + 1)(x^2 + x + 2) - 2x - 3$

Since $x^2 - x + 1 = 0$,

$$x^4 + 2x^2 - 3x - 1 = -2x - 3 = -2\left(\frac{1+\sqrt{3}i}{2}\right) - 3 = -1 - \sqrt{3}i - 3 = -4 - \sqrt{3}i$$

(8) When $x : y : z = 2 : 3 : 4$, find the value of $\dfrac{x+y+z}{2x+3y-4z}$.

Let $x = 2k$, $y = 3k$, $z = 4k$ $(k \neq 0)$.

Then $\dfrac{x+y+z}{2x+3y-4z} = \dfrac{2k+3k+4k}{4k+9k-16k} = \dfrac{9k}{-3k} = -3$

#10 When $x + y + z = 1$, factor the polynomial $x^2 - 3xy + 2y^2 - 4x - y - 2z - 1$.

Since $x + y + z = 1$, $z = 1 - x - y$

$x^2 - 3xy + 2y^2 - 4x - y - 2z - 1$

$= x^2 - 3xy + 2y^2 - 4x - y - 2(1 - x - y) - 1$

$= x^2 - 3xy + 2y^2 - 2x + y - 3 = x^2 - (3y + 2)x + 2y^2 + y - 3$

$= x^2 - (3y + 2)x + (2y + 3)(y - 1) = \big(x - (2y + 3)\big)\big(x - (y - 1)\big)$

$= (x - 2y - 3)(x - y + 1)$

#11 Find the value of k that will make each polynomial perfect square form.

(1) $x^2 + 5x + k = x^2 + 2 \cdot x \cdot \frac{5}{2} + \left(\frac{5}{2}\right)^2 = \left(x + \frac{5}{2}\right)^2 \quad \therefore k = \frac{25}{4}$

OR $x^2 + 5x + k = \left(x + \frac{5}{2}\right)^2 = x^2 + 5x + \left(\frac{5}{2}\right)^2 \quad \therefore k = \frac{25}{4}$

(2) $9x^2 - 12x + k = (3x)^2 - 2 \cdot 3x \cdot 2 + (2)^2 = (3x - 2)^2 \quad \therefore k = 4$

OR $9x^2 - 12x + k = 9\left(x^2 - \frac{12}{9}x + \frac{k}{9}\right) = 9\left(x^2 - \frac{4}{3}x + \frac{k}{9}\right) = 9\left(x - \frac{1}{2}\cdot\frac{4}{3}\right)^2 = 9\left(x - \frac{2}{3}\right)^2$

$= 9\left(x^2 - \frac{4}{3}x + \frac{4}{9}\right) = 9x^2 - 12x + 4 \quad \therefore k = 4$

(3) $2x^2 - 6x + k = 2\left(x^2 - 3x + \frac{k}{2}\right) = 2\left(x - \frac{1}{2}\cdot 3\right)^2 = 2\left(x - \frac{3}{2}\right)^2 = 2\left(x^2 - 3x + \frac{9}{4}\right)$

$= 2x^2 - 6x + \frac{9}{2} \quad \therefore k = \frac{9}{2}$

(4) $25x^2 + 4x + k = (5x)^2 + 2 \cdot 5x \cdot \frac{2}{5} + \left(\frac{2}{5}\right)^2 = \left(5x + \frac{2}{5}\right)^2 \quad \therefore k = \frac{4}{25}$

OR $25x^2 + 4x + k = 25\left(x^2 + \frac{4}{25}x + \frac{k}{25}\right) = 25\left(x + \frac{1}{2}\cdot\frac{4}{25}\right)^2 = 25\left(x + \frac{2}{25}\right)^2$

$= 25\left(x^2 + \frac{4}{25}x + \left(\frac{2}{25}\right)^2\right) = 25x^2 + 4x + \frac{4}{25} \quad \therefore k = \frac{4}{25}$

(5) $\frac{1}{25}x^2 + kx + 4 = \left(\frac{1}{5}x \pm 2\right)^2 = \frac{1}{25}x^2 \pm \frac{4}{5}x + 4 \quad \therefore k = \frac{4}{5}$ or $k = -\frac{4}{5}$

(6) $4x^2 - kx + 25 = (2x \pm 5)^2 = 4x^2 \pm 20x + 25 \quad \therefore -k = \pm 20 \quad \therefore k = -20$ or $k = 20$

(7) $2x^2 + kx + 8 = 2\left(x^2 + \frac{k}{2}x + 4\right) = 2\left(x + \frac{k}{4}\right)^2 = 2\left(x^2 + \frac{k}{2}x + \frac{k^2}{16}\right) = 2x^2 + kx + \frac{k^2}{8}$

$\therefore \frac{k^2}{8} = 8 \quad \therefore k^2 = 64 \quad \therefore k = 8$ or $k = -8$

OR $2\left(x^2 + \frac{k}{2}x + 4\right) = 2(x \pm 2)^2 = 2(x^2 \pm 4x + 4) \quad \therefore \frac{k}{2} = \pm 4 \quad \therefore k = \pm 8$

(8) $9x^2 + (2k - 4)x + 4 = (3x \pm 2)^2 = 9x^2 \pm 12x + 4$

$\therefore 2k - 4 = \pm 12 \quad \therefore 2k = 16$ or $2k = -8 \quad \therefore k = 8$ or $k = -4$

(9) $4x^2 + (k + 5)xy + 9y^2 = (2x \pm 3y)^2 = 4x^2 \pm 12xy + 9y^2$

$\therefore k + 5 = \pm 12 \quad \therefore k = 12 - 5 = 7$ or $k = -12 - 5 = -7$

(10) $k - \frac{1}{4}xy + \frac{1}{4}y^2 = \frac{1}{4}y^2 - \frac{1}{4}xy + k = \left(\frac{1}{2}y\right)^2 - 2\left(\frac{1}{2}y\right)\left(\frac{1}{4}x\right) + \left(\frac{1}{4}x\right)^2 = \left(\frac{1}{2}y - \frac{1}{4}x\right)^2$

$= \frac{1}{4}y^2 - \frac{1}{4}xy + \frac{1}{16}x^2 \quad \therefore k = \frac{1}{16}x^2$

(11) $9x^2 + (k - 1)xy + 25y^2 = (3x \pm 5y)^2 = 9x^2 \pm 30xy + 25y^2$

$\therefore k - 1 = \pm 30 \quad \therefore k = 31$ or $k = -29$

(12) $kx^2 + 3x + 9 = k\left(x^2 + \frac{3}{k}x + \frac{9}{k}\right) = k\left(x + \frac{3}{2k}\right)^2 = k\left(x^2 + \frac{3}{k}x + \frac{9}{4k^2}\right) = kx^2 + 3x + \frac{9}{4k}$

$\therefore \ \frac{9}{4k} = 9 \quad \therefore \ 4k = 1 \quad \therefore \ k = \frac{1}{4}$

(13) $(x+3)(x-4) - k = x^2 - x - 12 - k = x^2 - 2 \cdot x \cdot \frac{1}{2} + \left(\frac{1}{2}\right)^2 = \left(x - \frac{1}{2}\right)^2$

$\therefore \ \frac{1}{4} = -12 - k \quad \therefore \ k = -12 - \frac{1}{4} = -12\frac{1}{4}$

(14) $(2x+1)(2x-4) + k = 4x^2 + 2x - 8x - 4 + k = 4x^2 - 6x - 4 + k$

$$= (2x)^2 - 2 \cdot 2x \cdot \frac{6}{4} + \left(\frac{6}{4}\right)^2 = \left(2x - \frac{6}{4}\right)^2$$

$\therefore \ \left(\frac{6}{4}\right)^2 = -4 + k \quad \therefore \ k = \frac{9}{4} + 4 = 6\frac{1}{4}$

(15) $(x-1)(x-2)(x+4)(x+5) + k = \underline{(x-1)(x+4)}\,\underline{(x-2)(x+5)} + k$

$= (x^2 + 3x - 4)(x^2 + 3x - 10) + k = (A - 4)(A - 10) + k \quad \text{Letting } x^2 + 3x = A$

$= A^2 - 14A + 40 + k = (A - 7)^2 = A^2 - 14A + 49 \quad \therefore \ 40 + k = 49 \quad \therefore \ k = 9$

#12 Find the value of a for the following polynomials. Each polynomial has a given factor.

(1) $x^2 + 2x + a$ has the factor $(x + 3)$.

Let $x^2 + 2x + a = (x + 3)(x + b)$

$\Rightarrow x^2 + 2x + a = x^2 + (3 + b)x + 3b$

Since $3 + b = 2$ and $3b = a$, $b = -1$ and $a = -3 \quad \therefore \ a = -3$

(OR Since $x + 3$ is a factor of $x^2 + 2x + a$, $(-3)^2 + 2(-3) + a = 0$.

So $9 - 6 + a = 0 \quad \therefore \ a = -3$)

(2) $3x^2 + ax - 8$ has the factor $(x - 2)$.

Let $3x^2 + ax - 8 = (x - 2)(3x + b)$

$\Rightarrow 3x^2 + ax - 8 = 3x^2 + (b - 6)x - 2b$

Since $b - 6 = a$ and $-2b = -8$, $b = 4$ and $a = -2 \quad \therefore \ a = -2$

(OR Since $x - 2$ is a factor of $3x^2 + ax - 8$, $3(2)^2 + 2a - 8 = 0$.

So $4 + 2a = 0 \quad \therefore \ a = -2$)

(3) $4x^2 + ax - 6$ has the factor $(3 - 2x)$.

Let $4x^2 + ax - 6 = (3 - 2x)(b - 2x)$

$\Rightarrow 4x^2 + ax - 6 = 4x^2 + (-2b - 6)x + 3b$

Since $-2b - 6 = a$ and $3b = -6$, $b = -2$ and $a = -2 \quad \therefore \ a = -2$

(OR Since $3 - 2x$ is a factor of $4x^2 + ax - 6$, $4\left(\frac{3}{2}\right)^2 + a\left(\frac{3}{2}\right) - 6 = 0$.

So $3 + \frac{3}{2}a = 0 \quad \therefore \ a = -2$)

(4) $2x^2 + (3a - 1)x - 15$ **has the factor** $(2x + 3)$.

Let $2x^2 + (3a - 1)x - 15 = (2x + 3)(x + b)$

$\Rightarrow 2x^2 + (3a - 1)x - 15 = 2x^2 + (3 + 2b)x + 3b$

Since $3 + 2b = 3a - 1$ and $3b = -15$, $b = -5$ and $a = -2$ $\therefore a = -2$

(OR Since $2x + 3$ is a factor of $2x^2 + (3a - 1)x - 15$,

$$2\left(-\frac{3}{2}\right)^2 + (3a - 1)\left(-\frac{3}{2}\right) - 15 = 0.$$

So $\frac{9}{2} - \frac{9}{2}a + \frac{3}{2} - 15 = 0$ $\therefore 9 - 9a + 3 - 30 = 0$ $\therefore 9a = -18$ $\therefore a = -2$)

(5) $2ax^2 - 5x + 2$ **has the factor** $(3x - 2)$.

Let $2ax^2 - 5x + 2 = (3x - 2)(bx - 1)$

$\Rightarrow 2ax^2 - 5x + 2 = 3bx^2 + (-2b - 3)x + 2$

Since $3b = 2a$ and $-2b - 3 = -5$, $b = 1$ and $a = \frac{3}{2}$ $\therefore a = \frac{3}{2}$

(OR Since $3x - 2$ is a factor of $2ax^2 - 5x + 2$, $2a\left(\frac{2}{3}\right)^2 - 5\left(\frac{2}{3}\right) + 2 = 0$.

So $\frac{8a}{9} - \frac{10}{3} + 2 = 0$ $\therefore 8a - 30 + 18 = 0$ $\therefore 8a = 12$ $\therefore a = \frac{3}{2}$)

#13 Find the value of $a + b$ for any constants a and b.

(1) $3ax^2 - 6x + ab$ **has the factor** $(3x - 1)^2$.

$3ax^2 - 6x + ab = c(3x - 1)^2 = 9cx^2 - 6cx + c$

$\therefore 3a = 9c, \ -6 = -6c,$ and $ab = c$

$\therefore c = 1, \ a = 3, \ b = \frac{1}{3}$

Therefore, $a + b = 3\frac{1}{3}$

(2) $ax^2 + 8x + 4b$ **has two factors** $(3x + 2)$ **and** $(x - 2)$.

$ax^2 + 8x + 4b = c(3x + 2)(x - 2) = 3cx^2 - 4cx - 4c$

$\therefore a = 3c, \ 8 = -4c,$ and $4b = -4c$

$\therefore c = -2, \ b = 2, \ a = -6$

Therefore, $a + b = -4$

(3) $(4x - 3)^2 - (3x - 2)^2$ **has two factors** $(ax + 5)$ **and** $(b - x)$.

$(4x - 3)^2 - (3x - 2)^2 = (4x - 3 + 3x - 2)(4x - 3 - 3x + 2) = (7x - 5)(x - 1)$

$= -(-7x + 5)(x - 1) = (-7x + 5)(1 - x) = (ax + 5)\,(b - x)$

$\therefore a = -7$ and $b = 1$

Therefore, $a + b = -6$

(4) $2x^2 + ax - 4$ **and** $bx^2 - x - 2$ **have the same factor** $(2x + 1)$.

$2x^2 + ax - 4 = (2x + 1)(x + c) = 2x^2 + (1 + 2c)x + c$

$\therefore a = 1 + 2c$ and $-4 = c$ $\therefore a = -7$

$bx^2 - x - 2 = (2x + 1)(dx - 2) = 2dx^2 + (d - 4)x - 2$

$\therefore b = 2d$ and $-1 = d - 4$

$\therefore d = 3$ and $b = 6$ Therefore, $a + b = -1$

(OR $2x + 1 = 0$ $\therefore x = -\dfrac{1}{2}$

$\quad 2x^2 + ax - 4 = 2\left(-\dfrac{1}{2}\right)^2 + a\left(-\dfrac{1}{2}\right) - 4 = 0$ $\therefore \dfrac{1}{2} - \dfrac{1}{2}a - 4 = 0$ $\therefore a = -7$

$\quad bx^2 - x - 2 = b\left(-\dfrac{1}{2}\right)^2 - \left(-\dfrac{1}{2}\right) - 2 = 0$ $\therefore \dfrac{1}{4}b + \dfrac{1}{2} - 2 = 0$ $\therefore b = 6$

Therefore, $a + b = -1$)

#14 Factor the polynomial given that $P(k) = 0$.

(1) $P(x) = 2x^3 + 3x^2 + x + 6$ **;** $k = -2$

Since $P(-2) = 0$, $x + 2$ is a factor of $P(x)$.

Using synthetic division to find the other factors, we have

$$
\begin{array}{r|rrrr}
-2 & 2 & 3 & 1 & 6 \\
   &   & -4 & 2 & -6 \\
\hline
   & 2 & -1 & 3 & 0 \\
\end{array}
$$

$\therefore\ P(x) = (x + 2)(2x^2 - x + 3)$

(2) $P(x) = x^3 - 3x - 2$ **;** $k = -1$

Since $P(-1) = 0$, $x + 1$ is a factor of $P(x)$.

$$
\begin{array}{r|rrrr}
-1 & 1 & 0 & -3 & -2 \\
   &   & -1 & 1 & 2 \\
\hline
   & 1 & -1 & -2 & 0 \\
\end{array}
$$

$\therefore\ P(x) = (x + 1)(x^2 - x - 2) = (x + 1)(x - 2)(x + 1) = (x + 1)^2(x - 2)$

(3) $P(x) = 6x^4 - x^3 - 7x^2 + x + 1$ **;** $k = 1$

Since $P(1) = 0$, $x - 1$ is a factor of $P(x)$.

$$
\begin{array}{r|rrrrr}
1 & 6 & -1 & -7 & 1 & 1 \\
  &   & 6 & 5 & -2 & -1 \\
\hline
  & 6 & 5 & -2 & -1 & 0 \\
\end{array}
$$

$\therefore\ P(x) = (x-1)(6x^3 + 5x^2 - 2x - 1)$

Let $f(x) = 6x^3 + 5x^2 - 2x - 1$

Then $f(-1) = 6(-1)^3 + 5(-1)^2 - 2(-1) - 1 = -6 + 5 + 2 - 1 = 0$.

Thus $x + 1$ is a factor of $f(x)$.

$$\begin{array}{r|rrrr} -1 & 6 & 5 & -2 & -1 \\ & & -6 & 1 & 1 \\ \hline & 6 & -1 & -1 & 0 \end{array}$$

$\therefore\ f(x) = (x+1)(6x^2 - x - 1)$

Therefore, $P(x) = (x-1)(x+1)(6x^2 - x - 1) = (x-1)(x+1)(2x-1)(3x+1)$

(4) $P(x) = 2x^3 - x^2 - 7x + 6$; $k = -2$

Since $P(-2) = 0$, $x + 2$ is a factor of $P(x)$.

$$\begin{array}{r|rrrr} -2 & 2 & -1 & -7 & 6 \\ & & -4 & 10 & -6 \\ \hline & 2 & -5 & 3 & 0 \end{array}$$

$\therefore\ P(x) = (x+2)(2x^2 - 5x + 3) = (x+2)(x-1)(2x-3)$

#15 Find all real zeros of each polynomial.

(1) $P(x) = x^4 + 4x^3 - 2x^2 - 12x + 9$

Consider the possible rational zeros of the polynomial $P(x)$.

Since the leading coefficient is 1 and the constant term is 9,

the possible rational zeros are $x = \pm\dfrac{1}{1},\ \pm\dfrac{3}{1}, \pm\dfrac{9}{1}$.

To determine which of these possible zeros are actual zeros of $P(x)$, use synthetic division.

$$\begin{array}{r|rrrrr} 1 & 1 & 4 & -2 & -12 & 9 \\ & & 1 & 5 & 3 & -9 \\ \hline & 1 & 5 & 3 & -9 & 0 \end{array}$$

Since $x = 1$ is a zero of $P(x)$, $p(x) = (x-1)(x^3 + 5x^2 + 3x - 9)$.

Let $f(x) = x^3 + 5x^2 + 3x - 9$

$$\begin{array}{r|rrrr} 1 & 1 & 5 & 3 & -9 \\ & & 1 & 6 & 9 \\ \hline & 1 & 6 & 9 & 0 \end{array}$$

Since $x = 1$ is a zero of $f(x)$, $f(x) = (x - 1)(x^2 + 6x + 9)$.

$\therefore\ P(x) = (x - 1)(x - 1)(x^2 + 6x + 9) = (x - 1)^2(x + 3)^2$

Therefore, the real zeros of $p(x)$ are 1 and -3.

(2) $P(x) = 2x^3 + 3x^2 - 6x + 2$

Since factors of constant term 2 are ± 1, ± 2 and factors of leading coefficient 2 are ± 1,

± 2, the possible rational zeros of $P(x)$ are $x = \pm 1$, ± 2, $\pm \dfrac{1}{2}$

Since $P\left(\dfrac{1}{2}\right) = 2 \cdot \dfrac{1}{8} + 3 \cdot \dfrac{1}{4} - 6 \cdot \dfrac{1}{2} + 2 = \dfrac{1}{4} + \dfrac{3}{4} - 3 + 2 = 0$,

use synthetic division with the factor $x = \dfrac{1}{2}$.

$$
\begin{array}{c|rrrr}
\frac{1}{2} & 2 & 3 & -6 & 2 \\
 & & 1 & 2 & -2 \\
\hline
 & 2 & 4 & -4 & 0
\end{array}
$$

$\therefore\ P(x) = \left(x - \dfrac{1}{2}\right)(2x^2 + 4x - 4) = (2x - 1)(x^2 + 2x - 2)$

Using the quadratic formula, the zeros of the factor $x^2 + 2x - 2$ can be found.

Note that : Quadratic Formula I

$$
\boxed{\begin{array}{l}
ax^2 + bx + c = 0, \ \ a \neq 0 \\[4pt]
\Rightarrow\ \ x = \dfrac{-b \pm \sqrt{b^2 - 4ac}}{2a}\ ,\ \ b^2 - 4ac \geq 0
\end{array}}
$$

Therefore, the real zeros of $p(x)$ are $\dfrac{1}{2}, -1 + \sqrt{3}$, and $-1 - \sqrt{3}$.

(3) $P(x) = 4x^4 - 4x^3 + x^2 - 2x + 1$

The possible rational zeros of $P(x)$ are $x = \pm \dfrac{1}{1}$, $\pm \dfrac{1}{2}$, $\pm \dfrac{1}{4}$

Since $P(1) = 4 - 4 + 1 - 2 + 1 = 0$ and

$$P\left(\dfrac{1}{2}\right) = 4 \cdot \dfrac{1}{16} - 4 \cdot \dfrac{1}{8} + \dfrac{1}{4} - 2 \cdot \dfrac{1}{2} + 1 = \dfrac{1}{4} - \dfrac{1}{2} + \dfrac{1}{4} - 1 + 1 = 0,$$

use synthetic division with the factor $x = 1$ and $x = \dfrac{1}{2}$

$$
\begin{array}{c|rrrrr}
1 & 4 & -4 & 1 & -2 & 1 \\
 & & 4 & 0 & 1 & -1 \\
\hline
\frac{1}{2} & 4 & 0 & 1 & -1 & 0 \\
 & & 2 & 1 & 1 & \\
\hline
 & 4 & 2 & 2 & 0 &
\end{array}
$$

$$\therefore P(x) = (x-1)\left(x - \tfrac{1}{2}\right)(4x^2 + 2x + 2)$$

$$= (x-1)(2x-1)(2x^2 + x + 1)$$

Using the quadratic formula,

$$2x^2 + x + 1 \text{ has zeros}: x = \frac{-1 \pm \sqrt{1 - 4 \cdot 2}}{2 \cdot 2} = \frac{-1 \pm \sqrt{-7}}{4} = \frac{-1 \pm \sqrt{7}\,i}{4}$$

Therefore, the real zeros of $p(x)$ are 1, $\dfrac{1}{2}$, $\dfrac{-1 + \sqrt{7}\,i}{4}$, and $\dfrac{-1 - \sqrt{7}\,i}{4}$.

#16 Find each polynomial $P(x)$.

(1) Let the quotient of $P(x)$ be $2x - 1$.

If $P(x)$ is divided by $x^2 - 2x + 3$, then the remainder is $-3x + 1$.

$$P(x) = (x^2 - 2x + 3)(2x - 1) - 3x + 1$$

$$= 2x^3 - 4x^2 + 6x - x^2 + 2x - 3 - 3x + 1$$

$$= 2x^3 - 5x^2 + 5x - 2$$

(2) Let $P(x) = x^2 - a$ ($a \neq 0$). Then $P(x^2) = P(x)Q(x)$ for a polynomial $Q(x)$.

Since $P(x) = x^2 - a$, $P(x^2) = (x^2)^2 - a = x^4 - a$.

Since $P(x^2) = P(x)Q(x)$, $Q(x) = x^2 + bx + c$ (degree of 2)

$$\underrightarrow{P(x^2)} = \underrightarrow{P(x)}Q(x)$$

Degree of 4   degree of 2

$$\therefore x^4 - a = (x^2 - a)(x^2 + bx + c) = x^4 + bx^3 + cx^2 - ax^2 - abx - ac$$

$$= x^4 + bx^3 + (c - a)x^2 - abx - ac$$

$\therefore b = 0$, $c - a = 0$, $ab = 0$, $ac = a$

Since $a = c$, $a^2 = a$ $\therefore a(a - 1) = 0$

Since $a \neq 0$, $a = 1$ $\therefore a = 1$, $b = 0$, $c = 1$ Therefore, $P(x) = x^2 - 1$

(3) For constants a and b,

$$P(x) = x^3 + 4x^2 + ax + b \text{ is divided by } (x+1)^2 \text{ with no remainder.}$$

Let $Q(x)$ be the quotient of $\dfrac{P(x)}{(x+1)^2}$. Then, $x^3 + 4x^2 + ax + b = (x + 1)^2 Q(x)$.

Substituting $x = -1$ into both sides, $-1 + 4 - a + b = 0$ $\therefore b = a - 3$

$\therefore x^3 + 4x^2 + ax + (a - 3) = (x + 1)^2 Q(x)$

Let $f(x) = x^3 + 4x^2 + ax + (a - 3)$. Then $f(-1) = -1 + 4 - a + a - 3 = 0$

$$
\begin{array}{r|rrrr}
-1 & 1 & 4 & a & a-3 \\
   &   & -1 & -3 & -a+3 \\
\hline
   & 1 & 3 & a-3 & 0
\end{array}
$$

$$\therefore \ f(x) = (x+1)(x^2 + 3x + a - 3)$$

$$\therefore \ (x+1)(x^2 + 3x + a - 3) = (x+1)^2 Q(x)$$

$$\therefore \ x^2 + 3x + a - 3 = (x+1)Q(x)$$

Substituting $x = -1$ into both sides, $1 - 3 + a - 3 = 0$

$$\therefore a = 5, \ b = 2$$

Therefore, $P(x) = x^3 + 4x^2 + 5x + 2$

#17 Using synthetic division, divide $2x^3 + 4x^2 - 3x - 5$ by

(1) $2x - 1$

(2) $2x + 3$

Find the quotient and remainder for each division.

(1)

$$
\begin{array}{r|rrrr}
\frac{1}{2} & 2 & 4 & -3 & -5 \\
 & \downarrow & 1 & \frac{5}{2} & -\frac{1}{4} \\
\hline
 & 2 & 5 & -\frac{1}{2} & -\frac{21}{4}
\end{array}
$$

$$\therefore \ 2x^3 + 4x^2 - 3x - 5 = \left(x - \frac{1}{2}\right)\left(2x^2 + 5x - \frac{1}{2}\right) - \frac{21}{4}$$

$$= (2x - 1)\left(x^2 + \frac{5}{2}x - \frac{1}{4}\right) - \frac{21}{4}$$

Therefore, the quotient is $x^2 + \frac{5}{2}x - \frac{1}{4}$ and remainder is $-\frac{21}{4}$.

(2)

$$
\begin{array}{r|rrrr}
-\frac{3}{2} & 2 & 4 & -3 & -5 \\
 & \downarrow & -3 & -\frac{3}{2} & \frac{27}{4} \\
\hline
 & 2 & 1 & -\frac{9}{2} & \frac{7}{4}
\end{array}
$$

$$\therefore \ 2x^3 + 4x^2 - 3x - 5 = \left(x + \frac{3}{2}\right)\left(2x^2 + x - \frac{9}{2}\right) + \frac{7}{4}$$

$$= (2x + 3)\left(x^2 + \frac{1}{2}x - \frac{9}{4}\right) + \frac{7}{4}$$

Therefore, the quotient is $x^2 + \frac{1}{2}x - \frac{9}{4}$ and remainder is $\frac{7}{4}$.

#18 Obtain the remainder.

(1) When a polynomial $P(x) = 3x^4 - 2x^3 - x + 1$ is divided by a polynomial $x^2 - 1$, find the remainder.

Let $Q(x)$ be the quotient of $\frac{P(x)}{x^2-1}$.

Since the degree of the divisor is 2, the degree of the remainder is less than 2.

Let $R(x) = ax + b$ be the remainder.

Then $P(x) = 3x^4 - 2x^3 - x + 1 = (x^2 - 1)Q(x) + ax + b$

$\therefore$ $P(1) = 3 - 2 - 1 + 1 = 1 = a + b$ and

$P(-1) = 3 + 2 + 1 + 1 = 7 = -a + b$

$\therefore$ $a = -3$, $b = 4$

Therefore, the remainder is $-3x + 4$

(2) When $x + 1$ is a factor of a polynomial $P(x) = x^4 - 2x + a$, divide $P(x)$ by $x - 1$ and find the remainder.

Since $x + 1$ is a factor of a polynomial $P(x)$, $P(-1) = 0$

$\therefore$ $1 + 2 + a = 0$ $\therefore$ $a = -3$ $\therefore$ $P(x) = x^4 - 2x - 3$

Therefore, the remainder of $\frac{P(x)}{x-1}$ is $P(1) = 1 - 2 - 3 = -4$.

(3) For all real number x, a polynomial $P(x)$ satisfies $P(1 + x) = P(1 - x)$ and $P(0) = -4$

Find the remainder of $\frac{P(x)}{x(x-2)}$.

Substitute $x = 1$ into $P(1 + x) = P(1 - x)$.

Then, $P(1 + 1) = P(1 - 1) = P(0) = -4$ $\therefore$ $P(2) = -4$

Let $Q(x)$ and $R(x) = ax + b$ be the quotient and remainder of $\frac{P(x)}{x(x-2)}$, respectively.

Then $P(x) = x(x - 2)Q(x) + ax + b$

Since $P(0) = -4$, $b = -4$

Since $P(2) = -4$, $2a - 4 = -4$; i.e., $a = 0$

Therefore, the remainder of $\frac{P(x)}{x(x-2)}$ is -4.

(4) The remainder of $\frac{P(x)}{(x+1)^2}$ is $-3x+1$ and the remainder of $\frac{P(x)}{x-2}$ is 4.

   Find the remainder of $\frac{P(x)}{(x+1)^2(x-2)}$.

   Let Q_1 be the quotient of $\frac{P(x)}{(x+1)^2}$. Then $P(x) = (x+1)^2 Q_1(x) - 3x + 1$

   Let Q_2 be the quotient of $\frac{Q_1(x)}{x-2}$ and R be the remainder of $\frac{Q_1(x)}{x-2}$.

   Then $Q_1(x) = \underline{(x-2)Q_2(x)} + \underline{R}$

   Divisor : degree of 1   Remainder : degree of 0 (constant)

   $\therefore\ P(x) = (x+1)^2\big((x-2)Q_2(x) + R\big) - 3x + 1$

   Since the remainder of $\frac{P(x)}{x-2}$ is 4, $P(2) = 4$

   $\therefore\ 4 = (2+1)^2\big((2-2)Q_2(2) + R\big) - 3\cdot 2 + 1 = 9R - 5 \quad \therefore\ R = 1$

   $\therefore\ P(x) = (x+1)^2(x-2)Q_2(x) + (x+1)^2 - 3x + 1$

   Therefore, the remainder of $\frac{P(x)}{(x+1)^2(x-2)}$ is $(x+1)^2 - 3x + 1 = x^2 - x + 2$.

(5) When $P(x)$ is divided by $(x-1)(x-2)$, the remainder is $-5x+2$.

   Find the remainder when the polynomial $P(2x)$ is divided by $x-1$.

   Let $Q(x)$ be the quotient of $\frac{P(x)}{(x-1)(x-2)}$.

   Then $P(x) = (x-1)(x-2)Q(x) - 5x + 2$

   $\therefore\ P(2x) = (2x-1)(2x-2)Q(2x) - 10x + 2$

   $\therefore$ The remainder of $\frac{P(2x)}{x-1}$ is

   $P(2\cdot 1) = (2\cdot 1 - 1)(2\cdot 1 - 2)Q(2\cdot 1) - 10\cdot 1 + 2 = -10 + 2 = -8$

(6) When $P(x)$ is divided by $x-1$, the remainder is 1, and when divided by
   $(x-2)(x-3)$, the remainder is 5. Find the remainder when $P(x)$ is divided by
   $(x-1)(x-2)(x-3)$.

   Since $P(x) = (x-1)Q_1(x) + 1 \quad \therefore P(1) = 1$

   Since $P(x) = (x-2)(x-3)Q_2(x) + 5 \quad \therefore P(2) = P(3) = 5$

   Now, let $P(x) = \underline{(x-1)(x-2)(x-3)Q_3(x)} + \underline{ax^2 + bx + c}$

   Divisor : Degree of 3   Remainder : Degree of 2

   Then $P(1) = a + b + c = 1 \ \cdots\cdots\ ①$

   $\qquad P(2) = 4a + 2b + c = 5 \ \cdots\cdots\ ②$

   $\qquad P(3) = 9a + 3b + c = 5 \ \cdots\cdots\ ③$

$$2a + 2b + 2c = 2 \quad \cdots\cdots ① \times 2$$

$$-)\ 4a + 2b + c = 5 \quad \cdots\cdots ②$$

$$\overline{\qquad\qquad\qquad\qquad}$$

$$-2a + c = -3 \qquad \therefore\ c = 2a - 3$$

From ①, $a + b + (2a - 3) = 1 \quad \therefore b = -3a + 4$

From ③, $9a + 3(-3a + 4) + (2a - 3) = 5 \quad \therefore a = -2,\ b = 10,\ c = -7$

Therefore, the remainder is $-2x^2 + 10x - 7$.

#19 When a polynomial $P(x) = x^5 - 2ax + b$ is divided by $(x - 1)^2$, the remainder is -3.

Find the values of real numbers a and b.

Let $Q(x)$ be the quotient of $\dfrac{P(x)}{(x-1)^2}$.

Then $P(x) = x^5 - 2ax + b = (x - 1)^2 Q(x) - 3$.

$\therefore\ x^5 - 2ax + b + 3 = (x - 1)^2 Q(x)$

Substituting $x = 1$ into both sides of the equal sign, we have

$1 - 2a + b + 3 = 0$

$\therefore\ b = 2a - 4$

$\therefore\ x^5 - 2ax + (2a - 4) + 3 = (x - 1)^2 Q(x) \quad \cdots\cdots ①$

Let $f(x) = x^5 - 2ax + (2a - 4) + 3$.

Then $f(1) = 1 - 2a + (2a - 4) + 3 = 0$

$\therefore\ f(x)$ has a factor $x - 1$.

1	1	0	0	0	$-2a$	$2a - 1$
		1	1	1	1	$-2a + 1$
	1	1	1	1	$-2a + 1$	0

$\therefore\ f(x) = (x - 1)(x^4 + x^3 + x^2 + x - 2a + 1)$

From ①, $(x - 1)(x^4 + x^3 + x^2 + x - 2a + 1) = (x - 1)^2 Q(x)$

$\therefore\ (x^4 + x^3 + x^2 + x - 2a + 10) = (x - 1)Q(x) \quad \cdots\cdots ②$

Substituting $x = 1$ into ②, we have $4 - 2a + 1 = 0$

$\therefore\ a = \dfrac{5}{2}$ and $b = 2a - 4 = 1$

#20 For a triangle with lengths $a, b,$ and c, evaluate each triangle.

(1) $a^3 - ac^2 + a^2 b - bc^2 = 0$

$a^3 - ac^2 + a^2 b - bc^2$

$= a(a^2 - c^2) + b(a^2 - c^2)$

$= (a^2 - c^2)(a + b)$

$= (a + c)(a - c)(a + b) = 0$

Since $a + b > 0$ and $a + c > 0$, $a - c = 0$

$\therefore \ a = c$

Therefore, the triangle is an isosceles triangle.

(2) $a^3 + b^3 + c^3 = 3abc$

$a^3 + b^3 + c^3 = 3abc \ \Rightarrow \ a^3 + b^3 + c^3 - 3abc = 0$

$\therefore (a + b + c)(a^2 + b^2 + c^2 - ab - bc - ca) = 0$

Since $a + b + c > 0$, $a^2 + b^2 + c^2 - ab - bc - ca = 0$

Note that $a^2 + b^2 + c^2 \pm ab \pm bc \pm ca = \frac{1}{2}\{(a \pm b)^2 + (b \pm c)^2 + (c \pm a)^2\}$

$\therefore \ \frac{1}{2}\{(a - b)^2 + (b - c)^2 + (c - a)^2\} = 0$

$\therefore \ a = b, \ b = c, \ c = a$

$\therefore \ a = b = c$

Therefore, the triangle is a right triangle.

#21 Evaluate each expression using factorization.

(1) $99^2 - 1$ $= (99 + 1)(99 - 1) = 100 \cdot 98 = 9800$

(2) $99^2 - 89^2$ $= (99 + 89)(99 - 89) = 188 \cdot 10 = 1880$

(3) $49^2 - 51^2$ $= (49 + 51)(49 - 51) = 100 \cdot -2 = -200$

(4) $3^8 - 1$ $= (3^4)^2 - 1 = (3^4 + 1)(3^4 - 1) = (3^4 + 1)(3^2 + 1)(3^2 - 1)$

$\qquad = (3^4 + 1)(3^2 + 1)(3 + 1)(3 - 1) = 82 \cdot 10 \cdot 4 \cdot 2 = 6560$

(5) $6^2 - 5^2 + 4^2 - 3^2 + 2^2 - 1$ $= (6 + 5)(6 - 5) + (4 + 3)(4 - 3) + (2 + 1)(2 - 1)$

$\qquad = 11 \cdot 1 + 7 \cdot 1 + 3 \cdot 1 = 21$

(6) $\left(1 - \frac{1}{2^2}\right)\left(1 - \frac{1}{3^2}\right)\left(1 - \frac{1}{4^2}\right)\cdots\cdots\left(1 - \frac{1}{99^2}\right)\left(1 - \frac{1}{100^2}\right)$

$\qquad = \left(1 - \frac{1}{2}\right)\left(1 + \frac{1}{2}\right)\left(1 - \frac{1}{3}\right)\left(1 + \frac{1}{3}\right)\left(1 - \frac{1}{4}\right)\left(1 + \frac{1}{4}\right)\cdots\left(1 - \frac{1}{99}\right)\left(1 + \frac{1}{99}\right)\left(1 - \frac{1}{100}\right)\left(1 + \frac{1}{100}\right)$

$\qquad = \left(\frac{1}{2} \cdot \frac{3}{2}\right)\left(\frac{2}{3} \cdot \frac{4}{3}\right)\left(\frac{3}{4} \cdot \frac{5}{4}\right)\cdots\cdots\left(\frac{98}{99} \cdot \frac{100}{99}\right)\left(\frac{99}{100} \cdot \frac{101}{100}\right) = \frac{1}{2} \cdot \frac{101}{100} = \frac{101}{200}$

(7) $3\left(2^2 + 1\right)\left(2^4 + 1\right)\left(2^8 + 1\right) + 1$ $= (2^2 - 1)(2^2 + 1)(2^4 + 1)(2^8 + 1) + 1$

$\qquad = (2^4 - 1)(2^4 + 1)(2^8 + 1) + 1 = (2^8 - 1)(2^8 + 1) + 1 = (2^{16} - 1) + 1 = 2^{16}$

(8) $\dfrac{99 \times 101 + 99 \times 2}{101^2 - 4} = \dfrac{99\,(101 + 2)}{(101 + 2)(101 - 2)} = \dfrac{99}{(101 - 2)} = \dfrac{99}{99} = 1$

(9) $36 \times 34 - 35 \times 34 = 34(36 - 35) = 34 \cdot 1 = 34$

(10) $87 \times 56 + 87 \times 44 = 87(56 + 44) = 87 \cdot 100 = 8700$

(11) $65^2 - 2 \times 65 \times 35 + 35^2 = (65 - 35)^2 = 30^2 = 900$

(12) $25^2 + 30 \times 25 + 15^2 = 25^2 + 2 \cdot 25 \cdot 15 + 15^2 = (25 + 15)^2 = 40^2 = 1600$

(13) $\dfrac{1000^3 + 1}{1000^2 - 999}$

Let $x = 1000$.

Then, $\dfrac{1000^3 + 1}{1000^2 - 999} = \dfrac{x^3 + 1}{x^2 - (x-1)} = \dfrac{(x+1)(x^2 - x + 1)}{x^2 - x + 1} = x + 1 = 1001$

(14) $(3 + 1)(3^2 + 1)(3^4 + 1)(3^8 + 1) = \dfrac{(3-1)(3+1)(3^2+1)(3^4+1)(3^8+1)}{(3-1)}$

$$= \dfrac{(3^2-1)(3^2+1)(3^4+1)(3^8+1)}{2} = \dfrac{(3^4-1)(3^4+1)(3^8+1)}{2}$$

$$= \dfrac{(3^8-1)(3^8+1)}{2} = \dfrac{3^{16}-1}{2}$$

(15) $\left(\dfrac{97}{100}\right)^3 + \left(\dfrac{3}{100}\right)^3 - 1$

Let $x = \dfrac{97}{100}$ and $y = \dfrac{3}{100}$.

Then, $\left(\dfrac{97}{100}\right)^3 + \left(\dfrac{3}{100}\right)^3 - 1 = x^3 + y^3 - 1$

$$= (x + y)^3 - 3xy(x + y) - 1$$

$$= \left(\dfrac{97}{100} + \dfrac{3}{100}\right)^3 - 3 \cdot \dfrac{97}{100} \cdot \dfrac{3}{100}\left(\dfrac{97}{100} + \dfrac{3}{100}\right) - 1$$

$$= 1 - \dfrac{3^2 \cdot 97}{10^4} - 1$$

$$= -\dfrac{3^2 \cdot 97}{10^4}$$

#22 When a polynomial $P(x) = x^3 + ax^2 + bx - 2$ is divided by $x + 2$ and $x - 1$, respectively, there are no remainders. For the constants a and b, find the value of $a^2 - b^2$.

Since $\dfrac{P(x)}{x+2}$ and $\dfrac{P(x)}{x-1}$ have no remainders, $P(-2) = 0$ and $P(1) = 0$.

$\therefore -8 + 4a - 2b - 2 = 0$ and $1 + a + b - 2 = 0$

$\therefore 2a - b = 5$ and $a + b = 1$

$\therefore a = 2$ and $b = -1$

Therefore, $a^2 - b^2 = 4 - 1 = 3$

#23 Simplify each expression.

(1) $\dfrac{\frac{1}{x+y} + \frac{1}{x-y}}{\frac{1}{x+y} - \frac{1}{x-y}} = \dfrac{\frac{x-y+x+y}{(x+y)(x-y)}}{\frac{x-y-x-y}{(x+y)(x-y)}} = \dfrac{\frac{2x}{(x+y)(x-y)}}{\frac{-2y}{(x+y)(x-y)}} = \dfrac{2x}{-2y} = -\dfrac{x}{y}$

(2) $1 - \dfrac{1}{1-\dfrac{1}{1-x}} = 1 - \dfrac{1}{\frac{1-x-1}{1-x}} = 1 - \dfrac{1}{\frac{-x}{1-x}} = 1 + \dfrac{1-x}{x} = \dfrac{x+1-x}{x} = \dfrac{1}{x}$

(3) $\dfrac{1 - \dfrac{x-y}{x+y}}{1 + \dfrac{x-y}{x+y}} = \dfrac{\frac{x+y-x+y}{x+y}}{\frac{x+y+x-y}{x+y}} = \dfrac{\frac{2y}{x+y}}{\frac{2x}{x+y}} = \dfrac{2y(x+y)}{2x(x+y)} = \dfrac{y}{x}$

(4) $1 + \dfrac{1}{1 + \dfrac{1}{1 + \dfrac{1}{1+x}}} = 1 + \dfrac{1}{1 + \dfrac{1}{\frac{1+x+1}{1+x}}} = 1 + \dfrac{1}{1 + \dfrac{x+1}{x+2}}$

$= 1 + \dfrac{1}{\frac{x+2+x+1}{x+2}} = 1 + \dfrac{1}{\frac{2x+3}{x+2}} = 1 + \dfrac{x+2}{2x+3} = \dfrac{2x+3+x+2}{2x+3} = \dfrac{3x+5}{2x+3}$

#24 For two polynomials A and B whose leading coefficients are 1, G and L are GCF and LCM of A and B, respectively. Find each polynomial.

(1) When $G = x - 1$, $L = (x-1)^2(x-3)$,

Let $A = (x-1)a$ and $B = (x-1)b$ where a and b are disjoint.

Since $AB = LG$, $(x-1)^2ab = (x-1)^3(x-3)$

∴ $ab = (x-1)(x-3)$

∴ $(a = x-1, b = x-3)$ or $(a = x-3, b = x-1)$

∴ $A = (x-1)^2$, $B = (x-1)(x-3)$ or $A = (x-1)(x-3)$, $B = (x-1)^2$

(2) When $G = x - 1$, $L = x^3 - 7x + 6$,

Let $A = (x-1)a$ and $B = (x-1)b$ where a and b are disjoint.

Since $AB = LG$, $(x-1)^2ab = (x^3 - 7x + 6)(x-1)$

Let $f(x) = x^3 - 7x + 6$. Then $f(1) = 0$

$$
\begin{array}{c|rrrr}
1 & 1 & 0 & -7 & 6 \\
  &   & 1 & 1 & -6 \\
\hline
  & 1 & 1 & -6 & 0 \\
\end{array}
$$

∴ $f(x) = (x-1)(x^2 + x - 6) = (x-1)(x+3)(x-2)$

∴ $(x-1)^2ab = (x-1)^2(x+3)(x-2)$

∴ $ab = (x+3)(x-2)$

∴ $A = (x-1)(x+3)$, $B = (x-1)(x-2)$

Or $A = (x-1)(x-2)$, $B = (x-1)(x+3)$

(3) When $G = x + 1$, $AB = x^4 - 4x^3 - 3x^2 + 10x + 8$,

Let $A = (x+1)a$ and $B = (x+1)b$ where a and b are disjoint.

Then $AB = (x+1)^2ab = x^4 - 4x^3 - 3x^2 + 10x + 8$

Consider the possible zeros. $x = \pm1, \ \pm2, \ \pm4, \ \pm8$

$$
\begin{array}{r|rrrrr}
-1 & 1 & -4 & -3 & 10 & 8 \\
   &   & -1 & 5 & -2 & -8 \\
\hline
2 & 1 & -5 & 2 & 8 & 0 \\
  &   & 2 & -6 & -8 & \\
\hline
-1 & 1 & -3 & -4 & 0 & \\
   &   & -1 & 4 & & \\
\hline
  & 1 & -4 & 0 & &
\end{array}
$$

$\therefore \; x^4 - 4x^3 - 3x^2 + 10x + 8 = (x+1)(x-2)(x+1)(x-4)$

$$= (x+1)^2(x-2)(x-4)$$

$\therefore \; (x+1)^2 ab = (x+1)^2(x-2)(x-4) \quad \therefore \; ab = (x-2)(x-4)$

$\therefore \quad A = (x+1)(x-2), \; B = (x+1)(x-4)$

Or $A = (x+1)(x-4), \; B = (x+1)(x-2)$

#25 For three polynomials $x^2 - 1$, $x^2 - 6x + 5$, and $x^2 + 3x + 2a$, the GCF is a polynomial of degree 1. Find the value of a.

Since $x^2 - 1 = (x+1)(x-1)$ and $x^2 - 6x + 5 = (x-5)(x-1)$,

the GCF of three polynomials is $x - 1$.

Let $P(x) = x^2 + 3x + 2a$. Then, $P(1) = 0$

$\therefore \; 1 + 3 + 2a = 0 \quad \therefore \; a = -2$

#26 For two polynomials $f(x) = x^3 - ax^2 + 2x + 4$ and $g(x) = x^2 + bx + 6$, $R(x)$ is the remainder of $\frac{f(x)}{g(x)}$. When the GCF of $g(x)$ and $R(x)$ is $x + 2$, find the value of $R(-1)$.

Since $x + 2$ is the GCF of $g(x)$ and $R(x)$, $x + 2$ is the GCF of $f(x)$ and $g(x)$.

$\therefore \; f(-2) = 0$ and $g(-2) = 0$

Thus, $-8 - 4a - 4 + 4 = 0$ and $4 - 2b + 6 = 0$

$\therefore \; a = -2$ and $b = 5$

$\therefore \; f(x) = x^3 + 2x^2 + 2x + 4$ and $g(x) = x^2 + 5x + 6$

$$
\begin{array}{r}
x - 3 \\
x^2 + 5x + 6 \overline{) x^3 + 2x^2 + 2x + 4} \\
\underline{x^3 + 5x^2 + 6x} \qquad \longleftarrow x(x^2 + 5x + 6) \\
-3x^2 - 4x + 4 \\
\underline{-3x^2 - 15x - 18} \longleftarrow -3(x^2 + 5x + 6) \\
11x + 22 \longleftarrow \text{Remainder}
\end{array}
$$

∴ The remainder of $\frac{f(x)}{g(x)}$ is $R(x) = 11x + 22$.

Therefore, $R(-1) = 11(-1) + 22 = 11$.

#27 **For two polynomials $f(x)$ and $g(x)$ whose leading coefficients are 1, the GCF and LCM of $f(x)$ and $g(x)$ are $x - 2$ and $x^3 - 3x^2 - 4x + 12$, respectively. Find the value of $|f(0) - g(0)|$.**

Since the GCF $x - 2$ is a factor of the LCM, $x^3 - 3x^2 - 4x + 12$ has a factor $x - 2$.

Using synthetic division, we have

$$
\begin{array}{r|rrrr}
2 & 1 & -3 & -4 & 12 \\
  &   & 2 & -2 & -12 \\
\hline
  & 1 & -1 & -6 & \;\;0
\end{array}
$$

∴ $x^3 - 3x^2 - 4x + 12 = (x - 2)(x^2 - x - 6) = (x - 2)(x - 3)(x + 2)$

∴ Two polynomials with GCF $x - 2$ are $(x - 2)(x - 3)$ and $(x - 2)(x + 2)$.

Therefore, $|f(0) - g(0)| = |6 + 4| = 10$

#28 Simplify each expression.

(1) $\sqrt{7 + 2\sqrt{10}} = \sqrt{(5 + 2) + 2\sqrt{(5 \cdot 2)}} = \sqrt{5} + \sqrt{2}$

(2) $\sqrt{2 - \sqrt{3}} = \sqrt{\frac{4 - 2\sqrt{3}}{2}} = \frac{\sqrt{4 - 2\sqrt{3}}}{\sqrt{2}} = \frac{\sqrt{(3+1) - 2\sqrt{3 \cdot 1}}}{\sqrt{2}} = \frac{\sqrt{3} - 1}{\sqrt{2}} = \frac{\sqrt{6} - \sqrt{2}}{2}$

(3) $\sqrt{4 + \sqrt{7}} - \sqrt{4 - \sqrt{7}} = \sqrt{\frac{8 + 2\sqrt{7}}{2}} - \sqrt{\frac{8 - 2\sqrt{7}}{2}} = \sqrt{\frac{(7+1) + 2\sqrt{7 \cdot 1}}{2}} - \sqrt{\frac{(7+1) - 2\sqrt{7 \cdot 1}}{2}}$

$$= \frac{\sqrt{7} + 1}{\sqrt{2}} - \frac{\sqrt{7} - 1}{\sqrt{2}} = \frac{2}{\sqrt{2}} = \frac{2\sqrt{2}}{2} = \sqrt{2}$$

(4) $\sqrt{9 + \sqrt{72}} - \sqrt{9 - \sqrt{72}} = \sqrt{9 + 2\sqrt{18}} - \sqrt{9 - 2\sqrt{18}}$

$$= \sqrt{(6 + 3) + 2\sqrt{(6 \cdot 3)}} - \sqrt{(6 + 3) - 2\sqrt{(6 \cdot 3)}}$$

$$= \sqrt{6} + \sqrt{3} - \left(\sqrt{6} - \sqrt{3}\right) = 2\sqrt{3}$$

(5) $\frac{\sqrt{x+2} - \sqrt{x-2}}{\sqrt{x+2} + \sqrt{x-2}} = \frac{\left(\sqrt{x+2} - \sqrt{x-2}\right)^2}{\left(\sqrt{x+2} + \sqrt{x-2}\right)\left(\sqrt{x+2} - \sqrt{x-2}\right)} = \frac{(x+2) - 2\sqrt{x+2} \cdot \sqrt{x-2} + (x-2)}{(x+2) - (x-2)}$

$$= \frac{2x - 2\sqrt{(x+2)(x-2)}}{4} = \frac{x - \sqrt{x^2 - 4}}{2}$$

(6) $\dfrac{1-\sqrt{2}-\sqrt{3}}{1+\sqrt{2}+\sqrt{3}} = \dfrac{\left(1-(\sqrt{2}+\sqrt{3})\right)^2}{\{1+(\sqrt{2}+\sqrt{3})\}\{1-(\sqrt{2}+\sqrt{3})\}} = \dfrac{1-2(\sqrt{2}+\sqrt{3})+(\sqrt{2}+\sqrt{3})^2}{1-(\sqrt{2}+\sqrt{3})^2} = \dfrac{1-2\sqrt{2}-2\sqrt{3}+2+2\sqrt{6}+3}{1-(2+2\sqrt{6}+3)}$

$$= \dfrac{6+2(\sqrt{6}-\sqrt{2}-\sqrt{3})}{-4-2\sqrt{6}} = \dfrac{(3+\sqrt{6}-\sqrt{2}-\sqrt{3})(-2+\sqrt{6})}{(-2-\sqrt{6})(-2+\sqrt{6})} = \dfrac{-6-2\sqrt{6}+2\sqrt{2}+2\sqrt{3}+3\sqrt{6}+6-\sqrt{12}-\sqrt{18}}{4-6}$$

$$= \dfrac{-2\sqrt{6}+2\sqrt{2}+2\sqrt{3}+3\sqrt{6}-2\sqrt{3}-3\sqrt{2}}{-2} = \dfrac{\sqrt{6}-\sqrt{2}}{-2} = \dfrac{\sqrt{2}-\sqrt{6}}{2}$$

#29 Evaluate each expression.

(1) When $x = 2 + \sqrt{3}$, $\dfrac{1}{1+\sqrt{x+2}} + \dfrac{1}{1-\sqrt{x+2}}$

$$\dfrac{1}{1+\sqrt{x+2}} + \dfrac{1}{1-\sqrt{x+2}} = \dfrac{1-\sqrt{x+2}+1+\sqrt{x+2}}{(1+\sqrt{x+2})(1-\sqrt{x+2})} = \dfrac{2}{1-(x+2)} = \dfrac{2}{-x-1} = -\dfrac{2}{(2+\sqrt{3})+1}$$

$$= -\dfrac{2}{3+\sqrt{3}} = -\dfrac{2(3-\sqrt{3})}{(3+\sqrt{3})(3-\sqrt{3})} = -\dfrac{6-2\sqrt{3}}{9-3} = -\dfrac{6-2\sqrt{3}}{6} = -1+\dfrac{\sqrt{3}}{3}$$

(2) When $x = \sqrt{3+\sqrt{8}}$, $x^3 + \dfrac{1}{x^3}$

Since $x = \sqrt{3+\sqrt{8}} = \sqrt{3+2\sqrt{2}} = \sqrt{(2+1)+2\sqrt{(2\cdot 1)}} = \sqrt{2}+1,$

$$x + \dfrac{1}{x} = \sqrt{2}+1+\dfrac{1}{\sqrt{2}+1} = \sqrt{2}+1+\dfrac{1(\sqrt{2}-1)}{(\sqrt{2}+1)(\sqrt{2}-1)}$$

$$= \sqrt{2}+1+\dfrac{\sqrt{2}-1}{2-1} = \sqrt{2}+1+\sqrt{2}-1 = 2\sqrt{2}.$$

$$\therefore\ x^3 + \dfrac{1}{x^3} = \left(x+\dfrac{1}{x}\right)^3 - 3\left(x+\dfrac{1}{x}\right) = (2\sqrt{2})^3 - 3(2\sqrt{2}) = 16\sqrt{2} - 6\sqrt{2} = 10\sqrt{2}$$

(3) When $x = \sqrt{5+2\sqrt{6}}$ **and** $y = \sqrt{5-2\sqrt{6}}$, $\dfrac{\sqrt{x}-\sqrt{y}}{\sqrt{x}+\sqrt{y}}$

Since $x = \sqrt{5+2\sqrt{6}} = \sqrt{(3+2)+2\sqrt{(3\cdot 2)}} = \sqrt{3}+\sqrt{2}$ and

$$y = \sqrt{5-2\sqrt{6}} = \sqrt{(3+2)-2\sqrt{(3\cdot 2)}} = \sqrt{3}-\sqrt{2}\ (\because \sqrt{3} > \sqrt{2}),$$

$$\dfrac{\sqrt{x}-\sqrt{y}}{\sqrt{x}+\sqrt{y}} = \dfrac{(\sqrt{x}-\sqrt{y})^2}{(\sqrt{x}+\sqrt{y})(\sqrt{x}-\sqrt{y})} = \dfrac{x-2\sqrt{xy}+y}{x-y} = \dfrac{2\sqrt{3}-2\sqrt{(3-2)}}{2\sqrt{2}} = \dfrac{2\sqrt{3}-2}{2\sqrt{2}} = \dfrac{\sqrt{3}-1}{\sqrt{2}} = \dfrac{\sqrt{6}-\sqrt{2}}{2}$$

(4) When $x = \sqrt{2}$, $\dfrac{1}{\sqrt{x+1+2\sqrt{x}}} - \dfrac{1}{\sqrt{x+1-2\sqrt{x}}}$

Since $\sqrt{x+1+2\sqrt{x}} = \sqrt{(x+1)+2\sqrt{(x\cdot 1)}} = \sqrt{x}+\sqrt{1} = \sqrt{x}+1$ and

$$\sqrt{x+1-2\sqrt{x}} = \sqrt{(x+1)-2\sqrt{(x\cdot 1)}} = \sqrt{x}-\sqrt{1} = \sqrt{x}-1\ (\because x = \sqrt{2} > 1),$$

$$\frac{1}{\sqrt{x+1+2\sqrt{x}}} - \frac{1}{\sqrt{x+1-2\sqrt{x}}} = \frac{1}{\sqrt{x}+1} - \frac{1}{\sqrt{x}-1} = \frac{\sqrt{x}-1-(\sqrt{x}+1)}{(\sqrt{x}+1)(\sqrt{x}-1)} = \frac{-2}{x-1}$$

$$= \frac{-2}{\sqrt{2}-1} = \frac{-2(\sqrt{2}+1)}{(\sqrt{2}-1)(\sqrt{2}+1)} = \frac{-2\sqrt{2}-2}{2-1} = -2\sqrt{2}-2$$

#30 Simplify the expression : $\dfrac{1}{x^2+x} + \dfrac{1}{x^2+3x+2} + \dfrac{1}{x^2+5x+6} + \cdots\cdots + \dfrac{1}{x^2+19x+90}$

$$\frac{1}{x^2+x} + \frac{1}{x^2+3x+2} + \frac{1}{x^2+5x+6} + \cdots\cdots + \frac{1}{x^2+19x+90}$$

$$= \frac{1}{x(x+1)} + \frac{1}{(x+1)(x+2)} + \frac{1}{(x+2)(x+3)} + \cdots\cdots + \frac{1}{(x+9)(x+10)}$$

$$= \left(\frac{1}{x} - \frac{1}{x+1}\right) + \left(\frac{1}{x+1} - \frac{1}{x+2}\right) + \left(\frac{1}{x+2} - \frac{1}{x+3}\right) + \cdots\cdots + \left(\frac{1}{x+9} - \frac{1}{x+10}\right)$$

$$= \frac{1}{x} - \frac{1}{x+10} = \frac{x+10-x}{x(x+10)} = \frac{10}{x(x+10)}$$

#31 Find the value of k such that $\dfrac{x+y}{3} = \dfrac{2y-z}{4} = \dfrac{z}{5} = \dfrac{3x+7y-4z}{k}$

$$\frac{x+y}{3} = \frac{2y-z}{4} = \frac{z}{5} = \frac{3(x+y)+2(2y-z)+(-2)z}{3\cdot3+2\cdot4+(-2)\cdot5} = \frac{3x+7y-4z}{7}. \quad \therefore \ k = 7$$

Alternative Method :

Let $\dfrac{x+y}{3} = \dfrac{2y-z}{4} = \dfrac{z}{5} = \dfrac{3x+7y-4z}{7} = m \ (m \neq 0)$.

Then $x + y = 3m, \ 2y - z = 4m, \ z = 5m$

Since $2y - 5m = 4m, \ y = \dfrac{9}{2}m$

Since $x + \dfrac{9}{2}m = 3m, \ x = -\dfrac{3}{2}m$

$$\therefore \ \frac{3x+7y-4z}{k} = \frac{-\frac{9}{2}m + \frac{63}{2}m - 20m}{k} = \frac{7m}{k} = m$$

$$\therefore \ k = 7$$

#32 When a is the integer and b is the decimal part of an irrational number $\sqrt{11+\sqrt{72}}$,

(1) Find the value of $a + \dfrac{1}{b-1}$.

Since $\sqrt{11+\sqrt{72}} = \sqrt{11+2\sqrt{18}} = \sqrt{(9+2)+2\sqrt{(9\cdot2)}} = \sqrt{9}+\sqrt{2} = 3+\sqrt{2}$

and $3 + \sqrt{2} = 4.\times\times\times$, $a = 4$ and $b = (3+\sqrt{2}) - 4 = \sqrt{2} - 1$.

$$\therefore \ a + \frac{1}{b-1} = 4 + \frac{1}{(\sqrt{2}-1)-1} = 4 + \frac{1}{\sqrt{2}-2} = 4 + \frac{1(\sqrt{2}+2)}{(\sqrt{2}-2)(\sqrt{2}+2)}$$

$$= 4 + \frac{\sqrt{2}+2}{2-4} = \frac{-8+\sqrt{2}+2}{-2} = \frac{-6+\sqrt{2}}{-2}$$

$$= \frac{6-\sqrt{2}}{2}$$

(2) Find the value of $\sqrt{\dfrac{b+1-\sqrt{b^2+2b}}{b+1+\sqrt{b^2+2b}}}$.

Since $b = \sqrt{2} - 1$,

$b + 1 = \sqrt{2}$ and $b^2 + 2b = (\sqrt{2} - 1)^2 + 2(\sqrt{2} - 1) = 2 - 2\sqrt{2} + 1 + 2\sqrt{2} - 2 = 1$.

$$\therefore \sqrt{\dfrac{b+1-\sqrt{b^2+2b}}{b+1+\sqrt{b^2+2b}}} = \sqrt{\dfrac{\sqrt{2}-\sqrt{1}}{\sqrt{2}+\sqrt{1}}} = \sqrt{\dfrac{\sqrt{2}-1}{\sqrt{2}+1}} = \sqrt{\dfrac{(\sqrt{2}-1)^2}{(\sqrt{2}+1)(\sqrt{2}-1)}}$$

$$= \sqrt{\dfrac{2-2\sqrt{2}+1}{2-1}} = \sqrt{3 - 2\sqrt{2}} = \sqrt{2} - 1$$

(3) For a polynomial $P(x) = x^4 + 2x^3 + x^2 + 4x + 1$, find the value of $P(b)$.

Since $b = \sqrt{2} - 1$, $b + 1 = \sqrt{2}$.

$\therefore (b + 1)^2 = (\sqrt{2})^2$ $\therefore b^2 + 2b + 1 = 2$

$\therefore b^2 + 2b - 1 = 0$ $(2b^2 + 4b - 2 = 0)$

$\therefore P(b) = b^4 + 2b^3 + b^2 + 4b + 1$

$$= b^4 + 2b^3 - b^2 + (2b^2 + 4b - 2) + 3$$

$$= b^2(b^2 + 2b - 1) + (2b^2 + 4b - 2) + 3$$

$$= b^2 \cdot 0 + 0 + 3 = 3$$

#33 When a is the decimal part of $\sqrt{5}$, find the value of $\sqrt{\left(a - \dfrac{1}{2}\right)^2} + \sqrt{\left(a - \dfrac{1}{5}\right)^2}$.

Since $2 = \sqrt{4}$ and $3 = \sqrt{9}$, $2 < \sqrt{5} < 3$

$\therefore \sqrt{5} = 2.\times\times\times$

$\therefore$ The integer part of $\sqrt{5}$ is 2 and the decimal part of $\sqrt{5}$ is $\sqrt{5} - 2$.

$a - \dfrac{1}{2} = (\sqrt{5} - 2) - \dfrac{1}{2} = \sqrt{5} - \dfrac{5}{2} = \dfrac{2\sqrt{5}-5}{2} = \dfrac{\sqrt{20}-\sqrt{25}}{2} < 0$

$a - \dfrac{1}{5} = (\sqrt{5} - 2) - \dfrac{1}{5} = \sqrt{5} - \dfrac{11}{5} = \dfrac{5\sqrt{5}-11}{5} = \dfrac{\sqrt{125}-\sqrt{121}}{2} > 0$

$$\therefore \sqrt{\left(a - \dfrac{1}{2}\right)^2} + \sqrt{\left(a - \dfrac{1}{5}\right)^2} = \left|a - \dfrac{1}{2}\right| + \left|a - \dfrac{1}{5}\right|$$

$$= -\left(a - \dfrac{1}{2}\right) + \left(a - \dfrac{1}{5}\right)$$

$$= \dfrac{1}{2} - \dfrac{1}{5} = \dfrac{3}{10}$$

#34 **For a real number** $a\,(0 < a < 1)$, $x = a^2 + \frac{1}{a^2}$.

Express $\sqrt{x+2} - \sqrt{x-2}$ **as a polynomial in** a.

$$\sqrt{x+2} - \sqrt{x-2} = \sqrt{a^2 + \frac{1}{a^2} + 2} - \sqrt{a^2 + \frac{1}{a^2} - 2}$$

$$= \sqrt{\left(a + \frac{1}{a}\right)^2} - \sqrt{\left(a - \frac{1}{a}\right)^2}$$

$$= \left|a + \frac{1}{a}\right| - \left|a - \frac{1}{a}\right|$$

Since $0 < a < 1$, $\frac{1}{a} > 1$

$\therefore\ a + \frac{1}{a} > 0$ and $a - \frac{1}{a} < 0$

Therefore, $\sqrt{x+2} - \sqrt{x-2} = \left|a + \frac{1}{a}\right| - \left|a - \frac{1}{a}\right| = \left(a + \frac{1}{a}\right) + \left(a - \frac{1}{a}\right) = 2a$

#35 **When** $\frac{1}{4} < x < 1$, **simplify the expression** $\sqrt{4x + 1 - 4\sqrt{x}} + \left|\sqrt{x} - 1\right|$.

Since $\frac{1}{4} < x < 1$, $1 < 4x < 4$

$\therefore\ \sqrt{1} < \sqrt{4x} < \sqrt{4}$ $\therefore\ 1 < \sqrt{4x} < 2$ and $\sqrt{x} < 1$

$\therefore\ \sqrt{4x} - 1 > 0$ and $\sqrt{x} - 1 < 0$

$\therefore\ \sqrt{4x + 1 - 4\sqrt{x}} + \left|\sqrt{x} - 1\right| = \sqrt{4x + 1 - 2\sqrt{4x}} + \left|\sqrt{x} - 1\right|$

$$= \left|\sqrt{4x} - 1\right| + \left|\sqrt{x} - 1\right|$$

$$= \left(\sqrt{4x} - 1\right) - \left(\sqrt{x} - 1\right) = \sqrt{4x} - \sqrt{x} = 2\sqrt{x} - \sqrt{x} = \sqrt{x}$$

Chapter 3. Equations and Inequalities

#1 Solve the following equations for x :

(1) $3x - 2 = 7$ $3x = 9$; $x = 3$

(2) $2x + 3 = 3x - 2$ $2x - 3x = -2 - 3$; $-x = -5$; $x = 5$

(3) $5x - 2 = \frac{1}{2}x - 1\frac{1}{4}$ $5x - \frac{1}{2}x = -1\frac{1}{4} + 2$; $\frac{9}{2}x = \frac{3}{4}$; $\frac{18}{4}x = \frac{3}{4}$; $18x = 3$; $x = \frac{1}{6}$

(4) $0.2x - 0.3 = 0.4x - 0.5$ $0.2x - 0.4x = -0.5 + 0.3$; $-0.2x = -0.2$; $x = 1$

(5) $\frac{3}{4}\left(x - \frac{1}{3}\right) = \frac{1}{2}\left(\frac{1}{5} + 4x\right)$ $\frac{3}{4}x - \frac{1}{4} = \frac{1}{10} + 2x$; $\frac{3}{4}x - 2x = \frac{1}{10} + \frac{1}{4}$; $-\frac{5}{4}x = \frac{7}{20}$; $x = -\frac{7}{25}$

(6) $\frac{x-3}{2} - 1 = \frac{x}{4} - 3$ $2(x - 3) - 4 = x - 12$; $2x - x = -12 + 6 + 4$; $x = -2$

(7) $3(1 - 2x) + 7 = -2x - 2$ $3 - 6x + 7 = -2x - 2$; $-6x + 2x = -2 - 10$

$$-4x = -12 \; ; \; x = 3$$

(8) $3 - \frac{2x-1}{3} = 5x - \frac{x-2}{6}$ $18 - 2(2x - 1) = 30x - (x - 2)$; $-4x - 29x = -18$

$$-33x = -18 \; ; x = \frac{6}{11}$$

#2 Find the value.

(1) For any constants a, b, the solution of the equation $3x - 2 = ax - 4$ is $x = -1$ and the solution of the equation $\frac{1}{2}x + b = ax + 3$ is $x = -2$. Find $a \cdot b$

$3x - 2 = ax - 4 \Rightarrow 3(-1) - 2 = a(-1) - 4 \Rightarrow -5 = -a - 4 \Rightarrow a = 1$

$\frac{1}{2}x + b = ax + 3 \Rightarrow \frac{1}{2}(-2) + b = a(-2) + 3 \Rightarrow 2a + b = 4 \Rightarrow 2(1) + b = 4 \Rightarrow b = 2$

$\therefore a \cdot b = 1 \cdot 2 = 2$

(2) The solution of the equation $2ax + 5 = -3$ is half of the solution of the equation $x - 5 = 3x + 7$. Find the value of $3a - 4$.

$x - 5 = 3x + 7 \Rightarrow 2x = -12 \Rightarrow x = -6$

So, the solution x of an equation $2ax + 5 = -3$ is $x = -3$.

So, $2a(-3) + 5 = -3 \Rightarrow -6a + 5 = -3 \Rightarrow -6a = -8 \Rightarrow a = \frac{4}{3}$

Therefore, $3a - 4 = 3 \cdot \frac{4}{3} - 4 = 0$.

(3) For any constants a and b, $\frac{1}{a} - \frac{1}{b} = 3$ $(ab \neq 0)$. **Find the value of** $\frac{5a-3ab-5b}{a-b}$.

$$\frac{1}{a} - \frac{1}{b} = 3 \ \Rightarrow \ \frac{b-a}{ab} = 3 \ \Rightarrow \ b - a = 3ab \ \Rightarrow \ a - b = -3ab$$

$$\therefore \ \frac{5a-3ab-5b}{a-b} = \frac{5(a-b)-3ab}{a-b} = \frac{5(-3ab)-3ab}{-3ab} = \frac{-18ab}{-3ab} = 6$$

(4) $\begin{cases} \textcircled{1} \ \frac{a+3}{4} - \frac{2x-2}{3} = 1 \\ \textcircled{2} \ \frac{3a-2}{2} - \frac{2a-x}{3} = 1 \end{cases}$

When the ratio of the solution of $\textcircled{1}$ **to the solution of** $\textcircled{2}$ **is** $1 : 4$**, find the value of** a.

$\textcircled{1}$ $\quad \frac{a+3}{4} - \frac{2x-2}{3} = 1$

$$\Rightarrow \ \frac{3a+9-8x+8}{12} = 1 \ \Rightarrow \ 3a + 17 - 8x = 12 \ \Rightarrow \ 8x = 3a + 5 \ \Rightarrow \ x = \frac{3a+5}{8}.$$

$\textcircled{2}$ $\quad \frac{3a-2}{2} - \frac{2a-x}{3} = 1$

$$\Rightarrow \ \frac{9a-6-4a+2x}{6} = 1 \ \Rightarrow \ 5a - 6 + 2x = 6 \ \Rightarrow \ 2x = 12 - 5a \ \Rightarrow \ x = \frac{12-5a}{2}$$

$$\therefore \ \frac{3a+5}{8} : \frac{12-5a}{2} = 1 : 4 \ ; \ 1 \cdot \frac{12-5a}{2} = 4 \cdot \frac{3a+5}{8} \ ; \ \frac{12-5a}{2} = \frac{3a+5}{2}$$

$$12 - 5a = 3a + 5 \ ; \ 8a = 7$$

Therefore, $a = \frac{7}{8}$.

(5) For any x, the equation $3x - 5a = 2bx + 6$, where a and b are constants, is always true. Find the value of $\frac{a}{2b}$.

$$3x - 5a = 2bx + 6 \ \Rightarrow \ (3 - 2b)x = 5a + 6 \ \Rightarrow \ (3 - 2b)x + 0 = 0 \cdot x + 5a + 6$$

$$\therefore \ 3 - 2b = 0, \ 5a + 6 = 0 \ \therefore \ b = \frac{3}{2}, \ a = -\frac{6}{5}$$

Therefore, $\frac{a}{2b} = \frac{-\frac{6}{5}}{2 \cdot \frac{3}{2}} = -\frac{6}{15} = -\frac{2}{5}$

(6) The solution of an equation $\frac{2x-5a}{3} + x + 4 = 8$ **is a negative integer.**

Find the greatest value of a.

$\frac{2x-5a}{3} + x + 4 = 8$

$$\Rightarrow \ 2x - 5a + 3x + 12 = 24 \ \Rightarrow \ 5x - 5a = 12 \ \Rightarrow \ 5x = 5a + 12 \ \Rightarrow \ x = \frac{5a+12}{5}$$

Since $x = \frac{5a+12}{5} < 0$, $\frac{5a+12}{5} = -1, -2, -3, \cdots$ So $\frac{5a+12}{5} = -1$ to get the greatest value of a.

$\frac{5a+12}{5} = -1 \ \Rightarrow \ 5a + 12 = -5 \ \Rightarrow \ 5a = -17 \ \Rightarrow \ a = -\frac{17}{5}$

Therefore, the greatest value of a is $a = -\frac{17}{5} = -3\frac{2}{5}$.

(7) $a@b = ab^2 + a^2b$

When $\dfrac{1}{a} = 2$, $\dfrac{1}{b} = -3$, find the value of $b@a$.

$b@a = ba^2 + b^2a = -\dfrac{1}{3} \cdot \dfrac{1}{4} + \dfrac{1}{9} \cdot \dfrac{1}{2} = -\dfrac{1}{12} + \dfrac{1}{18} = \dfrac{-3+2}{36} = -\dfrac{1}{36}$

(8) A quadratic equation $x^2 - x + 1 = 0$ has two solutions α and β.

Find the value of $(1 - \alpha)(1 - \beta) + (2 - \alpha)(2 - \beta)$.

Since $\alpha + \beta = 1$ and $\alpha\beta = 1$,

$(1 - \alpha)(1 - \beta) + (2 - \alpha)(2 - \beta) = 1 - (\alpha + \beta) + \alpha\beta + 4 - 2(\alpha + \beta) + \alpha\beta$

$$= 5 - 3(\alpha + \beta) + 2\alpha\beta = 4$$

#3 Solve the equation involving absolute value.

(1) $|x - 3| = 5x + 2$

If $x - 3 \geq 0$ ($x \geq 3$) $\Rightarrow$ $|x - 3| = x - 3 = 5x + 2$ $\Rightarrow$ $4x = -5$ $\Rightarrow$ $x = -\dfrac{5}{4}$

Since $x \geq 3$, $x = -\dfrac{5}{4}$ is not a solution.

If $x - 3 < 0$ ($x < 3$) $\Rightarrow$ $|x - 3| = -(x - 3) = 5x + 2$ $\Rightarrow$ $6x = 1$ $\Rightarrow x = \dfrac{1}{6}$

Since $x < 3$, $x = \dfrac{1}{6}$ is a solution.

Therefore, the solution of $|x - 3| = 5x + 2$ is $x = \dfrac{1}{6}$

(2) $|x - 4| + |x + 2| = 10$

If $x - 4 = 0$ $\Rightarrow x = 4$

If $x + 2 = 0$ $\Rightarrow x = -2$

$\therefore$ Consider $x < -2$, $-2 \leq x < 4$, and $x \geq 4$

If $x < -2$ $\Rightarrow |x - 4| + |x + 2| = -(x - 4) - (x + 2) = 10$

$\Rightarrow -2x = 8 \Rightarrow x = -4$; A solution

If $-2 \leq x < 4$ $\Rightarrow |x - 4| + |x + 2| = -(x - 4) + (x + 2) = 10$

$\Rightarrow 0 \cdot x = 4$; Undefined

If $x \geq 4$ $\Rightarrow |x - 4| + |x + 2| = (x - 4) + (x + 2) = 10$

$\Rightarrow 2x = 12 \Rightarrow x = 6$; A solution

Therefore, the solutions are $x = -4$ and $x = 6$.

(3) $|x - 2| - |5 - x| = 0$

$|x - 2| - |5 - x| = 0$ $\Rightarrow |x - 2| = |5 - x|$ $\Rightarrow x - 2 = \pm(5 - x)$

If $x - 2 = 5 - x$ $\Rightarrow 2x = 7$ $\Rightarrow x = \dfrac{7}{2}$

If $x - 2 = -(5 - x)$ $\Rightarrow 0 \cdot x = -3$; Undefined   Therefore, the solution is $x = \dfrac{7}{2}$.

#4 Solve each equation for x by using the factorization.

    (1) $x^2 - 2x - 3 = 0$ $(x-3)(x+1) = 0$ $\therefore x = 3$ or $x = -1$

    (2) $2x^2 - 7x + 5 = 0$ $(x-1)(2x-5) = 0$ $\therefore x = 1$ or $x = \frac{5}{2}$

    (3) $-3x^2 + 6x = 0$ $-3x(x-2) = 0$ $\therefore x = 0$ or $x = 2$

    (4) $2x^2 + 2x - 4 = 0$ $2(x^2 + x - 2) = 0$; $2(x+2)(x-1) = 0$ $\therefore x = -2$ or $x = 1$

    (5) $x(x+5) = 6$ $(x+6)(x-1) = 0$ $\therefore x = -6$ or $x = 1$

    (6) $x^2 = \frac{x+1}{2}$ $2x^2 - x - 1 = 0$; $(x-1)(2x+1) = 0$ $\therefore x = 1$ or $x = -\frac{1}{2}$

#5 Solve each equation for x by using the square root.

    (1) $2x^2 = 8$ $x^2 = 4$ $\therefore x = \pm 2$

    (2) $9x^2 - 5 = 0$ $9x^2 = 5$; $x^2 = \frac{5}{9}$ $\therefore x = \pm\sqrt{\frac{5}{9}} = \pm\frac{\sqrt{5}}{3}$

    (3) $3(x-1)^2 = 15$ $(x-1)^2 = 5$; $x - 1 = \pm\sqrt{5}$ $\therefore x = 1 \pm \sqrt{5}$

    (4) $(2x+5)^2 - 3 = 0$ $(2x+5)^2 = 3$; $2x + 5 = \pm\sqrt{3}$ $\therefore x = \frac{-5\pm\sqrt{3}}{2}$

    (5) $4(x-2)^2 - 1 = 0$ $4(x-2)^2 = 1$; $(x-2)^2 = \frac{1}{4}$; $x - 2 = \pm\sqrt{\frac{1}{4}}$ $\therefore x = 2 \pm \frac{1}{2}$

#6 Solve each quadratic equation by using perfect squares.

    (1) $x^2 - 3x - 3 = 0$

$$\left(x - \frac{3}{2}\right)^2 - \frac{9}{4} - 3 = 0 \ ; \ \left(x - \frac{3}{2}\right)^2 = \frac{9}{4} + 3 = \frac{9+12}{4} = \frac{21}{4} \ \therefore x = \frac{3}{2} \pm \frac{\sqrt{21}}{2}$$

    (2) $2x^2 + 5x = 7$

$$2\left(x^2 + \frac{5}{2}x\right) = 7 \ ; \ x^2 + \frac{5}{2}x = \frac{7}{2} \ ; \ \left(x + \frac{5}{4}\right)^2 - \frac{25}{16} = \frac{7}{2} \ ; \ \left(x + \frac{5}{4}\right)^2 = \frac{25+56}{16} = \frac{81}{16}$$

$$\therefore x = -\frac{5}{4} \pm \frac{9}{4} \ \therefore x = 1 \text{ or } -\frac{7}{2}$$

    (3) $-x^2 - 3x + 5 = 0$

$$x^2 + 3x - 5 = 0 \ ; \left(x + \frac{3}{2}\right)^2 - \frac{9}{4} - 5 = 0 \ ; \ \left(x + \frac{3}{2}\right)^2 = \frac{9}{4} + 5 = \frac{29}{4}$$

$$\therefore x = -\frac{3}{2} \pm \frac{\sqrt{29}}{2}$$

    (4) $3x^2 - 4x + 1 = 0$

$$3\left(x^2 - \frac{4}{3}x + \frac{1}{3}\right) = 0 \ ; \ x^2 - \frac{4}{3}x + \frac{1}{3} = 0 \ ; \ \left(x - \frac{2}{3}\right)^2 - \frac{4}{9} + \frac{1}{3} = 0$$

$$\left(x - \frac{2}{3}\right)^2 = \frac{4}{9} - \frac{1}{3} = \frac{4-3}{9} = \frac{1}{9} \quad \therefore x = \frac{2}{3} \pm \frac{1}{3} \quad \therefore x = 1 \text{ or } x = \frac{1}{3}$$

#7 Find the constant k for each quadratic equation with a double root.

(1) $(3x - 4)^2 - k^2 = 0$

By formula, $k^2 = 0$; $k = 0$

OR $9x^2 - 24x + 16 - k^2 = 0$; $x^2 - \frac{8}{3}x + \frac{16-k^2}{9} = 0$; $\left(x - \frac{4}{3}\right)^2 - \frac{16}{9} + \frac{16-k^2}{9} = 0$

$\therefore \frac{16}{9} = \frac{16-k^2}{9}$ $\therefore k^2 = 0$ $\therefore k = 0$

(2) $x^2 - kx + 5 = 0$

By formula, $5 = \left(-\frac{k}{2}\right)^2 = \frac{k^2}{4}$; $k^2 = 20$ $\therefore k = \pm 2\sqrt{5}$

OR $\left(x - \frac{k}{2}\right)^2 - \frac{k^2}{4} + 5 = 0$; $\left(x - \frac{k}{2}\right)^2 = \frac{k^2}{4} - 5$; $\frac{k^2}{4} - 5 = 0$; $\frac{k^2}{4} = 5$; $k^2 = 20$

$\therefore k = \pm 2\sqrt{5}$

(3) $x^2 + 2x + k^2 = 0$

By formula, $k^2 = \left(\frac{2}{2}\right)^2 = 1$ $\therefore k = \pm 1$

OR $(x + 1)^2 - 1 + k^2 = 0$; $(x + 1)^2 = 1 - k^2$; $1 - k^2 = 0$; $k^2 = 1$ $\therefore k = \pm 1$

(4) $kx^2 + 3x + 2 = 0$

By formula, $\frac{2}{k} = \left(\frac{1}{2} \cdot \frac{3}{k}\right)^2 = \frac{9}{4k^2}$ $\therefore 2 = \frac{9}{4k}$ $\therefore k = \frac{9}{8}$

OR $k\left(x^2 + \frac{3}{k}x + \frac{2}{k}\right) = 0$; $\left(x + \frac{3}{2k}\right)^2 - \frac{9}{4k^2} + \frac{2}{k} = 0$,

since $k \neq 0$ ($\because$ It's a quadratic equation.)

$\left(x + \frac{3}{2k}\right)^2 = \frac{9}{4k^2} - \frac{2}{k}$ $\therefore \frac{9}{4k^2} = \frac{2}{k}$; $8k^2 = 9k$; $8k = 9$ $\therefore k = \frac{9}{8}$

(5) $2x^2 + 3x + k - 5 = 0$

By formula, $\frac{k-5}{2} = \left(\frac{1}{2} \cdot \frac{3}{2}\right)^2$ $\therefore k - 5 = \frac{9}{16} \cdot 2 = \frac{9}{8}$ $\therefore k = 5 + \frac{9}{8} = 6\frac{1}{8}$

OR $2(x^2 + \frac{3}{2}x + \frac{k-5}{2}) = 0$; $x^2 + \frac{3}{2}x + \frac{k-5}{2} = 0$; $\left(x + \frac{3}{4}\right)^2 - \frac{9}{16} + \frac{k-5}{2} = 0$

$\left(x + \frac{3}{4}\right)^2 = \frac{9}{16} - \frac{k-5}{2}$; $\frac{k-5}{2} = \frac{9}{16}$; $16(k - 5) = 18$; $8k = 49$ $\therefore k = \frac{49}{8} = 6\frac{1}{8}$

(6) $x^2 + kx + (k - 1) = 0$

By formula, $k - 1 = \left(\frac{k}{2}\right)^2 = \frac{k^2}{4}$; $k^2 - 4k + 4 = 0$; $(k - 2)^2 = 0$ $\therefore k = 2$

OR $\left(x + \frac{k}{2}\right)^2 - \frac{k^2}{4} + (k - 1) = 0$; $\frac{k^2}{4} = (k - 1)$; $k^2 - 4k + 4 = 0$; $(k - 2)^2 = 0$; $k = 2$

(7) $\frac{1}{3}x^2 + (k+1)x + 8 = 0$

By formula, $24 = \left(\frac{3(k+1)}{2}\right)^2 = \frac{9(k+1)^2}{4}$; $\frac{3(k+1)^2}{4} = 8$; $(k+1)^2 = \frac{32}{3}$

$\therefore$ $k = -1 \pm \sqrt{\frac{32}{3}} = -1 \pm \frac{4\sqrt{6}}{3}$

OR $x^2 + 3(k+1)x + 24 = 0$; $\left(x + \frac{3(k+1)}{2}\right)^2 - \frac{9(k+1)^2}{4} + 24 = 0$; $\frac{9(k+1)^2}{4} = 24$

$(k+1)^2 = 24 \cdot \frac{4}{9}$; $k+1 = \pm\sqrt{\frac{32}{3}} = \pm 4\sqrt{\frac{2}{3}} = \pm\frac{4\sqrt{6}}{3}$; $k+1 = \pm\frac{4\sqrt{6}}{3}$ $\therefore$ $k = -1 \pm \frac{4\sqrt{6}}{3}$

#8 Find the value of $p+q$ for each quadratic equation with the solution $x = p \pm \sqrt{q}$.

(1) $-2x^2 + 5x + 1 = 0$

$-2\left(x^2 - \frac{5}{2}x - \frac{1}{2}\right) = 0$; $x^2 - \frac{5}{2}x - \frac{1}{2} = 0$; $\left(x - \frac{5}{4}\right)^2 - \frac{25}{16} - \frac{1}{2} = 0$; $\left(x - \frac{5}{4}\right)^2 = \frac{25}{16} + \frac{1}{2} = \frac{33}{16}$

$\therefore$ $x = \frac{5}{4} \pm \sqrt{\frac{33}{16}}$

$\therefore$ $p = \frac{5}{4}$, $q = \frac{33}{16}$

Therefore, $p + q = \frac{5}{4} + \frac{33}{16} = \frac{53}{16}$

(2) $3(x-1)^2 = 4$

$(x-1)^2 = \frac{4}{3}$; $x = 1 \pm \sqrt{\frac{4}{3}}$

$\therefore$ $p = 1$, $q = \frac{4}{3}$

Therefore, $p + q = 1 + \frac{4}{3} = 2\frac{1}{3}$

(3) $-(x+1)^2 + 5 = 0$

$(x+1)^2 = 5$; $x = -1 \pm \sqrt{5}$

$\therefore$ $p = -1$, $q = 5$

Therefore, $p + q = -1 + 5 = 4$

#9 Find the constant a or the range of a for each quadratic equation with a condition.

(1) $(x+1)^2 = a + 2$ **has no real number solution.**

Since $(x+1)^2 \geq 0$, $a + 2 < 0$ $\therefore$ $a < -2$

(2) $x^2 + 3x + 3a = 0$ **has two different real number solutions.**

$\left(x + \frac{3}{2}\right)^2 - \frac{9}{4} + 3a = 0$; $\left(x + \frac{3}{2}\right)^2 = \frac{9}{4} - 3a$ $\therefore$ $\frac{9}{4} - 3a > 0$; $3a < \frac{9}{4}$ $\therefore$ $a < \frac{3}{4}$

(3) $ax^2 + x + 2 = 0$ **has one real number solution.**

$a(x^2 + \frac{1}{a}x + \frac{2}{a}) = 0$

Since $a \neq 0$, $x^2 + \frac{1}{a}x + \frac{2}{a} = 0$; $\left(x + \frac{1}{2a}\right)^2 - \frac{1}{4a^2} + \frac{2}{a} = 0$; $\frac{1}{4a^2} = \frac{2}{a}$; $\frac{1}{4a} = \frac{2}{1}$

$\therefore \ 8a = 1$; $a = \frac{1}{8}$

(4) $3x^2 - x + a = 0$ **has no real number solution.**

$3(x^2 - \frac{1}{3}x + \frac{a}{3}) = 0$; $x^2 - \frac{1}{3}x + \frac{a}{3} = 0$; $\left(x - \frac{1}{6}\right)^2 - \frac{1}{36} + \frac{a}{3} = 0$; $\left(x - \frac{1}{6}\right)^2 = \frac{1}{36} - \frac{a}{3}$

$\therefore \ \frac{1}{36} - \frac{a}{3} < 0$; $\frac{a}{3} > \frac{1}{36}$ $\quad \therefore \ a > \frac{1}{12}$

(5) $x^2 + (a+1)x + \frac{a+3}{2} = 0$ **has a double root.**

$\left(x + \frac{a+1}{2}\right)^2 - \frac{(a+1)^2}{4} + \frac{a+3}{2} = 0$; $\frac{(a+1)^2}{4} = \frac{a+3}{2}$; $(a+1)^2 = 2a + 6$; $a^2 + 2a + 1 = 2a + 6$

$\therefore \ a^2 = 5$; $a = \pm\sqrt{5}$

(6) $x^2 + x + a = 0$ **has two different real number solutions,** $x = 2$ **and** $x = b$.

$x^2 + x + a = (x - 2)(x - b) = x^2 + (-2 - b)x + 2b = 0$

$\therefore \ -2 - b = 1$ and $2b = a$

$\therefore \ b = -3$ and $a = -6$

(7) $x^2 + 2ax + b = 0$ **has a double root** $x = 3$.

$x^2 + 2ax + b = (x + a)^2 - a^2 + b = (x - 3)^2 = 0$

$\therefore \ a = -3$ and $-a^2 + b = 0$; $b = 9$

(8) $x^2 + ax + b = 0$ **has a real number solution** $x = 1 + \sqrt{2}$.

Since the other solution is $x = 1 - \sqrt{2}$,

$x^2 + ax + b = \left(x - \left(1 + \sqrt{2}\right)\right)\left(x - \left(1 - \sqrt{2}\right)\right)$

$\qquad = x^2 - \left(1 + \sqrt{2}\right)x - \left(1 - \sqrt{2}\right)x + (1 - 2) = x^2 - 2x - 1 = 0$

$\therefore \ a = -2$ and $b = -1$

OR $\left(1 + \sqrt{2}\right)^2 + a\left(1 + \sqrt{2}\right) + b = 0$; $1 + 2\sqrt{2} + 2 + a + a\sqrt{2} + b = 0$

$1 + 2 + a + b = 0$ and $2\sqrt{2} + a\sqrt{2} = 0$

$\therefore \ a = -2$ and $b = -1$

OR $\left(x + \frac{a}{2}\right)^2 - \frac{a^2}{4} + b = 0$; $x = -\frac{a}{2} \pm \sqrt{\frac{a^2 - 4b}{4}}$

$\therefore \ -\frac{a}{2} = 1$, $\frac{a^2 - 4b}{4} = 2$; $a = -2$ and $b = -1$

(9) $x^2 - ax - 2b^2 = 0$ has two different real number solutions $x = 4 \pm \sqrt{2a}$.

$$\left(x - \frac{a}{2}\right)^2 - \frac{a^2}{4} - 2b^2 = 0 \ ; \ \left(x - \frac{a}{2}\right)^2 = \frac{a^2}{4} + 2b^2 \ ; \ x = \frac{a}{2} \pm \sqrt{\frac{a^2 + 8b^2}{4}}$$

$$\therefore \ \frac{a}{2} = 4 \ , \frac{a^2 + 8b^2}{4} = 2a \ ; \ a = 8, \ a^2 + 8b^2 = 8a \ ; \ 8b^2 = 64 - 64 = 0 \ ; \ b = 0$$

$$\therefore \ \ a = 8 \text{ and } b = 0$$

#10 Solve each quadratic equation by using quadratic formulas.

(1) $x^2 - 2x - 4 = 0$

$$x = \frac{1 \pm \sqrt{1+4}}{1} = 1 \pm \sqrt{5}$$

(2) $3x^2 + 5x - 1 = 0$

$$x = \frac{-5 \pm \sqrt{25 - 4 \cdot 3 \cdot (-1)}}{2 \cdot 3} = \frac{-5 \pm \sqrt{37}}{6}$$

(3) $5x^2 - 2x - 1 = 0$

$$x = \frac{1 \pm \sqrt{1+5}}{5} = \frac{1 \pm \sqrt{6}}{5}$$

(4) $-2x^2 + 3x + 5 = 0$

$$x = \frac{-3 \pm \sqrt{9 - 4 \cdot (-2) \cdot 5}}{2 \cdot (-2)} = \frac{-3 \pm \sqrt{9+40}}{-4} = \frac{-3 \pm 7}{-4} \ ; \ x = -1 \text{ or } x = \frac{5}{2}$$

(5) $\frac{1}{2}x^2 - 3x + 2 = 0$

$$x^2 - 6x + 4 = 0 \ ; \ x = \frac{3 \pm \sqrt{9-4}}{1} = 3 \pm \sqrt{5}$$

(6) $\frac{1}{6}x^2 - 0.5x + \frac{1}{4} = 0$

$$2x^2 - 6x + 3 = 0 \ ; x = \frac{3 \pm \sqrt{9-6}}{2} = \frac{3 \pm \sqrt{3}}{2}$$

(7) $(x+1)^2 = 3(x+2)$

$$x^2 + 2x + 1 - 3x - 6 = 0 \ ; \ x^2 - x - 5 = 0 \ ; x = \frac{1 \pm \sqrt{1+20}}{2 \cdot 1} = \frac{1 \pm \sqrt{21}}{2}$$

(8) $(x+2)^2 + 3(x+2) - 2 = 0$

Letting $A = x + 2$, $A^2 + 3A - 2 = 0$

$$A = \frac{-3 \pm \sqrt{9+8}}{2} = \frac{-3 \pm \sqrt{17}}{2} \ ; \ x + 2 = \frac{-3 \pm \sqrt{17}}{2} \ ; \ x = \frac{-4 - 3 \pm \sqrt{17}}{2} = \frac{-7 \pm \sqrt{17}}{2}$$

(9) $-\frac{(x-1)^2}{2} + x = 0.4(x+1)$

$$-5(x-1)^2 + 10x = 4(x+1) ; \ 5(x-1)^2 + 4(x+1) - 10x = 0 \ ; \ 5x^2 - 16x + 9 = 0$$

$$x = \frac{8 \pm \sqrt{64-45}}{5} = \frac{8 \pm \sqrt{19}}{5}$$

(10) $(x+3)(2x+6) = 5$

Substituting $A = x+3$, $A \cdot 2A - 5 = 0$; $2A^2 - 5 = 0$

$A = \frac{-0 \pm \sqrt{0+40}}{2 \cdot 2} = \frac{\pm\sqrt{40}}{4} = \frac{\pm 2\sqrt{10}}{4} = \frac{\pm\sqrt{10}}{2}$; $x + 3 = \frac{\pm\sqrt{10}}{2}$; $x = -3 \pm \frac{\sqrt{10}}{2}$

OR $(x+3)(2x+6) - 5 = 2x^2 + 12x + 13 = 0$

$x = \frac{-6 \pm \sqrt{36-26}}{2} = \frac{-6 \pm \sqrt{10}}{2} = -3 \pm \frac{\sqrt{10}}{2}$

#11 Find the value of the given expression.

(1) $a+b$ when $(2x+1)^2 = 3$ has solutions $x = a \pm b\sqrt{3}$.

$4x^2 + 4x - 2 = 0$; $2x^2 + 2x - 1 = 0$; $x = \frac{-1 \pm \sqrt{1+2}}{2} = \frac{-1 \pm \sqrt{3}}{2}$

$\therefore a = \frac{-1}{2}, b = \frac{1}{2}$ $\therefore a + b = 0$

(2) $a-b$ when $2x^2 - 8x + 1 = 0$ has solutions $x = \frac{a \pm 3\sqrt{b}}{6}$.

$2x^2 - 8x + 1 = 0$; $x = \frac{4 \pm \sqrt{16-2}}{2} = \frac{4 \pm \sqrt{14}}{2} = \frac{12 \pm 3\sqrt{14}}{6}$

$\therefore a = 12, b = 14$ $\therefore a - b = -2$

(3) ab when $ax^2 + 5x + 2 = 0$ has solutions $x = \frac{-5 \pm 2\sqrt{b}}{4}$.

$x = \frac{-5 \pm \sqrt{25-8a}}{2a}$ $\therefore \frac{-5}{2a} = \frac{-5}{4}$; $a = 2$

$\therefore \frac{\sqrt{25-8a}}{2a} = \frac{2\sqrt{b}}{4}$; $\frac{\sqrt{25-16}}{4} = \frac{2\sqrt{b}}{4}$; $\sqrt{9} = 2\sqrt{b}$; $9 = 4b$; $b = \frac{9}{4}$

$\therefore ab = \frac{9}{2}$

(4) $\frac{b}{a}$ when $3x^2 - 5x + 1 = 0$ has solutions $x = a \pm \sqrt{b}$.

$x = \frac{5 \pm \sqrt{25-12}}{6} = \frac{5 \pm \sqrt{13}}{6}$

$\therefore a = \frac{5}{6}$, $b = \frac{13}{36}$

$\therefore \frac{b}{a} = \frac{13}{36} \cdot \frac{6}{5} = \frac{13}{30}$

(5) $\frac{a+b}{ab}$ when $x^2 - 3x + 1 = 0$ has solutions $x = \frac{a \pm 2\sqrt{b}}{2}$.

$x = \frac{3 \pm \sqrt{9-4}}{2} = \frac{3 \pm \sqrt{5}}{2}$

$\therefore a = 3$, $4b = 5$; $b = \frac{5}{4}$

Since $a + b = 3 + \frac{5}{4} = \frac{17}{4}$ and $ab = 3 \cdot \frac{5}{4} = \frac{15}{4}$, $\frac{a+b}{ab} = \frac{17}{4} \cdot \frac{4}{15} = \frac{17}{15}$

(6) $\frac{a-b}{a^2-b^2}$ when $ax^2 + 3x - 3b = 0$ has solutions $x = -1 \pm \sqrt{5}$.

$$x = \frac{-3 \pm \sqrt{9+12ab}}{2a} = -1 \pm \sqrt{5} \quad \therefore \quad \frac{-3}{2a} = -1 \; ; \; a = \frac{3}{2}$$

$$\frac{\sqrt{9+12ab}}{2a} = \frac{\sqrt{9+18b}}{3} = \sqrt{\frac{9+18b}{9}} = \sqrt{1+2b} = \sqrt{5} \; ; 1 + 2b = 5 \; ; \; b = 2$$

$$\therefore \quad \frac{a-b}{a^2-b^2} = \frac{a-b}{(a+b)(a-b)} = \frac{1}{(a+b)} = \frac{1}{\frac{3}{2}+2} = \frac{2}{7}$$

#12 State the number and type of solutions for each quadratic equation.

(1) $x^2 + 2x - 3 = 0$

$D = 4 - 4 \cdot 1 \cdot (-3) = 4 + 12 = 16 > 0 \quad \therefore \; 2$ different real number solutions.

(2) $-x^2 + x - 5 = 0$

$D = 1 - 4 \cdot (-1) \cdot (-5) = 1 - 20 = -19 < 0 \quad \therefore \;$ No real number solution.

(3) $4x^2 - 4x + 1 = 0$

$D = 16 - 4 \cdot 4 \cdot 1 = 0 \quad \therefore \;$ A double solution.

(4) $kx^2 - (k+5)x + 1 = 0$

$D = (k+5)^2 - 4 \cdot k \cdot 1 = k^2 + 6k + 25 = (k+3)^2 - 9 + 25$

$= (k+3)^2 + 16 > 0 \; (\because (k+3)^2 \geq 0) \quad \therefore \; 2$ different real number solutions.

(5) $3x^2 - x - k^2 = 0$

$D = 1 + 12k^2$

Since $k^2 \geq 0$, $D = 1 + 12k^2 > 0 \quad \therefore \; 2$ different real number solutions.

(6) $x^2 - 4kx + 5k^2 + 1 = 0$

$D = 16k^2 - 4(5k^2 + 1) = -4k^2 - 4 = -4(k^2 + 1) < 0 \; (\because k^2 + 1 > 0)$

$\therefore \;$ No real number solution.

#13 Determine the value of a or the range of the values of a at which the following quadratic equation will have a given condition.

(1) $x^2 + 5x + a = x + 2$ will have no real number solution.

$x^2 + 4x + a - 2 = 0$

$D = 16 - 4(a-2) < 0 \quad \therefore \; 4(a-2) > 16 \; ; \; a - 2 > 4 \; ; \; a > 6$

(2) $(a+3)x^2 - 2ax + a - 1 = 0$ will have two different real number solutions.

$D = 4a^2 - 4(a+3)(a-1) > 0 \; ; \; -8a + 12 > 0 \; ; \; a < \frac{3}{2}$

Since $a + 3 \neq 0$, $a \neq -3 \quad \therefore \; a < -3$ or $-3 < a < \frac{3}{2}$

(3) $x^2 + 3ax - 2a + 3 = 0$ **will have only one real number solution.**

$D = (3a)^2 - 4 \cdot 1 \cdot (-2a + 3) = 0$; $9a^2 + 8a - 12 = 0$

$\therefore$ $a = \dfrac{-4 \pm \sqrt{16 + 9 \cdot 12}}{9} = \dfrac{-4 \pm 2\sqrt{31}}{9}$

(4) $x^2 + ax + a + 2 = 0$ **will have a double root and** $x^2 + 4ax + (2a - 1)^2 = 0$ **will have two different real number solutions.**

Since $x^2 + ax + a + 2 = 0$ has a double root, $D = a^2 - 4(a + 2) = 0$

$a^2 - 4a - 8 = 0$; $(a - 2)^2 - 4 - 8 = 0$; $(a - 2)^2 = 12$

$\therefore$ $a = 2 + 2\sqrt{3}$ or $a = 2 - 2\sqrt{3}$

Since $x^2 + 4ax + (2a - 1)^2 = 0$ has two different real number solutions,

$D = (4a)^2 - 4 \cdot 1 \cdot (2a - 1)^2 > 0$

$16a^2 - 4(4a^2 - 4a + 1) > 0$; $16a - 4 > 0$; $a > \frac{1}{4}$

Therefore, $a = 2 + 2\sqrt{3}$

(5) $2x^2 + (2a - 1)x + a^2 + \frac{1}{4} = 0$ **will have real number solutions.**

$D \geq 0$

$D = (2a - 1)^2 - 4 \cdot 2 \cdot (a^2 + \frac{1}{4}) \geq 0$; $4a^2 - 4a + 1 - 8a^2 - 2 \geq 0$; $-4a^2 - 4a - 1 \geq 0$

$4a^2 + 4a + 1 \leq 0$; $(2a + 1)^2 \leq 0$

Since $(2a + 1)^2 \geq 0$, $(2a + 1)^2 = 0$

$\therefore$ $2a + 1 = 0$; $a = -\frac{1}{2}$

(6) $(a - 1)x^2 + 2(a - 1)x + (a + 1) = 0$ **will have real number solutions.**

Since this equation is quadratic, $a - 1 \neq 0$; $a \neq 1$

To have real number solutions, $D \geq 0$

$D = 4(a - 1)^2 - 4 \cdot (a - 1) \cdot (a + 1) \geq 0$; $a^2 - 2a + 1 - (a^2 - 1) \geq 0$

$-2a + 2 \geq 0$; $2a \leq 2$; $a \leq 1$

Since $a \neq 1$, $a < 1$

(7) The quadratic equation $x^2 - 3x + 2a = 0$ **will have two different positive real number solutions.**

Since the equation has two different real number solutions, $D > 0$

$\therefore$ $D = 9 - 4 \cdot 1 \cdot 2a > 0$; $8a < 9$; $a < \frac{9}{8}$

Let α, β be the solutions. Then, $\alpha + \beta = 3$, $\alpha\beta = 2a$

Since α and β are positive, $\alpha\beta = 2a > 0$; $a > 0$

Therefore, $0 < a < \frac{9}{8}$

(8) The quadratic equation $ax^2 + 2x + 3 = 0$ will have two different negative real number solutions.

Since the equation has two different real number solutions, $D > 0$

$\therefore D = 4 - 4 \cdot a \cdot 3 > 0$; $4 > 12a$; $a < \dfrac{1}{3}$

Let α, β be the solutions. Then, $\alpha + \beta = -\dfrac{2}{a}$, $\alpha\beta = \dfrac{3}{a}$

Since α and β are negative, $\alpha + \beta = -\dfrac{2}{a} < 0$; $a > 0$ and $\alpha\beta = \dfrac{3}{a} > 0$; $a > 0$

Therefore, $0 < a < \dfrac{1}{3}$

(9) The quadratic equation $x^2 - 4x + 3a = 0$ will have two different real number solutions α and β with opposite signs.

$\alpha\beta = 3a < 0$; $a < 0$

$\therefore D = 16 - 4 \cdot 1 \cdot 3a > 0$; $16 > 12a$; $a < \dfrac{4}{3}$

Therefore, $a < 0$

(10) The quadratic equation $(a + 3)x^2 - 2ax + a - 1 = 0$ will have real number solutions.

Since the equation is quadratic, $a + 3 \neq 0$; $a \neq -3$

$D = (-2a)^2 - 4 \cdot (a + 3) \cdot (a - 1) \geq 0$; $4a^2 - 4(a^2 + 2a - 3) \geq 0$

$-8a + 12 \geq 0$; $a \leq \dfrac{3}{2}$

Since $a \neq -3$, $a < -3$ or $-3 < a \leq \dfrac{3}{2}$

(11) The quadratic equation $x^2 + (a^2 + a - 6)x - a - 1 = 0$ will have two different real number solutions α and β with $|\alpha| = |\beta|$.

Since $\alpha \neq \beta$ and $|\alpha| = |\beta|$, $\alpha\beta < 0$ and $\alpha + \beta = 0$

$\therefore -a - 1 < 0$ and $-(a^2 + a - 6) = 0$

$\therefore a > -1$ and $(a + 3)(a - 2) = 0$

$\therefore a > -1$ and $(a = -3$ or $a = 2$)

Therefore, $a = 2$

(12) The quadratic equation $x^2 + 2(3 - a)x + a^2 = 0$ will have two different real number solutions and $2x^2 - x + a = 0$ will have no real number solution.

Let D_1 be the discriminant of the equation $x^2 + 2(3 - a)x + a^2 = 0$.

Then $D_1 = 4(3 - a)^2 - 4a^2 > 0$ $\quad \therefore (3 - a)^2 - a^2 > 0$ $\quad \therefore -6a + 9 > 0$ $\quad \therefore a < \dfrac{3}{2}$

Let D_2 be the discriminant of the equation $2x^2 - x + a = 0$.

Then $D_2 = 1 - 8a < 0$ $\therefore a > \dfrac{1}{8}$

Therefore, $\dfrac{1}{8} < a < \dfrac{3}{2}$

(13) The quadratic equation $x^2 - (a+5)x - a - 1 = 0$ will have roots α and β which are integers.

By the roots-coefficients relationship, we have

$\alpha + \beta = a + 5$ $\cdots\cdots$ ①

$\alpha\beta = -a - 1$ $\cdots\cdots$ ②

① + ② : $\alpha + \beta + \alpha\beta = 4$

$\therefore (\alpha + 1)(\beta + 1) - 1 = 4$ $\therefore (\alpha + 1)(\beta + 1) = 5$

Since $\alpha + 1$ and $\beta + 1$ are integers,

$\alpha + 1$	1	5	-1	-5
$\beta + 1$	5	1	-5	-1

$\therefore$ It gives :

α	0	4	-2	-6
β	4	0	-6	-2

Since $a = \alpha + \beta - 5$, $a = -1$ or $a = -13$

#14 Find the value of the given expression.

(1) A quadratic equation $3x^2 + 5x - 2 = 0$ has two solutions α and β.

1) $\alpha + \beta$

$\alpha + \beta = -\dfrac{5}{3}$

2) $\alpha^2 + \beta^2$

$\alpha^2 + \beta^2 = (\alpha + \beta)^2 - 2\alpha\beta = \left(-\dfrac{5}{3}\right)^2 - 2\left(-\dfrac{2}{3}\right) = \dfrac{37}{9}$

3) $\alpha - \beta$

$(\alpha - \beta)^2 = (\alpha + \beta)^2 - 4\alpha\beta = \left(-\dfrac{5}{3}\right)^2 - 4\left(-\dfrac{2}{3}\right) = \dfrac{49}{9}$; $\alpha - \beta = \pm\dfrac{7}{3}$

4) $\alpha^2 - \beta^2$

$\alpha^2 - \beta^2 = (\alpha + \beta)(\alpha - \beta) = \begin{cases} -\dfrac{35}{9} & \text{when } \alpha - \beta = \dfrac{7}{3} \\ \dfrac{35}{9} & \text{when } \alpha - \beta = -\dfrac{7}{3} \end{cases}$

5) $\dfrac{1}{\alpha} + \dfrac{1}{\beta}$

$$\dfrac{1}{\alpha} + \dfrac{1}{\beta} = \dfrac{\alpha + \beta}{\alpha\beta} = \dfrac{-\frac{5}{3}}{-\frac{2}{3}} = \dfrac{5}{2}$$

(2) A quadratic equation $x^2 + 3kx + 2k^2 - 4k - 1 = 0$ has two solutions α and β.

Find the value of the given expressions in terms of k

1) $\alpha + \beta$; $\alpha + \beta = -\dfrac{3k}{1} = -3k$

2) $\alpha\beta$; $\alpha\beta = \dfrac{2k^2 - 4k - 1}{1} = 2k^2 - 4k - 1$

3) $\alpha^2 + \beta^2$; $\alpha^2 + \beta^2 = (\alpha + \beta)^2 - 2\alpha\beta = 9k^2 - 4k^2 + 8k + 2 = 5k^2 + 8k + 2$

4) $(\alpha - \beta)^2$; $(\alpha - \beta)^2 = (\alpha + \beta)^2 - 4\alpha\beta = 9k^2 - 8k^2 + 16k + 4 = k^2 + 16k + 4$

5) $\dfrac{\beta}{\alpha} + \dfrac{\alpha}{\beta}$; $\dfrac{\beta}{\alpha} + \dfrac{\alpha}{\beta} = \dfrac{\alpha^2 + \beta^2}{\alpha\beta} = \dfrac{5k^2 + 8k + 2}{2k^2 - 4k - 1}$

(3) $x^2 + ax + b = 0$ has two solutions -2 and -3.

Find the value of $\alpha^2 + \beta^2$ for $x^2 - bx - a = 0$ which has solutions α and β.

$-2 + (-3) = -\dfrac{a}{1}$; $a = 5$ and $-2 \cdot (-3) = \dfrac{b}{1}$; $b = 6$

(OR $(x + 2)(x + 3) = 0$; $x^2 + 5x + 6 = 0$ $\therefore a = 5,\ b = 6$)

$\therefore\ \ x^2 - bx - a = x^2 - 6x - 5 = 0$; $\alpha + \beta = 6,\ \alpha\beta = -5$

$\therefore\ \ \alpha^2 + \beta^2 = (\alpha + \beta)^2 - 2\alpha\beta = 6^2 + 10 = 46$

(4) $x^2 - x - 1 = 0$ has two solutions α and β. Find the value of $(\alpha^4 + 1) + (\beta^4 + 1)$.

Since α and β are roots of $x^2 - x - 1 = 0$, $\alpha^2 - \alpha - 1 = 0$ and $\beta^2 - \beta - 1 = 0$

$\therefore\ \alpha^2 = \alpha + 1,\ \ \beta^2 = \beta + 1$

$\therefore\ \alpha^4 = (\alpha + 1)^2 = \alpha^2 + 2\alpha + 1 = (\alpha + 1) + 2\alpha + 1 = 3\alpha + 2$,

$\quad \beta^4 = (\beta + 1)^2 = \beta^2 + 2\beta + 1 = (\beta + 1) + 2\beta + 1 = 3\beta + 2$

Note that $\alpha + \beta = 1$, $\alpha\beta = -1$ by roots-coefficients relationship.

$\therefore\ \ (\alpha^4 + 1) + (\beta^4 + 1) = (3\alpha + 3)(3\beta + 3) = 9(\alpha + 1)(\beta + 1) = 9(\alpha\beta + \alpha + \beta + 1)$

$$= 9(-1 + 1 + 1) = 9$$

(5) $(1 - i)x^2 + 2(1 + i)x + 3(1 - i) = 0$ has two solutions α and β.

Find the value of $(\alpha - \beta)^2$.

By roots-coefficients relationship, $\alpha + \beta = \frac{-2(1+i)}{1-i}$ and $\alpha\beta = \frac{3(1-i)}{1-i}$

So $\alpha + \beta = \frac{-2(1+i)}{1-i} = \frac{-2(1+i)^2}{(1-i)(1+i)} = \frac{-2(1+2i-1)}{1+1} = -2i$ and $\alpha\beta = 3$

$\therefore \ (\alpha - \beta)^2 = (\alpha + \beta)^2 - 4\alpha\beta = (-2i)^2 - 4\cdot 3 = -4 - 12 = -16$

(6) When $x^2 - |x - 1| - 1 = 0$ and $x^2 + ax + b = 0$ have the same solution,

find the value of $a^2 + b^2$.

Case 1. $x < 1 \ \Rightarrow \ x^2 - |x - 1| - 1 = x^2 + (x - 1) - 1$

$$= x^2 + x - 2 = (x + 2)(x - 1) = 0$$

$\therefore \ x = -2$ or $x = 1 \quad$ Since $x < 1$, $x = -2$

Case 2. $x \geq 1 \ \Rightarrow \ x^2 - |x - 1| - 1 = x^2 - (x - 1) - 1$

$$= x^2 - x = x(x - 1) = 0$$

$\therefore \ x = 0$ or $x = 1 \quad$ Since $x \geq 1$, $x = 1$

Thus $x = -2$ and $x = 1$ are solutions of $x^2 + ax + b = 0$.

Thus $4 - 2a + b = 0$ and $1 + a + b = 0 \quad \therefore a = 1, \ b = -2$

Therefore, $a^2 + b^2 = 1 + 4 = 5$.

(7) When $x^2 - 4xy + ay^2 + 2y - 1$ is factorized as a product of two linear factors,

find the value of a.

Consider $x^2 - 4xy + ay^2 + 2y - 1$

$$= x^2 - 4yx + (ay^2 + 2y - 1) = 0 : \text{Desending order of powers for } x.$$

Then $x = \frac{2y \pm \sqrt{(-2y)^2 - (ay^2 + 2y - 1)}}{1} = 2y \pm \sqrt{(4 - a)y^2 - 2y + 1}$

Since the given equation is in the form of $(x + \alpha)(x + \beta)$,

the discriminant $D = (4 - a)y^2 - 2y + 1$ must be in the form of a perfect square for y.

Thus, consider $(4 - a)y^2 - 2y + 1 = 0$

$\Rightarrow$ The discriminant $D' = (-2)^2 - 4(4 - a) = -12 + 4a = 0$

Therefore, $a = 3$

(8) For real numbers a and b, $x^2 + ax + b = 0$ has a solution $\frac{1}{1+i}$. Find the value of ab.

$\frac{1}{1+i} = \frac{1-i}{(1+i)(1-i)} = \frac{1-i}{1+1} = \frac{1}{2} - \frac{1}{2}i$

Since $\frac{1}{2} - \frac{1}{2}i$ is a solution of the given equation, $\frac{1}{2} + \frac{1}{2}i$ is the other solution.

$\therefore \ \left(\frac{1}{2} - \frac{1}{2}i\right) + \left(\frac{1}{2} + \frac{1}{2}i\right) = -a$ and $\left(\frac{1}{2} - \frac{1}{2}i\right)\left(\frac{1}{2} + \frac{1}{2}i\right) = b$

$\therefore$ $a = -1$ and $b = \dfrac{1}{2}$

Therefore, $ab = -\dfrac{1}{2}$

(9) $x^2 + 2(3 + i)x + (a - 4i) = 0$ **has real number solutions. Find the value of** a.

Let α be the real number solution. Then, $\alpha^2 + 2(3 + i)\alpha + (a - 4i) = 0$.

$\therefore$ $(\alpha^2 + 6\alpha + a) + (2\alpha - 4)i = 0$

$\therefore$ $\alpha^2 + 6\alpha + a = 0$ and $2\alpha - 4 = 0$

$\therefore$ $\alpha = 2$, $4 + 12 + a = 0$

Therefore, $a = -16$

(10) $x^2 - 2x + 2 = 0$ **has two roots** α **and** β. **Find the value of** $\dfrac{\beta}{\alpha^2 - 3\alpha + 2} + \dfrac{\alpha}{\beta^2 - 3\beta + 2}$.

Since α and β are roots, $\alpha^2 - 2\alpha + 2 = 0$ and $\beta^2 - 2\beta + 2 = 0$

By roots-coefficients relationship, $\alpha + \beta = 2$ and $\alpha\beta = 2$

$\therefore$ $\dfrac{\beta}{\alpha^2 - 3\alpha + 2} + \dfrac{\alpha}{\beta^2 - 3\beta + 2} = \dfrac{\beta}{(\alpha^2 - 2\alpha + 2) - \alpha} + \dfrac{\alpha}{(\beta^2 - 2\beta + 2) - \beta}$

$\qquad = \dfrac{\beta}{-\alpha} + \dfrac{\alpha}{-\beta} = -\dfrac{\alpha^2 + \beta^2}{\alpha\beta} = -\dfrac{(\alpha + \beta)^2 - 2\alpha\beta}{\alpha\beta} = -\dfrac{4 - 4}{2} = 0$

(11) $ax^2 - 2x + b = 0$, $a \neq 0$, **has a solution** $1 + 2i$. **Find the value of** $a + b$.

Since $1 + 2i$ is a root of the quadratic equation, $a(1 + 2i)^2 - 2(1 + 2i) + b = 0$

$\therefore$ $a(1 + 4i - 4) - 2 - 4i + b = 0$

$\therefore$ $(-3a - 2 + b) + (4a - 4)i = 0$

$\therefore$ $-3a - 2 + b = 0$ and $4a - 4 = 0$ $\quad \therefore$ $a = 1$, $b = 5$ $\quad$ Therefore, $a + b = 6$

(12) **For rational numbers** a **and** b, $x^2 - ax + b = 0$ **has a solution** $\sqrt{4 - 2\sqrt{3}}$.
Find the value of $a^2 - b^2$.

$\sqrt{4 - 2\sqrt{3}} = \sqrt{(3 + 1) - 2\sqrt{(3 \cdot 1)}} = \sqrt{3} - \sqrt{1} = \sqrt{3} - 1 = -1 + \sqrt{3}$

Since a and b are rational numbers, $-1 - \sqrt{3}$ is the other solution of $x^2 - ax + b = 0$.

$\therefore$ $\left(-1 + \sqrt{3}\right) + \left(-1 - \sqrt{3}\right) = a$ and $\left(-1 + \sqrt{3}\right)\left(-1 - \sqrt{3}\right) = b$

$\therefore$ $a = -2$ and $b = -2$

Therefore, $a^2 - b^2 = 4 - 4 = 0$

#15 Find the real number solution for the quadratic equation $ax^2 + (b-1)x + 4 = 0$,

(1) When the quadratic equation $2x^2 + (a-1)x + b = 0$ has two solutions $\frac{1}{2}$ and $\frac{1}{3}$.

$\frac{1}{2} + \frac{1}{3} = -\frac{a-1}{2}$; $\frac{5}{6} = -\frac{3(a-1)}{6}$; $a - 1 = -\frac{5}{3}$; $a = -\frac{2}{3}$

$\frac{1}{2} \cdot \frac{1}{3} = \frac{b}{2}$; $b = \frac{1}{3}$

(OR $2\left(x - \frac{1}{2}\right)\left(x - \frac{1}{3}\right) = 0$; $2\left(x^2 - \frac{5}{6}x + \frac{1}{6}\right) = 0$

$2x^2 - \frac{5}{3}x + \frac{1}{3} = 0$; $a - 1 = -\frac{5}{3}$; $a = -\frac{2}{3}, b = \frac{1}{3}$)

$\therefore\ ax^2 + (b-1)x + 4 = -\frac{2}{3}x^2 - \frac{2}{3}x + 4 = 0$; $x^2 + x - 6 = 0$; $(x+3)(x-2) = 0$

$\therefore\ x = -3$ or $x = 2$

(2) When the quadratic equation $3ax^2 + 8bx + 3 = 0$ has a double root -2.

$-2 + (-2) = -\frac{8b}{3a}$, $-2 \cdot (-2) = \frac{3}{3a}$

$\therefore\ -4 = -\frac{8b}{3a}$, $4 = \frac{1}{a}$; $a = \frac{1}{4}, b = \frac{3}{8}$

(OR $3a(x+2)^2 = 0$; $3a(x^2 + 4x + 4) = 0$; $3ax^2 + 12ax + 12a = 0$

$8b = 12a$, $3 = 12a$; $a = \frac{1}{4}, b = \frac{3}{8}$)

$\therefore\ ax^2 + (b-1)x + 4 = \frac{1}{4}x^2 - \frac{5}{8}x + 4 = 0$; $2x^2 - 5x + 32 = 0$

Since $D = 25 - 4 \cdot 2 \cdot 32 < 0$, no real number solution.

(3) When the quadratic equation $ax^2 + 3ax - 4 = 0$ has two solutions, b and $b+1$.

$b + (b+1) = -\frac{3a}{a} = -3$; $2b = -4$; $b = -2$

$b(b+1) = \frac{-4}{a}$; $-2(-2+1) = \frac{-4}{a}$; $2a = -4$; $a = -2$

(OR $a(x-b)\big(x - (b+1)\big) = 0$; $a\big(x^2 - (b+b+1)x + b(b+1)\big) = 0$

; $ax^2 - (2b+1)ax + ab(b+1) = 0$ $\therefore\ -(2b+1) = 3, ab(b+1) = -4$

$-2b = 4$; $b = -2$ and $ab(b+1) = -4$; $a = -2$)

$\therefore\ ax^2 + (b-1)x + 4 = -2x^2 - 3x + 4$

$\therefore\ x = \frac{3 \pm \sqrt{9 - 4 \cdot (-2) \cdot 4}}{2 \cdot (-2)} = \frac{3 \pm \sqrt{41}}{-4}$

(4) When the quadratic equation $ax^2 + 3x + b = 0$ has two solutions, α and β, which satisfy the conditions $\alpha + \beta = -2$ and $\alpha\beta = 4$.

$\alpha + \beta = -\dfrac{3}{a} = -2$; $a = \dfrac{3}{2}$

$\alpha\beta = \dfrac{b}{a} = 4$; $b = 4a = 4 \cdot \dfrac{3}{2} = 6$

(OR $ax^2 + 3x + b = a(x^2 - (\alpha + \beta)x + \alpha\beta) = a(x^2 - (-2)x + 4) = 0$

$3 = -a(-2)$, $b = a \cdot (4)$; $a = \dfrac{3}{2}, b = 6$)

$\therefore ax^2 + (b - 1)x + 4 = \dfrac{3}{2}x^2 + 5x + 4 = 0$

$3x^2 + 10x + 8 = 0$ $\quad \therefore x = \dfrac{-5 \pm \sqrt{25 - 24}}{3} = \dfrac{-5 \pm 1}{3}$ $\quad \therefore x = -\dfrac{4}{3}$ or $x = -2$

(5) When the quadratic equation $x^2 + ax + 3 = 0$ has two different solutions. The one of the solutions is $x = -2 + 3\sqrt{b}$.

The other solution is $x = -2 - 3\sqrt{b}$.

$\therefore -\dfrac{a}{1} = \left(-2 + 3\sqrt{b}\right) + \left(-2 - 3\sqrt{b}\right) = -4$; $a = 4$

and $\dfrac{3}{1} = \left(-2 + 3\sqrt{b}\right) \cdot \left(-2 - 3\sqrt{b}\right) = 4 - 9b$; $b = \dfrac{1}{9}$

(OR $\left(x - (-2 + 3\sqrt{b})\right)\left(x - (-2 - 3\sqrt{b})\right) = 0$

$x^2 - \left(-2 + 3\sqrt{b} - 2 - 3\sqrt{b}\right)x + \left(-2 + 3\sqrt{b}\right)\left(-2 - 3\sqrt{b}\right) = 0$

$x^2 + 4x + 4 - 9b = 0$ $\therefore a = 4$, $3 = 4 - 9b$ $\therefore a = 4$, $b = \dfrac{1}{9}$)

Therefore, $ax^2 + (b - 1)x + 4 = 4x^2 - \dfrac{8}{9}x + 4 = 0$; $36x^2 - 8x + 36 = 0$

$9x^2 - 2x + 9 = 0$.

Since $D = 4 - 4 \cdot 9 \cdot 9 < 0$, no real number solution.

#16 The following quadratic equations have only one solution.

Find the solution (a double root) for each equation.

(1) $x^2 + kx + 2k - 3 = 0$

$D = k^2 - 4(2k - 3) = 0$; $k^2 - 8k + 12 = 0$; $(k - 6)(k - 2) = 0$

$\therefore k = 6$ or $k = 2$

If $k = 6$, then $x^2 + kx + 2k - 3 = x^2 + 6x + 9 = 0$

$\therefore (x + 3)^2 = 0$; $x = -3$ (A double root)

If $k = 2$, then $x^2 + kx + 2k - 3 = x^2 + 2x + 1 = 0$

$\therefore (x + 1)^2 = 0$; $x = -1$ (A double root)

(OR by formula, $x = \frac{-b \pm \sqrt{b^2 - 4ac}}{2a}$

Since $D = 0$, $b^2 - 4ac = 0$ $\therefore$ $x = \frac{-b}{2a}$

$\therefore x = \frac{-b}{2a} = \frac{-6}{2 \cdot 1} = -3$ (when $k = 6$) or $x = \frac{-b}{2a} = \frac{-2}{2 \cdot 1} = -1$ (when $k = 2$))

(2) $(k+2)x^2 - 2kx + k + 1 = 0$

$D = 4k^2 - 4(k+2)(k+1) = 0$; $4k^2 - 4k^2 - 12k - 8 = 0$ $\therefore k = -\frac{2}{3}$

$\therefore (k+2)x^2 - 2kx + k + 1 = \frac{4}{3}x^2 + \frac{4}{3}x + \frac{1}{3} = 0$; $4x^2 + 4x + 1 = 0$; $(2x+1)^2 = 0$

$\therefore x = -\frac{1}{2}$ (A double root)

(OR by formula, $x = \frac{-b}{2a} = \frac{-\frac{4}{3}}{2 \cdot \frac{4}{3}} = -\frac{1}{2}$)

(3) $x^2 + (k+2)x + k^2 - k + 2 = 0$

$D = (k+2)^2 - 4(k^2 - k + 2) = 0$; $k^2 + 4k + 4 - 4k^2 + 4k - 8 = 0$

$-3k^2 + 8k - 4 = 0$; $3k^2 - 8k + 4 = 0$; $(k-2)(3k-2) = 0$

$\therefore k = 2$ or $k = \frac{2}{3}$

If $k = 2$, then $x^2 + (k+2)x + k^2 - k + 2 = x^2 + 4x + 4 = 0$

$\therefore (x+2)^2 = 0$ $\therefore x = -2$ (A double root)

If $k = \frac{2}{3}$, then $x^2 + \frac{8}{3}x + \frac{4}{9} - \frac{2}{3} + 2 = 0$; $x^2 + \frac{8}{3}x + \frac{16}{9} = 0$; $9x^2 + 24x + 16 = 0$

$\therefore x = \frac{-12 \pm \sqrt{144 - 9 \cdot 16}}{9} = \frac{-12}{9} = -\frac{4}{3}$ (A double root)

#17 For $a \neq b$, a quadratic equation $x^2 + ax + b + 5 = 0$ has two solutions α, β and a quadratic equation $x^2 + bx + a + 5 = 0$ has two solutions α, γ.
Find the value of $\alpha + \beta + \gamma$.

Since α is the common solution of two quadratic equations, we have

$\alpha^2 + a\alpha + b + 5 = 0$ $\cdots\cdots$ ①

$\alpha^2 + b\alpha + a + 5 = 0$ $\cdots\cdots$ ②

① $-$ ② : $\alpha(a - b) + (b - a) = 0$ $\therefore (a - b)(\alpha - 1) = 0$

Since $a \neq b$, $\alpha = 1$

Since $\alpha = 1$ and β are solutions of $x^2 + ax + b + 5 = 0$, $1 + \beta = -a$ and $1 \cdot \beta = b + 5$

Since $\alpha = 1$ and γ are solutions of $x^2 + bx + a + 5$, $1 + \gamma = -b$ and $1 \cdot \gamma = a + 5$

Thus, $\beta + \gamma = (-a - 1) + (a + 5) = 4$

Therefore, $\alpha + \beta + \gamma = 1 + 4 = 5$

#18 Create a quadratic equation with leading coefficient 1.

(1) $x^2 + 2x - 3 = 0$ has two roots α and β.

Find a quadratic equation that has the roots $\alpha + \beta$ and $\alpha\beta$.

$\alpha + \beta = -2,\ \alpha\beta = -3$

$\therefore\ x^2 - (\alpha + \beta + \alpha\beta)x + (\alpha + \beta)\cdot(\alpha\beta) = 0 \quad \therefore\ x^2 + 5x + 6 = 0$

(OR $(x - (\alpha + \beta))(x - \alpha\beta) = 0$; $(x + 2)(x + 3) = 0 \quad \therefore\ x^2 + 5x + 6 = 0$)

(2) $x^2 + 2x + 3 = 0$ has two roots α and β.

Find a quadratic equation that has the roots $\dfrac{1}{\alpha}$ and $\dfrac{1}{\beta}$.

$\alpha + \beta = -2,\ \alpha\beta = 3$

Since $\dfrac{1}{\alpha} + \dfrac{1}{\beta} = \dfrac{\alpha + \beta}{\alpha\beta} = \dfrac{-2}{3}$ and $\dfrac{1}{\alpha}\cdot\dfrac{1}{\beta} = \dfrac{1}{\alpha\beta} = \dfrac{1}{3}$, $x^2 - \left(\dfrac{-2}{3}\right)x + \dfrac{1}{3} = 0$

$\therefore\ x^2 + \dfrac{2}{3}x + \dfrac{1}{3} = 0$

(3) Find a quadratic equation that has the roots $\sqrt{2 + \sqrt{3}}$ and $\sqrt{2 - \sqrt{3}}$.

$\sqrt{2 + \sqrt{3}} + \sqrt{2 - \sqrt{3}} = \sqrt{\dfrac{4 + 2\sqrt{3}}{2}} + \sqrt{\dfrac{4 - 2\sqrt{3}}{2}} = \dfrac{\sqrt{4 + 2\sqrt{3}}}{\sqrt{2}} + \dfrac{\sqrt{4 - 2\sqrt{3}}}{\sqrt{2}}$

$= \dfrac{\sqrt{3} + 1}{\sqrt{2}} + \dfrac{\sqrt{3} - 1}{\sqrt{2}} = \dfrac{\sqrt{6} + \sqrt{2}}{2} + \dfrac{\sqrt{6} - \sqrt{2}}{2} = \dfrac{2\sqrt{6}}{2} = \sqrt{6}$

$\sqrt{2 + \sqrt{3}}\cdot\sqrt{2 - \sqrt{3}} = \sqrt{(2 + \sqrt{3})(2 - \sqrt{3})} = \sqrt{4 - 3} = 1$

$\therefore\ x^2 - \sqrt{6}x + 1 = 0$

#19 Solve each quadratic equation.

(1) $x^2 + |x| + x - 1 = 0$

Case 1. $x \geq 0 \quad \Rightarrow \quad |x| = x$

$x^2 + |x| + x - 1 = x^2 + x + x - 1 = x^2 + 2x - 1 = 0$

$\therefore\ x = \dfrac{-1 \pm \sqrt{1+1}}{1} = -1 \pm \sqrt{2}$

Since $x \geq 0$, $x = -1 + \sqrt{2}$

Case 2. $x < 0 \quad \Rightarrow \quad |x| = -x$

$x^2 + |x| + x - 1 = x^2 - x + x - 1 = x^2 - 1 = 0$

$\therefore\ x^2 = 1 \quad \therefore x = \pm 1$

Since $x < 0$, $x = -1$

Therefore, $x = -1 + \sqrt{2}$ or $x = -1$

(2) $|x| - 1 = \sqrt{(2x-5)^2}$

$|x| - 1 = \sqrt{(2x-5)^2} = |2x - 5|$

Case 1. $x < 0$ $\Rightarrow$ $-x - 1 = -(2x - 5)$ $\therefore$ $x = 6$

Since $x < 0$, $x = 6$ cannot be a solution.

Case 2. $0 \le x < \frac{5}{2}$ $\Rightarrow$ $x - 1 = -(2x - 5)$ $\therefore$ $x = 2$

Case 3. $x \ge \frac{5}{2}$ $\Rightarrow$ $x - 1 = 2x - 5$ $\therefore$ $x = 4$

Therefore, $x = 2$ or $x = 4$

#20 **For rational numbers a, b, and c, $ax^2 + bx + c = 0$ has a solution which is the decimal part of $\sqrt{4 + \sqrt{12}}$. When α and β are solutions of $cx^2 + bx + a = 0$, find the value of $|\alpha - \beta|$.**

$\sqrt{4 + \sqrt{12}} = \sqrt{4 + 2\sqrt{3}} = \sqrt{3} + 1$

Since $1 < \sqrt{3} < 2$, $2 < \sqrt{3} + 1 < 3$ $\therefore$ $\sqrt{3} + 1 = 2.\times\times\times$

$\therefore$ The decimal part is $(\sqrt{3} + 1) - 2 = \sqrt{3} - 1$.

Since the coefficients of $ax^2 + bx + c = 0$ are rational numbers and $-1 + \sqrt{3}$ is one solution, $-1 - \sqrt{3}$ is the other solution.

$\therefore$ $(-1 + \sqrt{3}) + (-1 - \sqrt{3}) = -\frac{b}{a}$ $\therefore$ $\frac{b}{a} = 2$

$(-1 + \sqrt{3})(-1 - \sqrt{3}) = \frac{c}{a}$ $\therefore$ $\frac{c}{a} = -2$

Since $\alpha + \beta = -\frac{b}{c} = -\frac{\frac{b}{a}}{\frac{c}{a}} = -\frac{2}{-2} = 1$ and $\alpha\beta = \frac{a}{c} = \frac{\frac{a}{a}}{\frac{c}{a}} = \frac{1}{-2}$,

$|\alpha - \beta| = \sqrt{(\alpha - \beta)^2} = \sqrt{(\alpha + \beta)^2 - 4\alpha\beta} = \sqrt{1 + 2} = \sqrt{3}$

#21 Solve each equation.

(1) $x^3 - x^2 + x - 1 = 0$

$\Rightarrow$ $x^2(x - 1) + x - 1 = 0$ $\Rightarrow$ $(x - 1)(x^2 + 1) = 0$ $\Rightarrow$ $x = 1$ or $x^2 = -1$

Therefore, solutions are $x = 1$, $x = i$, and $x = -i$

(2) $2x^3 - 5x^2 + 5x - 2 = 0$

$\Rightarrow$ $2(x^3 - 1) - 5x(x - 1) = 0$ $\Rightarrow$ $2(x - 1)(x^2 + x + 1) - 5x(x - 1) = 0$

$\Rightarrow$ $(x - 1)(2x^2 + 2x + 2 - 5x) = 0$ $\Rightarrow$ $(x - 1)(2x^2 - 3x + 2) = 0$

$2x^2 - 3x + 2 \Rightarrow x = \dfrac{-(-3)\pm\sqrt{(-3)^2-4(2)(2)}}{2(2)} = \dfrac{3\pm\sqrt{-7}}{4} = \dfrac{3\pm\sqrt{7}\,i}{4}$

Therefore, solutions are $x = 1$, $x = \dfrac{3+\sqrt{7}\,i}{4}$, and $x = \dfrac{3-\sqrt{7}\,i}{4}$

(3) $x^3 - 4x^2 - 4x - 5 = 0$

Let $P(x) = x^3 - 4x^2 - 4x - 5$.

Since $P(5) = 5^3 - 4(5^2) - 4(5) - 5 = 5^2(5-4) - 25 = 0$, $P(x)$ has a factor $x - 5$.

$$
\begin{array}{c|cccc}
5 & 1 & -4 & -4 & -5 \\
  &   & 5 & 5 & 5 \\
\hline
  & 1 & 1 & 1 & 0
\end{array}
$$

$\therefore\ P(x) = (x - 5)(x^2 + x + 1)$

$x^2 + x + 1 = 0 \Rightarrow x = \dfrac{-1\pm\sqrt{1-4}}{2} = \dfrac{-1\pm\sqrt{-3}}{2} = \dfrac{-1\pm\sqrt{3}\,i}{2}$

Therefore, solutions are $x = 5$, $x = \dfrac{-1+\sqrt{3}\,i}{2}$, and $x = \dfrac{-1-\sqrt{3}\,i}{2}$

(4) $x^4 + x^3 - 2x^2 - x + 1 = 0$

Let $P(x) = x^4 + x^3 - 2x^2 - x + 1$

Since $P(1) = 1 + 1 - 2 - 1 + 1 = 0$, $P(x)$ has a factor $x - 1$.

$$
\begin{array}{c|ccccc}
1 & 1 & 1 & -2 & -1 & 1 \\
  &   & 1 & 2 & 0 & -1 \\
\hline
  & 1 & 2 & 0 & -1 & 0
\end{array}
$$

$\therefore\ P(x) = (x - 1)(x^3 + 2x^2 - 1)$

Let $f(x) = x^3 + 2x^2 - 1$

Since $f(-1) = -1 + 2 - 1 = 0$, $f(x)$ has a factor $x + 1$.

$$
\begin{array}{c|cccc}
-1 & 1 & 2 & 0 & -1 \\
   &   & -1 & -1 & 1 \\
\hline
   & 1 & 1 & -1 & 0
\end{array}
$$

$\therefore\ f(x) = (x + 1)(x^2 + x - 1)$

$\therefore\ P(x) = (x - 1)(x + 1)(x^2 + x - 1)$

$x^2 + x - 1 = 0 \Rightarrow x = \dfrac{-1\pm\sqrt{1+4}}{2} = \dfrac{-1\pm\sqrt{5}}{2}$

Therefore, solutions are $x = 1$, $x = -1$, $x = \dfrac{-1+\sqrt{5}}{2}$, and $x = \dfrac{-1-\sqrt{5}}{2}$

(5) $2x^4 - 3x^3 - x^2 - 3x + 2 = 0$

Since $x \neq 0$, divide both sides by x^2.

Then $2x^2 - 3x - 1 - \dfrac{3}{x} + \dfrac{2}{x^2} = 0$

$\Rightarrow 2\left(x^2 + \dfrac{1}{x^2}\right) - 3\left(x + \dfrac{1}{x}\right) - 1 = 0$

$\Rightarrow 2\left(\left(x + \dfrac{1}{x}\right)^2 - 2\right) - 3\left(x + \dfrac{1}{x}\right) - 1 = 0$

$\Rightarrow 2(X^2 - 2) - 3X - 1 = 0$ Letting $X = x + \dfrac{1}{x}$

$\Rightarrow 2X^2 - 3X - 5 = 0$

$\Rightarrow (2X - 5)(X + 1) = 0$

$\Rightarrow X = -1$ or $X = \dfrac{5}{2}$

$\therefore x + \dfrac{1}{x} = -1$ or $x + \dfrac{1}{x} = \dfrac{5}{2}$

$x + \dfrac{1}{x} = -1 \Rightarrow x^2 + x + 1 = 0 \quad \therefore x = \dfrac{-1 \pm \sqrt{1-4}}{2} = \dfrac{-1 \pm \sqrt{3}\,i}{2}$

$x + \dfrac{1}{x} = \dfrac{5}{2} \Rightarrow 2x^2 - 5x + 2 = 0 \quad \therefore (2x - 1)(x - 2) = 0 \quad \therefore x = 2$ or $x = \dfrac{1}{2}$

Therefore, solutions are $x = 2$, $x = \dfrac{1}{2}$, $x = \dfrac{-1 + \sqrt{3}\,i}{2}$, and $x = \dfrac{-1 - \sqrt{3}\,i}{2}$

(6) $(x + 1)(x + 2)(x + 3)(x + 4) - 3 = 0$

$\Rightarrow \{(x + 1)(x + 4)\}\{(x + 2)(x + 3)\} - 3 = 0$

$\Rightarrow \{x^2 + 5x + 4\}\{x^2 + 5x + 6\} - 3 = 0$

$\Rightarrow (X + 4)(X + 6) - 3 = 0$ Letting $X = x^2 + 5x$

$\Rightarrow X^2 + 10X + 21 = 0$

$\Rightarrow (X + 3)(X + 7) = 0$

$\Rightarrow X = -3$ or $X = -7$

$\therefore x^2 + 5x + 3 = 0$ or $x^2 + 5x + 7 = 0$

$\therefore x = \dfrac{-5 \pm \sqrt{25-12}}{2} = \dfrac{-5 \pm \sqrt{13}}{2}$ or $x = \dfrac{-5 \pm \sqrt{25-28}}{2} = \dfrac{-5 \pm \sqrt{3}\,i}{2}$

Therefore, solutions are $x = \dfrac{-5 + \sqrt{13}}{2}$, $x = \dfrac{-5 - \sqrt{13}}{2}$, $x = \dfrac{-5 + \sqrt{3}\,i}{2}$, and $x = \dfrac{-5 - \sqrt{3}\,i}{2}$

(7) $x^3 + ax^2 - b = 0$ **with a root** $1 + i$

Substituting $x = 1 + i$, $(1 + i)^3 + a(1 + i)^2 - b = 0$

∴ $(1 + 3i - 3 - i) + a(1 + 2i - 1) - b = 0$

∴ $(-2 - b) + 2(1 + a)i = 0$

Since a and b are real numbers, $-2 - b = 0$ and $1 + a = 0$

∴ $a = -1$, $b = -2$

∴ The given equation is $x^3 - x^2 + 2 = 0$

Let $P(x) = x^3 - x^2 + 2$

Since $P(-1) = -1 - 1 + 2 = 0$, $P(x)$ has a factor $x + 1$.

$$
\begin{array}{r|rrrr}
-1 & 1 & -1 & 0 & 2 \\
   &   & -1 & 2 & -2 \\
\hline
   & 1 & -2 & 2 & 0
\end{array}
$$

∴ $P(x) = (x + 1)(x^2 - 2x + 2)$

$x^2 - 2x + 2 = 0 \Rightarrow x = \frac{1 \pm \sqrt{1-2}}{1} = 1 \pm \sqrt{-1} = 1 \pm i$

Therefore, solutions are $x = -1$, $x = 1 + i$, and $x = 1 - i$

#22 Find the value of each expression.

(1) A fourth-degree polynomial equation $x^4 - 2x^2 + 3x - 2 = 0$ **has a complex number**

solution α. **Find the value of** $\alpha + \frac{1}{\alpha}$.

Let $P(x) = x^4 - 2x^2 + 3x - 2$

Since $P(1) = 1 - 2 + 3 - 2 = 0$ and $P(-2) = 16 - 8 - 6 - 2 = 0$,

$P(x)$ has factors $x - 1$ and $x + 2$.

$$
\begin{array}{r|rrrrr}
1 & 1 & 0 & -2 & 3 & -2 \\
  &   & 1 & 1 & -1 & 2 \\
\hline
-2 & 1 & 1 & -1 & 2 & 0 \\
  &   & -2 & 2 & -2 & \\
\hline
  & 1 & -1 & 1 & 0 &
\end{array}
$$

∴ $P(x) = (x - 1)(x + 2)(x^2 - x + 1)$

Since α is a complex number solution, $\alpha^2 - \alpha + 1 = 0$.

Since $\alpha \neq 0$, divide both sides of the equation by α.

Then, $\alpha - 1 + \dfrac{1}{\alpha} = 0$

Therefore, $\alpha + \dfrac{1}{\alpha} = 1$

(2) A fourth-degree polynomial equation $x^4 - x^3 - 4x^2 - x + 1 = 0$ has a positive solution α. Find the value of $\alpha^3 + \dfrac{1}{\alpha^3}$.

Since α is a positive solution, $\alpha^4 - \alpha^3 - 4\alpha^2 - \alpha + 1 = 0$

Since $\alpha \neq 0$, divide both sides of the equation by α^2.

Then, $\alpha^2 - \alpha - 4 - \dfrac{1}{\alpha} + \dfrac{1}{\alpha^2} = 0$

$\Rightarrow \left(\alpha^2 + \dfrac{1}{\alpha^2}\right) - \left(\alpha + \dfrac{1}{\alpha}\right) - 4 = 0$

$\Rightarrow \left(\alpha + \dfrac{1}{\alpha}\right)^2 - 2 - \left(\alpha + \dfrac{1}{\alpha}\right) - 4 = 0$

$\Rightarrow X^2 - X - 6 = 0$ Letting $X = \alpha + \dfrac{1}{\alpha}$

$\Rightarrow (X - 3)(X + 2) = 0 \quad \Rightarrow \quad X = 3 \text{ or } X = -2$

$\Rightarrow \alpha + \dfrac{1}{\alpha} = 3 \text{ or } \alpha + \dfrac{1}{\alpha} = -2$

Since $\alpha > 0$, $\dfrac{1}{\alpha} > 0 \quad \therefore \quad \alpha + \dfrac{1}{\alpha} > 0 \quad \therefore \quad \alpha + \dfrac{1}{\alpha} = 3$

Therefore, $\alpha^3 + \dfrac{1}{\alpha^3} = \left(\alpha + \dfrac{1}{\alpha}\right)^3 - 3\left(\alpha + \dfrac{1}{\alpha}\right) = 27 - 9 = 18$

#23 For a cube polynomial equation $x^3 = 1$, ω is a complex number solution. Find the value of each expression.

(1) $\omega^2 + \omega + 1$

Since $\omega^3 = 1$, $\omega^3 - 1 = 0$

$\therefore \quad (\omega - 1)(\omega^2 + \omega + 1) = 0$

$\therefore \quad \omega^2 + \omega + 1 = 0$

(2) $\omega + \dfrac{1}{\omega} = \dfrac{\omega^2 + 1}{\omega} = \dfrac{-\omega}{\omega} = -1$

(3) $\omega^2 + \dfrac{1}{\omega^2} = \dfrac{\omega^4 + 1}{\omega^2} = \dfrac{\omega^3 \cdot \omega + 1}{\omega^2} = \dfrac{\omega + 1}{\omega^2} = \dfrac{-\omega^2}{\omega^2} = -1$

(4) $\omega + \dfrac{1}{\omega} + \bar{\omega} + \dfrac{1}{\bar{\omega}}$

Noth that : $\omega + \bar{\omega} = -1$, $\omega \cdot \bar{\omega} = 1$

$\therefore \quad \omega + \dfrac{1}{\omega} + \bar{\omega} + \dfrac{1}{\bar{\omega}} = (\omega + \bar{\omega}) + \left(\dfrac{1}{\omega} + \dfrac{1}{\bar{\omega}}\right) = (\omega + \bar{\omega}) + \left(\dfrac{\omega + \bar{\omega}}{\omega \, \bar{\omega}}\right) = -1 + (-1) = -2$

(5) $\dfrac{1}{1+\omega} + \dfrac{1}{1+\bar{\omega}} = \dfrac{1+\bar{\omega}+1+\omega}{(1+\omega)(1+\bar{\omega})} = \dfrac{2+\omega+\bar{\omega}}{1+(\omega+\bar{\omega})+\omega\cdot\bar{\omega}} = \dfrac{2-1}{1+(-1)+1} = 1$

#24 Find the value of given expression:

(1) For a cube polynomial equation $x^3 - 2x^2 - 5x + 3 = 0$ **with roots** α, β, **and** γ,

find the value of $(\alpha + \beta)(\beta + \gamma)(\gamma + \alpha)$.

By roots-coefficients relationship, $\alpha + \beta + \gamma = 2$, $\alpha\beta + \beta\gamma + \gamma\alpha = -5$, $\alpha\beta\gamma = -3$

$$\begin{aligned} \therefore \ (\alpha + \beta)(\beta + \gamma)(\gamma + \alpha) &= (2 - \gamma)(2 - \alpha)(2 - \beta) = (4 - 2(\alpha + \gamma) + \alpha\gamma)(2 - \beta) \\ &= 8 - 4(\alpha + \gamma) + 2\alpha\gamma - 4\beta + 2(\alpha + \gamma)\beta - \alpha\beta\gamma \\ &= 8 - 4(\alpha + \beta + \gamma) + 2(\alpha\beta + \beta\gamma + \gamma\alpha) - \alpha\beta\gamma \\ &= 8 - 4(2) + 2(-5) - (-3) = -7 \end{aligned}$$

(2) For a cube polynomial equation $x^3 + ax^2 + 4x + b = 0$ **with a root** $1 + i$,

find the value of $a + b$ **(**a, b **are real numbers).**

Since $1 + i$ is a root, $1 - i$ is also a root.

Let α be the other root. Then

① $(1 + i) + (1 - i) + \alpha = -a$ $\therefore\ 2 + \alpha = -a$

② $(1 + i)(1 - i) + (1 - i)\alpha + \alpha(1 + i) = 4$ $\therefore\ (1 + 1) + 2\alpha = 4$ $\therefore\ \alpha = 1$

③ $(1 + i)(1 - i)\alpha = -b$ $\therefore\ (1 + 1)\alpha = -b$

$\therefore\ a = -3$, $b = -2$

Therefore, $a + b = -5$

(3) For a cube polynomial equation $x^3 + ax + b = 0$ **with a root** α $(\alpha < 0)$ **such that**

$\alpha^2 = 6 - 4\sqrt{2}$, **find the value of** $a - b$ **(where** a **and** b **are rational numbers).**

Since $\alpha^2 = 6 - 4\sqrt{2}$, $\alpha = \pm\sqrt{6 - 4\sqrt{2}}$

Since $\alpha < 0$, $\alpha = -\sqrt{6 - 4\sqrt{2}} = -\sqrt{6 - 2\sqrt{8}} = -\sqrt{(4 + 2) - 2\sqrt{(4 \cdot 2)}}$

$$= -\left(\sqrt{4} - \sqrt{2}\right) = -\left(2 - \sqrt{2}\right) = -2 + \sqrt{2}$$

Since $-2 + \sqrt{2}$ is a solution, $-2 - \sqrt{2}$ is also a solution.

Let β be the other solution. Then,

① $(-2 + \sqrt{2}) + (-2 - \sqrt{2}) + \beta = 0$ $\therefore\ \beta = 4$

② $(-2 + \sqrt{2})(-2 - \sqrt{2}) + (-2 - \sqrt{2})\beta + \beta(-2 + \sqrt{2}) = a$ $\therefore\ 2 - 4\beta = a$

③ $(-2 + \sqrt{2})(-2 - \sqrt{2})\beta = -b$ $\therefore\ 2\beta = -b$

$\therefore\ a = -14$, $b = -8$

Therefore, $a - b = -6$

#25 Find the range of a real number a so that a fourth-degree quadratic equation

$x^4 - 2(a+1)x^2 + a^2 - 2a - 3 = 0$ **will have four different real number solutions.**

Let $x^2 = A$ $(A > 0)$

Then, $A^2 - 2(a+1)A + a^2 - 2a - 3 = 0 \cdots \cdots \text{①}$

If ① has 2 different real number solutions, then the original equation has 4 different real number solutions.

i) (The discriminant of ①) > 0

ii) (The sum of two roots of ①) > 0

iii) (The product of two roots of ①) > 0

∴ i) $\dfrac{D}{4} = (a+1)^2 - (a^2 - 2a - 3) = 4a + 4 > 0$ ∴ $a > -1$

ii) $2(a+1) > 0$ ∴ $a > -1$

iii) $a^2 - 2a - 3 = (a-3)(a+1) > 0$ ∴ $a > 3$, $a < -1$

Therefore, by i), ii), and iii), $a > 3$

#26 Find the value of ab for each system.

(1) $\begin{cases} ax - by = -2 \\ bx + 2y = a \end{cases}$ **with solution** $(3, 2)$

$\Rightarrow \begin{cases} 3a - 2b = -2 \\ 3b + 4 = a \end{cases}$ ∴ $3(3b + 4) - 2b = -2$; $9b + 12 - 2b = -2$; $7b = -14$; $b = -2$

∴ $a = -6 + 4 = -2$. Therefore, $ab = 4$

(2) $\begin{cases} x + 5y = -3 \\ 2x - by = 5 \end{cases}$ **with solution** $(a, -1)$

$\Rightarrow \begin{cases} a - 5 = -3 \\ 2a + b = 5 \end{cases}$ $\Rightarrow \begin{cases} a = 2 \\ 2a + b = 5 \end{cases}$ ∴ $b = 5 - 2a = 5 - 4 = 1$

Therefore, $ab = 2$

(3) $\begin{cases} -2x + y = 5 \\ x - 2y = -1 \end{cases}$ **with solution** (a, b)

$\Rightarrow \begin{cases} -2x + y = 5 \\ 2x - 4y = -2 \end{cases}$

$\Rightarrow$ $-2x + y = 5$

 $+)\ \underline{\ 2x - 4y = -2\ }$

 $-3y = 3$; $y = -1$

∴ $x = 2y - 1 = -2 - 1 = -3$

$\therefore \ (x,y) = (a,b) = (-3,-1)$ Therefore, $ab = 3$

(4) $\begin{cases} 3x - by = 2 \\ ax + y = -2 \end{cases}$ with solution $(b-1, 2)$

$\Rightarrow \begin{cases} 3(b-1) - 2b = 2 \\ a(b-1) + 2 = -2 \end{cases} \Rightarrow \begin{cases} b = 5 \\ a(b-1) = -4 \end{cases}$

$\therefore \ 4a = -4 \ ; a = -1$

Therefore, $ab = -5$

#27 The system $\begin{cases} 2x + 3y = 5 \\ -x - 2y = -3 \end{cases}$ **has a solution** (a, b).

Find the solution for the system $\begin{cases} (3-a)x + 2y = -2 \\ 2x + 3y = 2b + 1 \end{cases}$.

$\begin{cases} 2x + 3y = 5 \\ -x - 2y = -3 \end{cases} \Rightarrow \begin{cases} 2x + 3y = 5 \\ -2x - 4y = -6 \end{cases}$

$\Rightarrow \quad 2x + 3y = 5$

$+) \ \underline{-2x - 4y = -6}$

$-y = -1 \ ; \ y = 1$

$\therefore \ 2x = 5 - 3 = 2 \quad \therefore \ x = 1 \quad \therefore \ (x,y) = (a,b) = (1,1)$

$\begin{cases} (3-a)x + 2y = -2 \\ 2x + 3y = 2b + 1 \end{cases} \Rightarrow \begin{cases} 2x + 2y = -2 \\ 2x + 3y = 3 \end{cases}$

$\Rightarrow \quad 2x + 2y = -2$

$-) \ \underline{2x + 3y = 3}$

$-y = -5 \ ; \ y = 5$

$\therefore \ 2x = 3 - 3y = 3 - 15 = -12 \ ; \ x = -6$

Therefore, $(x,y) = (-6, 5)$

#28 Find the value for the following :

(1) The system $\begin{cases} 2x - y = 3 \\ x + 3y = 5 \end{cases}$ **has a solution** $(a+1, b-1)$.

Find the value of $(a+b)^2 - (a-b)^2$.

Since $2x - y = 3$, $y = 2x - 3$; $x + 3(2x-3) = 5$; $7x = 14$; $x = 2$; $y = 1$

$\therefore \ a + 1 = 2, \ b - 1 = 1 \quad \therefore \ a = 1, \ b = 2$

$\therefore \ a + b = 3, \ a - b = -1$

Therefore, $(a+b)^2 - (a-b)^2 = 3^2 - (-1)^2 = 9 - 1 = 8$

(2) The solution of the system $\begin{cases} 3x - 2y = -2 \\ (k-1)x + y = -3 \end{cases}$

is the same as the solution of the equation $2x - y = 3$. **Find the constant** k.

Since the solution is the same, rearrange the system.

$$\begin{cases} 3x - 2y = -2 \\ 2x - y = 3 \end{cases} \Rightarrow \begin{cases} 3x - 2y = -2 \\ 4x - 2y = 6 \end{cases}$$

$$\Rightarrow \quad 3x - 2y = -2$$

$$-) \, \underline{4x - 2y = 6}$$

$$-x \quad\quad = -8 \; ; \; x = 8$$

Since $2x - y = 3$, $y = 2x - 3 = 16 - 3 = 13$

$\therefore \; (k-1)x + y = -3 \;\Rightarrow\; (k-1)8 + 13 = -3 \;;\; 8k - 8 + 13 = -3 \;;\; 8k = -8$

$\therefore \; k = -1$

Alternative approach :

$y = -3 - (k-1)x$ and $y = 2x - 3$ must be the same line.

$\therefore \; 2x - 3 = -3 - (k-1)x \;;\; 2x = (1-k)x$

$\therefore \; 2 = 1 - k \quad \therefore \; k = -1.$

(3) Two systems $\begin{cases} 2x + by = 4 \\ x + 2y = -3 \end{cases}$ **and** $\begin{cases} x - 3y = 2 \\ ax + 2y = -1 \end{cases}$ **have the same solution.**

Find the value of $a + b$.

Let (m, n) be the solution. Then, we have

$\begin{cases} 2m + bn = 4 \\ m + 2n = -3 \end{cases}$ and $\begin{cases} m - 3n = 2 \\ am + 2n = -1 \end{cases}$

Since m and n must satisfy both $m + 2n = -3$ and $m - 3n = 2$,

we can solve for m and n by solving the system $\begin{cases} m + 2n = -3 \\ m - 3n = 2 \end{cases}$

$$\Rightarrow \quad m + 2n = -3$$

$$-) \, \underline{m - 3n = 2}$$

$$5n = -5 \;;\; n = -1$$

Substitute $n = -1$ into $m - 3n = 2$. Then $m = -1$.

$\therefore \; (m, n) = (-1, -1)$

Substitute $(m, n) = (-1, -1)$ into $2m + bn = 4$. Then $-2 - b = 4 \;;\; b = -6$

Substitute $(m, n) = (-1, -1)$ into $am + 2n = -1$. Then $-a - 2 = -1 \;;\; a = -1$

$\therefore \; a + b = -1 - 6 = -7$

(4) The system $\begin{cases} \frac{a+1}{2}x - \frac{3}{4}y = -2 \\ 5x + \frac{b-1}{2}y = 4 \end{cases}$ has all real number solutions.

Find the value of $a + b$.

$$\frac{\frac{a+1}{2}}{5} = \frac{-\frac{3}{4}}{\frac{b-1}{2}} = \frac{-2}{4} = \frac{-1}{2}$$

$$\therefore \quad 2 \cdot \frac{a+1}{2} = -5 \ ; \ a+1 = -5 \ ; \ a = -6$$

$$-\frac{3}{4} \cdot 2 = -\frac{b-1}{2} \ ; \ -\frac{3}{2} = -\frac{b-1}{2} \ ; \ 3 = b-1 \ ; \ b = 4$$

$$\therefore \quad a+b = -6+4 = -2$$

(5) The system $\begin{cases} a(x-y) + \frac{y}{2} = -1 \\ -\frac{x}{2} - \frac{1}{a}y = 3 \end{cases}$ has no solution. Find the value of a.

$$\begin{cases} ax + \left(-a + \frac{1}{2}\right)y = -1 \\ -\frac{1}{2}x - \frac{1}{a}y = 3 \end{cases}$$

$$\therefore \quad \frac{a}{-\frac{1}{2}} = \frac{-a+\frac{1}{2}}{-\frac{1}{a}} \neq \frac{-1}{3}$$

$$\therefore \quad a \cdot -\frac{1}{a} = -\frac{1}{2}\left(-a + \frac{1}{2}\right) \ ; \ -1 = -\frac{1}{2}\left(-a + \frac{1}{2}\right) \ ; \ -a + \frac{1}{2} = 2 \ ; \ a = -\frac{3}{2}$$

(6) The system $\begin{cases} 2kx - (3x + y) = 2y \\ -(k-1)x + 2y = kx \end{cases}$ has a solution other than $(0, 0)$.

Find the value of the constant k.

$$\begin{cases} (2k-3)x - 3y = 0 \\ (-k+1-k)x + 2y = 0 \end{cases} \ ; \ \frac{2k-3}{-k+1-k} = -\frac{3}{2} \ ; \ 4k - 6 = 6k - 3 \ ; \ 2k = -3 \ ; \ k = -\frac{3}{2}$$

Note that : If two lines have more than 1 intersection,

they have infinite intersections (i.e., they are the same line).

(7) The system $\begin{cases} 3x - 2y = k \\ -2x + y = 3 \end{cases}$ has the solution (a, b) with the condition $a : b = 1 : 3$.

Find the constant k.

Since $a : b = 1 : 3$, $3a = b$

Since (a, b) is the solution, $\begin{cases} 3a - 2b = k \\ -2a + b = 3 \end{cases}$

$$\therefore \begin{cases} 3a - 2b = k \\ -2a + b = 3 \end{cases} \Rightarrow \begin{cases} 3a - 6a = k \\ -2a + 3a = 3 \end{cases} \Rightarrow \begin{cases} -3a = k \\ a = 3 \end{cases}$$

$$\therefore \quad k = -3a = -9$$

(8) Find the value of $\frac{1}{x} - \frac{1}{y}$ for variables x and y that satisfy the system

$$\begin{cases} 2x - xy - 2y - 3 = 0 \\ 3x + 2xy - 3y + 1 = 0 \end{cases}.$$

$$\begin{cases} 2x - xy - 2y - 3 = 0 \\ 3x + 2xy - 3y + 1 = 0 \end{cases} \Rightarrow \begin{cases} 2(x - y) - xy = 3 \\ 3(x - y) + 2xy = -1 \end{cases}$$

Let $x - y = A$ and $xy = B$. Then,

$$\Rightarrow \begin{cases} 2A - B = 3 \\ 3A + 2B = -1 \end{cases} \Rightarrow \begin{cases} 4A - 2B = 6 \\ 3A + 2B = -1 \end{cases}$$

$$\Rightarrow \qquad 4A - 2B = 6$$
$$+\,)\ \underline{\quad 3A + 2B = -1 \quad}$$
$$\qquad 7A \qquad = 5 \quad ; A = \frac{5}{7}$$

$$\therefore B = 2A - 3 = \frac{10}{7} - 3 = \frac{-11}{7}$$

Since $A = \frac{5}{7}$, $x - y = \frac{5}{7}$.

Since $B = \frac{-11}{7}$, $xy = \frac{-11}{7}$.

$$\therefore \frac{1}{x} - \frac{1}{y} = \frac{y - x}{xy} = \frac{-\frac{5}{7}}{\frac{-11}{7}} = \frac{5}{11}$$

OR $\begin{cases} 2(x - y) - xy = 3 \\ 3(x - y) + 2xy = -1 \end{cases} \Rightarrow \begin{cases} \dfrac{2(x-y)}{xy} - 1 = \dfrac{3}{xy} \\ \dfrac{3(x-y)}{xy} + 2 = \dfrac{-1}{xy} \end{cases} \Rightarrow \begin{cases} \dfrac{2(x-y)}{xy} - 1 = \dfrac{3}{xy} \\ \dfrac{9(x-y)}{xy} + 6 = \dfrac{-3}{xy} \end{cases}$

Adding the two equations gives :

$$\frac{11(x-y)}{xy} + 5 = 0 \quad \Rightarrow \quad \frac{(x-y)}{xy} = -\frac{5}{11} \quad \therefore \frac{1}{x} - \frac{1}{y} = \frac{y-x}{xy} = \frac{5}{11}$$

#29 Find the value of $x + y$ for variables x and y that satisfy the equations $2^x \cdot 8^y = 32$ and $3^{x+1} \cdot 9^{y-1} = 3^3$.

$$2^x \cdot 8^y = 32 \Rightarrow 2^x \cdot (2^3)^y = 2^5 \Rightarrow 2^{x+3y} = 2^5 \,; x + 3y = 5$$

$$3^{x+1} \cdot 9^{y-1} = 3^3 \Rightarrow 3^{x+1} \cdot (3^2)^{y-1} = 3^3 \Rightarrow 3^{x+1+2y-2} = 3^3$$

$$x + 1 + 2y - 2 = 3 \,; x + 2y = 4$$

$$\therefore \begin{cases} x + 3y = 5 \\ x + 2y = 4 \end{cases}$$

$$\Rightarrow \qquad x + 3y = 5$$
$$-\,)\ \underline{\quad x + 2y = 4 \quad}$$
$$\qquad y = 1 \;; x = 5 - 3y = 5 - 3 = 2$$

$$\therefore \ x + y = 2 + 1 = 3$$

#30 Solve the following systems:

(1) $\begin{cases} x^2 + xy + y^2 = 1 \\ x^2 + y^2 + x + y = 2 \end{cases}$

Consider $X^2 - uX + v$ where $u = x + y$, $v = xy$

Since $x^2 + y^2 = (x + y)^2 - 2xy = u^2 - 2v$, we have a new system:

$\begin{cases} u^2 - 2v + v = 1 & \cdots\cdots ① \\ u^2 - 2v + u = 2 & \cdots\cdots ② \end{cases}$

② − ① : $u - v = 1$; $u = v + 1$ $\cdots\cdots ③$

Substituting ③ into ①,

$(v + 1)^2 - v = 1$; $v^2 + v = 0$; $v(v + 1) = 0$ $\quad \therefore v = 0$ or $v = -1$

$\therefore (u, v) = (1, 0)$ or $(u, v) = (0, -1)$

$\therefore \begin{cases} x + y = 1 \\ xy = 0 \end{cases}$ or $\begin{cases} x + y = 0 \\ xy = -1 \end{cases}$

Therefore, $(x, y) = (1, 0)$, $(x, y) = (0, 1)$, $(x, y) = (1, -1)$, $(x, y) = (-1, 1)$

(2) $\begin{cases} x^2 + y^2 = 10 \\ xy = 3 \end{cases}$

Consider $X^2 - uX + v$ where $u = x + y$, $v = xy$

Since $x^2 + y^2 = (x + y)^2 - 2xy = u^2 - 2v$, we have a new system:

$\begin{cases} u^2 - 2v = 10 & \cdots\cdots ① \\ v = 3 & \cdots\cdots ② \end{cases}$

Substituting ② into ①,

$u^2 = 16$ $\quad \therefore u = \pm 4$

$\therefore (u, v) = (\pm 4, 3)$

$\therefore \begin{cases} x + y = 4 \\ xy = 3 \end{cases}$ or $\begin{cases} x + y = -4 \\ xy = 3 \end{cases}$

Therefore, $(x, y) = (1, 3)$, $(x, y) = (3, 1)$, $(x, y) = (-1, -3)$, $(x, y) = (-3, -1)$

#31 Determine the range.

(1) Find the range of x for each expression when $-1 \le x \le 1$.

① $2x + 1$ $\quad$; $-2 \le 2x \le 2$ $\quad \therefore -1 \le 2x + 1 \le 3$

② $-3x - 2$; $3 \ge -3x \ge -3$; $3 - 2 \ge -3x - 2 \ge -3 - 2$ $\quad \therefore -5 \le -3x - 2 \le 1$

③ $\frac{1}{4}x - 3$ $\quad$; $-\frac{1}{4} \le \frac{1}{4}x \le \frac{1}{4}$; $-\frac{1}{4} - 3 \le \frac{1}{4}x - 3 \le \frac{1}{4} - 3$ $\quad \therefore -3\frac{1}{4} \le \frac{1}{4}x - 3 \le -2\frac{3}{4}$

(2) When $y = \frac{4-2x}{3}$,

  ① **Find the range of** y **when** $1 < x < 5$.

$$-2 > -2x > -10 \ ; \ 4-2 > 4-2x > 4-10 \ ; \ \frac{2}{3} > \frac{4-2x}{3} > \frac{-6}{3} \quad \therefore \ -2 < y < \frac{2}{3}$$

  ② **Find the range of** y **when** $-3 < x < -1$.

$$6 > -2x > 2 \ ; \ 4+6 > 4-2x > 4+2 \ ; \ \frac{10}{3} > \frac{4-2x}{3} > \frac{6}{3} \ ; \ 2 < y < \frac{10}{3}$$

  ③ **Find the range of** x **when** $2 \le y \le 4$.

$$2 \le \frac{4-2x}{3} \le 4 \ ; \ 6 \le 4-2x \le 12 \ ; \ 6-4 \le -2x \le 12-4 \ ; \ \frac{2}{-2} \ge \frac{-2x}{-2} \ge \frac{8}{-2}$$

$$-1 \ge x \ge -4 \ ; \ -4 \le x \le -1$$

#32 Find the constant k **if :**

  (1) The inequality $\frac{1}{2}x - \frac{k}{3} < -1$ **has the solution** $x < 2$.

$$3x - 2k < -6 \ ; \ 3x < 2k - 6 \ ; \ x < \frac{2k-6}{3}$$

$$\therefore \ \frac{2k-6}{3} = 2 \ ; \ 2k-6 = 6 \ ; \ 2k = 12 \quad \therefore k = 6$$

  (2) The inequality $\frac{kx}{4} - \frac{1}{2} > 1$ **has the solution** $x < -1$.

$$kx - 2 > 4 \ ; \ kx > 6.$$

Since the direction of the symbol for the solution changes, $k < 0$ and $x < \frac{6}{k}$.

$$\therefore \ \frac{6}{k} = -1 \quad \therefore \ k = -6$$

  (3) The inequality $\frac{2-kx}{5} - 2 \le \frac{x}{2} + 1$ **has the solution** $x \le -4$.

$$2(2-kx) - 20 \le 5x + 10 \ ; \ (-2k-5)x \le 10+20-4 \ ; \ (-2k-5)x \le 26$$

$$x \le \frac{26}{-2k-5} . \quad \text{Since } x \le -4, \quad \frac{26}{-2k-5} = -4 .$$

$$\therefore \ 8k + 20 = 26 \ ; \ 8k = 6 \ ; \ k = \frac{3}{4}$$

  (4) Two inequalities $2(1-2x) - 3 \le x - 5$ **and** $\frac{3k-2x}{3} \le x + 2k$ **have the same solution.**

$$2(1-2x) - 3 \le x - 5 \ \Rightarrow -5x \le -4 \ ; \ x \ge \frac{4}{5}$$

$$\frac{3k-2x}{3} \le x + 2k \ \Rightarrow 3k - 2x \le 3x + 6k \ ; \ 5x \ge -3k \ ; \ x \ge -\frac{3k}{5}$$

$$\therefore \ -\frac{3k}{5} = \frac{4}{5} \ ; \ 3k = -4 \ \therefore \ k = -\frac{4}{3}$$

(5) The inequality $2 - kx < 2x + k$ has no solution.

$(2 + k)x > 2 - k$.

If $0 \cdot x \, (= 0) > 0$ (Positive number), then there is no solution.

Thus, $2 + k = 0$ and $2 - k > 0$ (Positive number)

$\therefore \ k = -2$ and $k < 2$ $\therefore k = -2$

(6) The inequality $-2kx + 5 > 6$ has the solution $x > 2$.

$-2kx + 5 > 6 \ \Rightarrow \ 2kx < -1$.

Since the solution is $x > 2$, $k < 0$ and $x > \frac{-1}{2k}$

$\therefore \ \frac{-1}{2k} = 2$; $4k = -1$; $k = -\frac{1}{4}$

(7) $1 - 5x \le 2x - 5k$ has -2 as minimum value of the solution.

$1 - 5x \le 2x - 5k \ \Rightarrow 7x \ge 1 + 5k \ \Rightarrow x \ge \frac{1+5k}{7}$

$\therefore \ \frac{1+5k}{7}$ is minimum value of x

$\therefore \ \frac{1+5k}{7} = -2$; $1 + 5k = -14$; $5k = -15$; $k = -3$

(8) The inequality $x - \left(3 + \frac{k}{2}\right) > 2x + k$ has no positive solution.

$x - \left(3 + \frac{k}{2}\right) > 2x + k \ \Rightarrow \ -x > k + 3 + \frac{k}{2} \ \Rightarrow \ x < -\left(\frac{3}{2}k + 3\right)$

$\therefore \ -\left(\frac{3}{2}k + 3\right) = 1$; $\frac{3}{2}k + 3 = -1$; $\frac{3}{2}k = -4$; $k = -4 \cdot \frac{2}{3} = -\frac{8}{3}$

#33 Solve each inequality.

(1) The inequality $(-a + 2b)x + b - 3a \le 0$ has the solution $x \le -1$.

Find the solution for the inequality $(a - b)x + a - 2b > 0$, where $b > 0$.

$(-a + 2b)x + b - 3a \le 0 \ \Rightarrow \ (-a + 2b)x \le -b + 3a$

Since the solution is $x \le -1$ (The direction of symbol does not change.),

$-a + 2b > 0$ and $x \le \frac{-b+3a}{-a+2b}$

$\therefore \ \frac{-b+3a}{-a+2b} = -1$; $-a + 2b = b - 3a$; $2a = -b$; $a = -\frac{1}{2}b$

Substitute $a = -\frac{1}{2}b$ into $(a - b)x + a - 2b > 0 \ \Rightarrow \ -\frac{3}{2}bx - \frac{5}{2}b > 0$; $-\frac{3}{2}bx > \frac{5}{2}b$

Since $b > 0$, $-\frac{3}{2}x > \frac{5}{2}$; $x < -\frac{5}{3}$

Note that you cannot use $b = -2a$ instead of $a = -\frac{1}{2}b$.

($\because$ Since $-a + 2b > 0$, $2b > a$. Since $b > 0$, $2b > 0$. So, we have $2b > a$.

Since $a > 0$ or $a < 0$, we cannot decide which condition is necessary for a.

Therefore, we have to use $a = -\frac{1}{2}b,\ $ not $\ b = -2a.$) $\qquad \therefore$ The solution is $x < -\frac{5}{3}.$

(2) The inequality $(a + b)x \geq 3a - b$ has the solution $x \leq \frac{3}{2}$.

Find the solution for the inequality $(a - 2b)x < 2a + 3b$.

Since $(a + b)x \geq 3a - b$ has the solution $x \leq \frac{3}{2}$ (The direction is changed.), $a + b < 0$.

$\therefore\ x \leq \frac{3a-b}{a+b}$ $\qquad \therefore\ \frac{3a-b}{a+b} = \frac{3}{2}$ $\qquad \therefore\ 6a - 2b = 3a + 3b$ $\qquad \therefore\ 3a = 5b$ $\qquad \therefore\ a = \frac{5b}{3}$

Since $a + b < 0,\ \frac{5b}{3} + b < 0$ $\qquad \therefore\ \frac{8b}{3} < 0$ $\qquad \therefore\ b < 0$

From $(a - 2b)x < 2a + 3b,\ \left(\frac{5b}{3} - 2b\right)x < \frac{10b}{3} + 3b$

$\therefore\ -\frac{b}{3}x < \frac{19b}{3}\ ;\ -bx < 19b$

Since $b < 0,\ -b > 0$ $\qquad \therefore\ x < \frac{19b}{-b}$ $\qquad$ Therefore, $x < -19$

#34 Solve each inequality for x and graph the solution :

(1) $|x - 2| \leq 0$

Since $|a| \geq 0$ for any $a > 0,\ x - 2 = 0$ has to be true to satisfy the inequality.

Therefore, the solution is only $x = 2$.

(2) $|3x + 9| > 0$

Since $|a| \geq 0$ for any $a > 0,\ 3x + 9 \neq 0$ (To satisfy the inequality).

$\therefore\ x \neq -3$

Therefore, the solutions are all real numbers except -3.

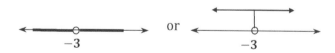

(3) $|x + 4| < 0$

Since $|a| \geq 0$ for any $a > 0,$ there is no solution to satisfy the inequality.

(4) $|-2x + 1| + 3 \le 6$

$|-2x + 1| \le 6 - 3$; $|-2x + 1| \le 3$

$\Rightarrow -3 \le -2x + 1 \le 3$ $\Rightarrow -4 \le -2x \le 2$ $\Rightarrow \frac{-4}{-2} \ge x \ge \frac{2}{-2}$ $\Rightarrow -1 \le x \le 2$

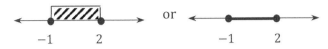

(5) $0 < |2 - 4x| < 8$

Case 1 . If $2 - 4x \ge 0$, then $x \le \frac{1}{2}$

$\Rightarrow 0 < 2 - 4x < 8$ $\Rightarrow -2 < -4x < 6$ $\Rightarrow \frac{-2}{-4} > x > \frac{6}{-4}$ $\Rightarrow -\frac{3}{2} < x < \frac{1}{2}$

 $\therefore -\frac{3}{2} < x < \frac{1}{2}$

Case 2. If $2 - 4x < 0$, then $x > \frac{1}{2}$

$\Rightarrow 0 < -(2 - 4x) < 8$ $\Rightarrow 0 < -2 + 4x < 8$ $\Rightarrow 2 < 4x < 10$ $\Rightarrow \frac{1}{2} < x < \frac{5}{2}$

 $\therefore \frac{1}{2} < x < \frac{5}{2}$

Therefore, the solution is $-\frac{3}{2} < x < \frac{1}{2}$ or $\frac{1}{2} < x < \frac{5}{2}$.

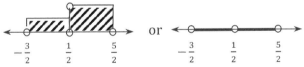

(6) $2 < |x + 1| < 3$

Case 1. If $x + 1 \ge 0$, then $x \ge -1$

$\Rightarrow 2 < x + 1 < 3$ $\Rightarrow 1 < x < 2$

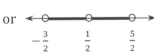 $\therefore 1 < x < 2$

Case 2. If $x + 1 < 0$, then $x < -1$

$\Rightarrow 2 < -(x + 1) < 3$ $\Rightarrow 3 < -x < 4$ $\Rightarrow -3 > x > -4$ $\Rightarrow -4 < x < -3$

$\therefore -4 < x < -3$

Therefore, the solutions are $1 < x < 2$ or $-4 < x < -3$.

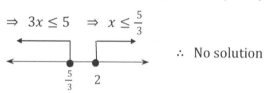

(7) $2|x + 1| + |x - 2| \leq 5$

Case 1. If $x < -1$, then $-2(x + 1) - (x - 2) \leq 5$

$\Rightarrow -3x \leq 5 \Rightarrow x \geq -\dfrac{5}{3}$

$\therefore -\dfrac{5}{3} \leq x < -1$

Case 2. If $-1 \leq x < 2$, then $2(x + 1) - (x - 2) \leq 5$

$\Rightarrow x \leq 1$

$\therefore -1 \leq x \leq 1$

Case 3. If $x \geq 2$, then $2(x + 1) + (x - 2) \leq 5$

$\Rightarrow 3x \leq 5 \Rightarrow x \leq \dfrac{5}{3}$

$\therefore$ No solution

Therefore, 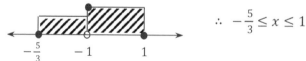 $\therefore -\dfrac{5}{3} \leq x \leq 1$

(8) $|x + 1| + \sqrt{x^2 - 4x + 4} < 4$

Since $\sqrt{x^2 - 4x + 4} = \sqrt{(x - 2)^2} = |x - 2|$, we have $|x + 1| + |x - 2| < 4$.

Case 1. If $x < -1$, then $-(x + 1) - (x - 2) < 4$

$\Rightarrow -2x < 3 \Rightarrow x > -\dfrac{3}{2}$

Since $x < -1$,

$\therefore -\dfrac{3}{2} < x < -1$

Case 2. If $-1 \leq x < 2$, then $(x + 1) - (x - 2) < 4$

$\Rightarrow 0 \cdot x \leq 1$; It is always true for all x.

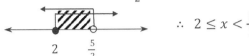 $\therefore\ -1 \le x < 2$

Case 3. If $x \ge 2$, then $(x + 1) + (x - 2) < 4$

$\Rightarrow\ 2x < 5\ \ \Rightarrow\ x < \dfrac{5}{2}$

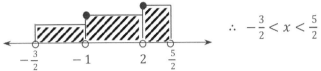 $\therefore\ 2 \le x < \dfrac{5}{2}$

Therefore, 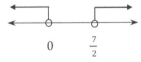 $\therefore\ -\dfrac{3}{2} < x < \dfrac{5}{2}$

(9) $|x| - 2|x - 3|\ > 1 - x$

Case 1. If $x < 0$, then $-x + 2(x - 3) > 1 - x$

$\Rightarrow\ 2x > 7\ \ \Rightarrow\ x > \dfrac{7}{2}$

Since $x < 0$,

$\therefore$ There is no solution.

Case 2. If $0 \le x < 3$, then $x + 2(x - 3) > 1 - x$

$\Rightarrow\ 4x > 7\ \ \Rightarrow\ x > \dfrac{7}{4}$

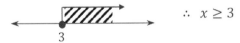 $\therefore\ \dfrac{7}{4} < x < 3$

Case 3. If $x \ge 3$, then $x - 2(x - 3) > 1 - x$

$\Rightarrow\ 0 \cdot x > -5\ \ $; This is always true for all x.

$\therefore\ x \ge 3$

Therefore, 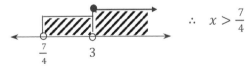 $\therefore\ x > \dfrac{7}{4}$

(10) $2|x + 1| - 3|x - 3| \ge 1$

Case 1. If $x < -1$, then $-2(x + 1) + 3(x - 3) \ge 1$

$\Rightarrow\ x \ge 12$

Since $x < -1$,

∴ There is no solution.

Case 2. If $-1 \le x < 3$, then $2(x+1) + 3(x-3) \ge 1$

$\Rightarrow 5x \ge 8 \quad \Rightarrow x \ge \dfrac{8}{5}$

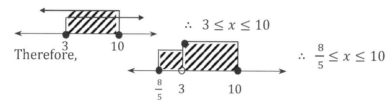

∴ $\dfrac{8}{5} \le x < 3$

Case 3. If $x \ge 3$, then $2(x+1) - 3(x-3) \ge 1$

$\Rightarrow -x \ge -10 \quad \Rightarrow x \le 10$

∴ $3 \le x \le 10$

Therefore,

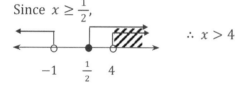

∴ $\dfrac{8}{5} \le x \le 10$

(11) $x^2 - x - 5 > |2x - 1|$

Case 1. If $x \ge \dfrac{1}{2}$, then $x^2 - x - 5 > 2x - 1$

$\Rightarrow x^2 - 3x - 4 > 0 \quad \Rightarrow (x-4)(x+1) > 0 \quad \Rightarrow x > 4 \text{ or } x < -1$

Since $x \ge \dfrac{1}{2}$,

∴ $x > 4$

Case 2. If $x < \dfrac{1}{2}$, then $x^2 - x - 5 > -(2x - 1)$

$\Rightarrow x^2 + x - 6 > 0 \quad \Rightarrow (x+3)(x-2) > 0 \quad \Rightarrow x > 2 \text{ or } x < -3$

∴ $x < -3$

Therefore,

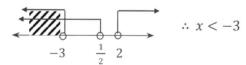

∴ $x < -3$ or $x > 4$

(12) $|x - 3| < 4$

Case 1. $x - 3 \ge 0 \;\; (x \ge 3)$

$\Rightarrow |x - 3| = x - 3 < 4 \;\; ; \; x < 7$

Since $x \ge 3$, $3 \le x < 7$

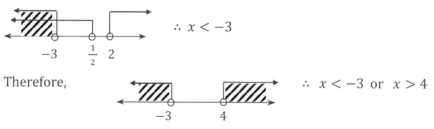

Case 2. $x - 3 < 0 \ (x < 3)$

$\Rightarrow |x - 3| = -(x - 3) < 4 \ ; \ x > -1$

Since $x < 3, \ -1 < x < 3$

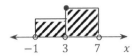

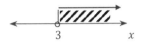

Therefore, the sum of all intervals is $-1 < x < 7$.

(13) $|x + 2| < 3x - 4$

Case 1. $x + 2 \geq 0 \ (x \geq -2)$

$\Rightarrow |x + 2| = x + 2 < 3x - 4 \ ; 2x > 6 \ ; \ x > 3$

Since $x \geq -2, \ x > 3$

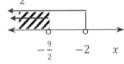

Case 2. $x + 2 < 0 \ (x < -2)$

$\Rightarrow |x + 2| = -(x + 2) < 3x - 4 \ ; \ 4x > 2 \ ; \ x > \dfrac{1}{2}$

Since $x < -2, \ x > \dfrac{1}{2}$ is not a solution.

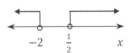

Therefore, $x > 3$

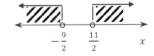

(14) $|x + 2| + |3 - x| > 10$

Since $(x + 2 = 0 \ \Rightarrow \ x = -2)$ and $(3 - x = 0 \ \Rightarrow \ x = 3)$,

consider $x < -2, -2 \leq x < 3$ and $x \geq 3$

Case 1. $x < -2$

$\Rightarrow -(x + 2) + (3 - x) > 10 \ ; \ -2x > 9 \ ; \ x < -\dfrac{9}{2}$

Since $x < -2, \ x < -\dfrac{9}{2}$

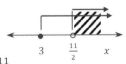

Case 2. $-2 \leq x < 3$

$\Rightarrow (x + 2) + (3 - x) > 10 \ ; \ 0 \cdot x > 5 \ ; \ $ False.

Case 3. $x \geq 3$

$\Rightarrow (x + 2) - (3 - x) > 10 ; \ 2x > 11 \ ; \ x > \dfrac{11}{2}$

Since $x \geq 3 , \ x > \dfrac{11}{2}$

Therefore, the sum of all intervals is $x < -\dfrac{9}{2}$ or $x > \dfrac{11}{2}$.

#35 When $1 \leq x \leq 2$, find maximum and minimum values of $\frac{2x}{x+1}$.

$$\frac{2x}{x+1} = \frac{2(x+1)-2}{x+1} = 2 - \frac{2}{x+1}$$

Since $1 \leq x \leq 2$, $2 \leq x + 1 \leq 3$

$\therefore \ \frac{1}{3} \leq \frac{1}{x+1} \leq \frac{1}{2}$ $\quad \therefore \ \frac{-2}{2} \leq \frac{-2}{x+1} \leq \frac{-2}{3}$ $\quad \therefore \ 1 \leq 2 - \frac{2}{x+1} \leq \frac{4}{3}$

Therefore, maximum value of $\frac{2x}{x+1}$ is $\frac{4}{3}$ and minimum value of $\frac{2x}{x+1}$ is 1.

#36 Find the value.

(1) When the inequality $|2x - a| > 4$ has solution: $x < b$ or $x > 3$, find the values of real numbers a and b.

$$|2x - a| > 4 \quad \Rightarrow \quad 2x - a > 4 \ \text{ or } \ 2x - a < -4$$

$\therefore \ x > \frac{a+4}{2} \ \text{ or } \ x < \frac{a-4}{2}$

Since $\frac{a+4}{2} = 3$ and $\frac{a-4}{2} = b$, $a = 2$ and $b = -1$

(2) When the inequality $|4x + 2| - 1 \leq a$ has solution $-3 \leq x \leq 2$, find the value of a.

$$\begin{aligned} |4x + 2| - 1 \leq a \quad &\Rightarrow \quad |4x + 2| \leq a + 1 \\ &\Rightarrow \quad -(a + 1) \leq 4x + 2 \leq a + 1 \\ &\Rightarrow \quad -a - 3 \leq 4x \leq a - 1 \\ &\Rightarrow \quad \frac{-a-3}{4} \leq x \leq \frac{a-1}{4} \end{aligned}$$

Since the solution is $-3 \leq x \leq 2$, $\dfrac{-a-3}{4} = -3$ and $\dfrac{a-1}{4} = 2$

Therefore, $a = 9$

#37 Find the value of k for the following conditions :

(1) The system $\begin{cases} x + 5 < 2k \\ 3x - 2 \geq 4 \end{cases}$ has the solution $2 \leq x < 5$.

$\Rightarrow \begin{cases} x < 2k - 5 \\ 3x \geq 6 \end{cases} \Rightarrow \begin{cases} x < 2k - 5 \\ x \geq 2 \end{cases}$

$\therefore \ 2 \leq x < 2k - 5$

$2k - 5 = 5 \ ; \ 2k = 10 \qquad \therefore \ k = 5$

(2) The system $\begin{cases} \frac{2x+1}{3} > \frac{x-3}{5} \\ 0.6x - 2.4 < kx - 0.8 \end{cases}$ **has the solution** $-2 < x < 2$.

$\Rightarrow \begin{cases} 5(2x+1) > 3(x-3) \\ 6x - 24 < 10kx - 8 \end{cases} \Rightarrow \begin{cases} 7x > -14 \\ (6-10k)x < 16 \end{cases} \Rightarrow \begin{cases} x > -2 \\ x < \frac{16}{6-10k} \end{cases}$

$\therefore \ -2 < x < \frac{16}{6-10k}$

$\frac{16}{6-10k} = 2 \ ; 12 - 20k = 16 \ ; 20k = -4$

$\therefore \ k = -\frac{1}{5}$

(3) The system $\begin{cases} -x + 2 \leq 0 \\ \frac{x}{2} + 3 \leq -k + 5 \end{cases}$ **has the solution** $x = 2$.

$\Rightarrow \begin{cases} x \geq 2 \\ \frac{x}{2} \leq -k + 2 \end{cases} \Rightarrow \begin{cases} x \geq 2 \\ x \leq 2(-k+2) \end{cases}$

$\therefore 2(-k+2) = 2 \ ; -k + 2 = 1 \quad \therefore \ k = 1$

(4) The system $\begin{cases} \frac{k-x}{2} \leq x + 5 \\ 3 - 2x < 3x - 2 \end{cases}$ **has the solution** $x \geq 3$.

$\Rightarrow \begin{cases} k - x \leq 2x + 10 \\ 3 - 2x < 3x - 2 \end{cases} \Rightarrow \begin{cases} 3x \geq k - 10 \\ 5x > 5 \end{cases} \Rightarrow \begin{cases} x \geq \frac{k-10}{3} \\ x > 1 \end{cases}$

$\therefore \ \frac{k-10}{3} = 3$

$\therefore \ k = 19$

(5) The system $\begin{cases} 2x + 3 \leq 4x - 5 \\ 3(x - k) \leq x + 3 \end{cases}$ **has only one solution.**

$\Rightarrow \begin{cases} 2x \geq 8 \\ 2x \leq 3 + 3k \end{cases} \Rightarrow \begin{cases} x \geq 4 \\ x \leq \frac{3+3k}{2} \end{cases}$

$\therefore \frac{3+3k}{2} = 4 \ ; \ 3 + 3k = 8 \ ; \ 3k = 5 \quad \therefore \ k = \frac{5}{3}$

(6) The system $3 < \dfrac{k-4x}{-2} < 5$ **has the solution** $1 < x < 2$.

$$3 < \frac{k-4x}{-2} < 5 \Rightarrow -6 > k - 4x > -10 \Rightarrow -6 - k > -4x > -10 - k \; ; \; \frac{-6-k}{-4} < x < \frac{-10-k}{-4}$$

$$\therefore \frac{-6-k}{-4} = 1 \text{ and } \frac{-10-k}{-4} = 2$$

$$\therefore \text{ if } \frac{-6-k}{-4} = 1 \Rightarrow -6 - k = -4 \; ; \; k = -2 \text{ and}$$

$$\text{if } \frac{-10-k}{-4} = 2 \Rightarrow -10 - k = -8 \; ; \; k = -2$$

Therefore, $k = -2$

#38 Find the range of k for the following conditions :

(1) The system $\begin{cases} 2x \le 5 - k \\ 3x - 3 \ge 2x - 1 \end{cases}$ **has no solution.**

$$\Rightarrow \begin{cases} x \le \frac{5-k}{2} \\ x \ge 2 \end{cases}$$

$$\therefore \frac{5-k}{2} < 2 \; ; \; 5 - k < 4 \; ; \; k > 1$$

(2) The system $\begin{cases} x - 3 \le 2x - 6 \\ 5x + k < 3x + 1 \end{cases}$ **has no solution.**

$$\Rightarrow \begin{cases} x \ge 3 \\ 2x < 1 - k \end{cases} \Rightarrow \begin{cases} x \ge 3 \\ x < \frac{1-k}{2} \end{cases}$$

If $\begin{cases} x \ge 3 \\ x < 3 \end{cases}$, then there is no solution.

Therefore, $\frac{1-k}{2} \le 3 \; ; \; 1 - k \le 6 \quad \therefore k \ge -5$

(3) The system $\begin{cases} 2x + 3 \le -5 \\ x + k > 1 \end{cases}$ **has only one integer in the solution.**

$$\Rightarrow \begin{cases} 2x \le -8 \\ x > 1 - k \end{cases} \Rightarrow \begin{cases} x \le -4 \\ x > 1 - k \end{cases}$$

If $1 - k = -4$, then $\begin{cases} x \le -4 \\ x > -4 \end{cases}$; No solution.

If $1 - k > -4$, then there is no solution.

If $1 - k = -5$, then $\begin{cases} x \le -4 \\ x > -5 \end{cases}$; $x = -4$ is the only one integer.

$\therefore -5 \le 1 - k < -4$; $-6 \le -k < -5$; $6 \ge k > 5$

$\therefore 5 < k \le 6$

(4) The system $\begin{cases} x - 3 \ge 0 \\ 3x + k \le 2x + 3 \end{cases}$ **has solutions.**

$\Rightarrow \begin{cases} x \ge 3 \\ x \le 3 - k \end{cases}$

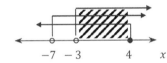

$\quad\quad 3 \quad\quad 3 - k \quad\quad x$

$\therefore 3 \le 3 - k$; $k \le 0$

#39 Find the sum of all integers that satisfy the following systems :

(1) $\begin{cases} 3x - 5 \le 7 \\ \frac{x-1}{2} < x + 3 \\ 2x - 5 < 5x + 4 \end{cases}$

$\Rightarrow \begin{cases} 3x \le 12 \\ x - 1 < 2x + 6 \\ 3x > -9 \end{cases} \Rightarrow \begin{cases} x \le 4 \\ x > -7 \\ x > -3 \end{cases}$

$\quad\quad -7 \quad -3 \quad\quad 4 \quad\quad x$

$\therefore$ The integers are $-2, -1, \ 0, \ 1, \ 2, \ 3, \ $ and $\ 4$.

$\therefore$ The sum is $-2 + (-1) + 0 + 1 + 2 + 3 + 4 = 7$.

(2) $\begin{cases} |x| \ \le 5 \\ |x| \ > 2 \end{cases}$

$\Rightarrow \begin{cases} -5 \le x \le 5 \\ x > 2 \text{ or } x < -2 \end{cases}$

$\quad\quad -5 \ -2 \quad\ 2 \quad 5 \quad x$

$\therefore -5 \le x < -2 \ $ or $\ 2 < x \le 5$

$\therefore$ The integers are $-5, -4, -3, \ 3, \ 4, \ 5$

$\therefore$ The sum is $-5 + (-4) + (-3) + 3 + 4 + 5 = 0$.

#40 The system $\begin{cases} x - 1 \geq 2x - 4 \\ \frac{x+k}{2} < 3x - 2 \end{cases}$ **has 5 integers in the solution.**

What is minimum value for k?

$$\Rightarrow \begin{cases} x \leq 3 \\ x + k < 6x - 4 \end{cases} \Rightarrow \begin{cases} x \leq 3 \\ 5x > k + 4 \end{cases} \Rightarrow \begin{cases} x \leq 3 \\ x > \frac{k+4}{5} \end{cases}$$

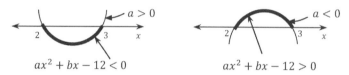

$$\therefore \; -2 \leq \frac{k+4}{5} < -1 \; ; \; -10 \leq k + 4 < -5 \; ; \; -14 \leq k < -9$$

∴ Minimum value for k is -14 .

#41 Find the value.

(1) Find the real numbers a and b so that a quadratic inequality $ax^2 + bx - 12 > 0$ will have solution $2 < x < 3$.

$a > 0$

$ax^2 + bx - 12 < 0$

$a < 0$

$ax^2 + bx - 12 > 0$

Since the solution is $2 < x < 3$, $a < 0$

If a quadratic inequality with leading coefficient 1 has a solution $2 < x < 3$, then the inequality is $(x - 2)(x - 3) < 0$; $x^2 - 5x + 6 < 0$.

Multiplying both sides by a ($a < 0$), $ax^2 - 5ax + 6a > 0$

Since this inequality must be equal to $ax^2 + bx - 12 > 0$, $-5a = b$ and $6a = -12$.

Therefore, $a = -2$, $b = 10$

Alternative Method :

Since the solution is $2 < x < 3$, $a < 0$

Consider a quadratic equation $ax^2 + bx - 12 = 0$ with roots 2 and 3.

By roots and coefficients relationship, $2 + 3 = -\frac{b}{a}$ and $2 \cdot 3 = -\frac{12}{a}$

∴ $a = -2$, $b = 10$

(2) When the solution of a quadratic inequality $x^2 - ax + 12 \leq 0$ is $\alpha \leq x \leq \beta$ and the solution of a quadratic inequality $x^2 - 5x + b \geq 0$ is $x \leq \alpha - 1$ or $x \geq \beta - 1$, find the value of $a + b$.

Since $x^2 - ax + 12 \leq 0$ has solution $\alpha \leq x \leq \beta$, $(x - \alpha)(x - \beta) \leq 0$

∴ $x^2 - (\alpha + \beta) + \alpha\beta \leq 0$

∴ $\alpha + \beta = a, \; \alpha\beta = 12$ ······ ①

Since $x^2 - 5x + b \geq 0$ has solution $x \leq \alpha - 1$ or $x \geq \beta - 1$,

$(x - (\alpha - 1))(x - (\beta - 1)) \geq 0$

$\therefore \ x^2 - (\alpha - 1 + \beta - 1)x + (\alpha - 1)(\beta - 1) \geq 0$

$\therefore \ x^2 - (\alpha + \beta - 2)x + \alpha\beta - (\alpha + \beta) + 1 \geq 0$

$\therefore \ \alpha + \beta - 2 = 5, \ \alpha\beta - (\alpha + \beta) + 1 = b \ \cdots\cdots ②$

By ① and ②, $a = 7$, $b = 6$ \qquad Therefore, $a + b = 13$

(3) When the solution of a quadratic inequality $x^2 + |x| - 6 \leq 0$ is $a \leq x \leq b$,

find the value of $a - b$.

Case 1. $x \geq 0 \ \Rightarrow \ x^2 + x - 6 \leq 0$

$\qquad\qquad\qquad \therefore \ (x + 3)(x - 2) \leq 0 \qquad \therefore -3 \leq x \leq 2$

$\qquad\qquad\qquad\qquad\qquad\qquad\qquad\qquad\qquad \therefore \ 0 \leq x \leq 2$

Case 2. $x < 0 \ \Rightarrow \ x^2 - x - 6 \leq 0$

$\qquad\qquad\qquad \therefore \ (x - 3)(x + 2) \leq 0 \qquad \therefore -2 \leq x \leq 3$

$\qquad\qquad\qquad\qquad\qquad\qquad\qquad\qquad\qquad \therefore \ -2 \leq x < 0$

Therefore,

$\qquad\qquad\qquad\qquad\qquad\qquad\qquad \therefore \ -2 \leq x \leq 2$

$\qquad\qquad\qquad\qquad\qquad\qquad\qquad\qquad \ \ \underset{a}{\uparrow} \qquad \underset{b}{\uparrow}$

Hence $a - b = (-2) - (2) = -4$

#42 Find the range of a real number a so that:

(1) A quadratic equation $x^2 - 2ax + 2a = 0$ will have two different real number

solutions.

Since the equation has two different real number solutions, $D > 0$.

$\therefore \ D = (-2a)^2 - 4 \cdot 2a = 4a^2 - 8a > 0$

$\therefore \ 4a(a - 2) > 0 \qquad \therefore \ a > 2$ or $a < 0$

(2) A quadratic inequality $ax^2 - x + a > 0$ is always true for all real number x.

\quad i) $a > 0$

\quad ii) Considering a quadratic equation $ax^2 - x + a = 0$, $D < 0$

$\qquad \therefore \ D = (-1)^2 - 4 \cdot a \cdot a = 1 - 4a^2 < 0$

$$\therefore \ 4a^2 - 1 > 0 \quad \therefore \ (2a-1)(2a+1) > 0 \quad \therefore \ a > \frac{1}{2} \ \text{or} \ a < -\frac{1}{2}$$

By i) and ii), 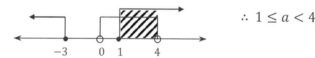 $\therefore \ a > \frac{1}{2}$

(3) A quadratic equation $x^2 + 2ax + 3 - 2a = 0$ will have real number solutions and a quadratic equation $x^2 - ax + a = 0$ will have two different complex number solutions.

$x^2 + 2ax + 3 - 2a = 0$

$\Rightarrow \ D \geq 0 \quad \therefore \ D = (2a)^2 - 4(3-2a) = 4a^2 + 8a - 12$

$$= 4(a^2 + 2a - 3) = 4(a+3)(a-1) \geq 0 \quad \therefore \ a \geq 1 \ \text{or} \ a \leq -3 \ \cdots \text{①}$$

$x^2 - ax + a = 0$

$\Rightarrow \ D < 0 \quad \therefore \ D = a^2 - 4a$

$$= a(a-4) < 0 \quad \therefore \ 0 < a < 4 \ \cdots \text{②}$$

By ① and ②, 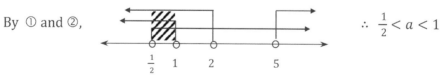 $\therefore \ 1 \leq a < 4$

(4) A quadratic equation $x^2 + 2(a-2)x + 2a - 1 = 0$ will have two different positive real number solutions.

Let α and β be the two different positive real number solutions.

Then, $\alpha + \beta > 0$ and $\alpha\beta > 0$

$\therefore \ \alpha + \beta = -2(a-2) > 0$ and $\alpha\beta = 2a - 1 > 0$

$\therefore \ a < 2$ and $a > \frac{1}{2} \ \cdots\cdots \text{①}$

Since $D > 0$, $D = 4(a-2)^2 - 4(2a-1) > 0$

$\therefore \ (a-2)^2 - (2a-1) > 0 \quad \therefore \ a^2 - 6a + 5 > 0$

$\therefore \ (a-1)(a-5) > 0 \quad \therefore \ a > 5 \ \text{or} \ a < 1 \ \cdots\cdots \text{②}$

By ① and ②, $\therefore \ \frac{1}{2} < a < 1$

(5) A quadratic inequality $ax^2 + 4x + a > 3$ will have all real number solutions.

$ax^2 + 4x + a > 3 \quad \Rightarrow \ ax^2 + 4x + a - 3 > 0$

① $a > 0$ (To have all real number solutions)

② Considering a quadratic equation $ax^2 + 4x + a - 3 = 0$, $D < 0$

$$\therefore \ D = 16 - 4a(a-3) = -4a^2 + 12a + 16 < 0$$

$$\therefore \ a^2 - 3a - 4 > 0 \qquad \therefore \ (a-4)(a+1) > 0 \qquad \therefore \ a > 4 \text{ or } a < -1$$

By ① and ②, 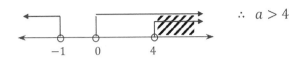 $\qquad \therefore \ a > 4$

(6) A quadratic inequality $a(2x^2 + 1) \leq (x-1)^2$ will have no real number solution.

$$a(2x^2 + 1) \leq (x-1)^2 \quad \Rightarrow \quad (2a-1)x^2 + 2x + a - 1 \leq 0$$

① $2a - 1 > 0 \quad \therefore \ a > \dfrac{1}{2}$ (To have no real number solution)

② Considering a quadratic equation $(2a-1)x^2 + 2x + a - 1 = 0, \ D < 0$

$$\therefore \ D = 4 - 4(2a-1)(a-1) = -8a^2 + 12a = -4a(2a-3) < 0$$

$$\therefore \ a(2a-3) > 0 \qquad \therefore \ a > \dfrac{3}{2} \text{ or } a < 0$$

By ① and ②, 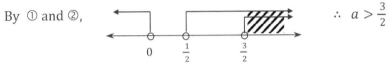 $\qquad \therefore \ a > \dfrac{3}{2}$

(7) A system $\begin{cases} x^2 - 5x < 0 \\ (x-4)(x-a) < 0 \end{cases}$ will have only one positive integer.

$$x^2 - 5x < 0 \quad \Rightarrow \quad x(x-5) < 0 \qquad \therefore \ 0 < x < 5$$

$$(x-4)(x-a) < 0 \quad \Rightarrow \quad a < x < 4 \text{ or } 4 < x < a$$

To have only one positive integer, $4 < x < a$ is impossible.   Consider $a < x < 4$

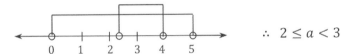 $\qquad \therefore \ 2 \leq a < 3$

#43 Solve each inequality.

Let $x = n + \alpha$ (n: integer, $0 \leq \alpha < 1$)

($[x]$ is the Maximum value of an integer which is not greater than x .)

(1) $2[x]^2 - 5[x] - 3 < 0$

Let $[x] = X$

Then, $2X^2 - 5X - 3 < 0 \qquad \therefore \ (2X + 1)(X - 3) < 0 \qquad \therefore \ -\dfrac{1}{2} < X < 3$

$\therefore \ -\dfrac{1}{2} < [x] < 3$

Since $[x]$ is an integer, $[x] = 0, \ 1, \ 2$

If $[x] = 0 \ \Rightarrow \ 0 \leq x < 1$

If $[x] = 1 \ \Rightarrow \ 1 \leq x < 2$

If $[x] = 2 \Rightarrow 2 \leq x < 3$

Therefore, $0 \leq x < 3$

(2) $[x-1]^2 + 3[x] - 3 < 0$

Let $x - 1 = n + \alpha$, where n is an integer and $0 \leq \alpha < 1$

Then, $x = n + \alpha + 1 = (n+1) + \alpha$

$\therefore$ $[x-1] = n$ and $[x] = n + 1$

$[x-1]^2 + 3[x] - 3 < 0 \Rightarrow n^2 + 3(n+1) - 3 < 0$

$$\Rightarrow n^2 + 3n < 0 \Rightarrow n(n+3) < 0 \qquad \therefore -3 < n < 0$$

Since n is an integer, $n = -2$ or $n = -1$

If $n = -2$, then $[x-1] = -2$ $\quad \therefore -2 \leq x - 1 < -1$ $\quad \therefore -1 \leq x < 0$

If $n = -1$, then $[x-1] = -1$ $\quad \therefore -1 \leq x - 1 < 0$ $\quad \therefore 0 \leq x < 1$

Therefore, $-1 \leq x < 1$

#44 Prove each inequality.

(1) $a^2 + b^2 > ab$ when $a + b = 1$

$a + b = 1 \Rightarrow b = 1 - a$

$a^2 + b^2 - ab = (a+b)^2 - 2ab - ab = (a+b)^2 - 3ab$

$$= 1 - 3a(1-a) = 3a^2 - 3a + 1 = 3(a^2 - a) + 1$$

$$= 3\left(\left(a - \tfrac{1}{2}\right)^2 - \left(\tfrac{1}{2}\right)^2\right) + 1 = 3\left(a - \tfrac{1}{2}\right)^2 + \tfrac{1}{4}$$

Since $3\left(a - \tfrac{1}{2}\right)^2 \geq 0$, $3\left(a - \tfrac{1}{2}\right)^2 + \tfrac{1}{4} > 0$

$\therefore$ $a^2 + b^2 - ab > 0$

Therefore, $a^2 + b^2 > ab$

(2) $a + b > ab$ when $a + b = 1$

Since $a + b = 1$, $a = 1 - b$

Show that : $1 > ab$

$$1 - ab = 1 - (1-b)b = b^2 - b + 1 = \left(b - \tfrac{1}{2}\right)^2 - \left(\tfrac{1}{2}\right)^2 + 1 = \left(b - \tfrac{1}{2}\right)^2 + \tfrac{3}{4}$$

Since $\left(b - \tfrac{1}{2}\right)^2 \geq 0$, $\left(b - \tfrac{1}{2}\right)^2 + \tfrac{3}{4} > 0$

$\therefore$ $1 - ab > 0$

$\therefore$ $1 > ab$

Therefore, $a + b > ab$

(3) $(a + b)(b + c)(c + a) \geq 8abc$ **when** $a > 0$, $b > 0$, **and** $c > 0$

Since $a > 0$, $b > 0$, and $c > 0$,

① $\frac{a+b}{2} \geq \sqrt{ab}$ ② $\frac{b+c}{2} \geq \sqrt{bc}$ and ③ $\frac{c+a}{2} \geq \sqrt{ca}$

∴ $\left(\frac{a+b}{2}\right)\left(\frac{b+c}{2}\right)\left(\frac{c+a}{2}\right) \geq \sqrt{a^2 b^2 c^2}$

∴ $\frac{(a+b)(b+c)(c+a)}{8} \geq abc$

Therefore, $(a + b)(b + c)(c + a) \geq 8abc$

(4) $a^3 + b^3 \geq ab(a + b)$ **when** $a > 0$ **and** $b > 0$

$$a^3 + b^3 - ab(a + b) = a^3 + b^3 - a^2 b - ab^2$$
$$= a^2(a - b) + b^2(b - a)$$
$$= (a - b)(a^2 - b^2)$$
$$= (a - b)(a + b)(a - b)$$
$$= (a - b)^2(a + b)$$

Since $(a - b)^2 \geq 0$ and $a + b > 0$, $(a - b)^2(a + b) \geq 0$.

∴ $a^3 + b^3 - ab(a + b) \geq 0$

Therefore, $a^3 + b^3 \geq ab(a + b)$

When $a = b$, LHS (Left Hand Side) $= RHS$ (Right Hand Side)

#45 Prove each inequality.

Then state for which values the *LHS* (Left Hand Side) equals the *RHS* (Right Hand Side).

(1) $a^2 + b^2 \geq ab$

$$a^2 + b^2 - ab = \left(a - \frac{1}{2}b\right)^2 - \left(\frac{1}{2}b\right)^2 + b^2 = \left(a - \frac{1}{2}b\right)^2 + \frac{3}{4}b^2$$

Since $\left(a - \frac{1}{2}b\right)^2 \geq 0$ and $\frac{3}{4}b^2 \geq 0$, $\left(a - \frac{1}{2}b\right)^2 + \frac{3}{4}b^2 \geq 0$

∴ $a^2 + b^2 - ab \geq 0$ ∴ $a^2 + b^2 \geq ab$

When $\left(a - \frac{1}{2}b\right)^2 = 0$ and $\frac{3}{4}b^2 = 0$, the equality becomes true.

∴ $LHS = RHS$ when $a = \frac{1}{2}b$ and $b = 0$

That is, $LHS = RHS$ when $a = b = 0$

(2) $a + \frac{9}{a} \geq 6$

Note that: $a > 0$, $b > 0 \Rightarrow \frac{2ab}{a+b} \leq \sqrt{ab} \leq \frac{a+b}{2}$

Since $\frac{a + \frac{9}{a}}{2} \geq \sqrt{a \cdot \frac{9}{a}}$, $\frac{a + \frac{9}{a}}{2} \geq 3$ ∴ $a + \frac{9}{a} \geq 6$

Multiplying both sides by a, $\quad a^2 - 6a + 9 \geq 0 \quad \therefore (a-3)^2 \geq 0$

When $a = 3$, $LHS = RHS$

(3) $\left(a + \dfrac{1}{b}\right)\left(b + \dfrac{1}{a}\right) \geq 4$ **when** $a > 0$ **and** $b > 0$

$\left(a + \dfrac{1}{b}\right)\left(b + \dfrac{1}{a}\right) = ab + 2 + \dfrac{1}{ab}$

Since $a > 0$ and $b > 0$, $\quad \dfrac{ab + \frac{1}{ab}}{2} \geq \sqrt{ab \cdot \dfrac{1}{ab}}$

$\therefore \quad ab + \dfrac{1}{ab} \geq 2 \quad \therefore \quad ab + \dfrac{1}{ab} + 2 \geq 4$

$\therefore \quad \left(a + \dfrac{1}{b}\right)\left(b + \dfrac{1}{a}\right) \geq 4$

When $ab = 1$, $\quad ab + \dfrac{1}{ab} + 2 = 4$

$\therefore$ That is, $LHS = RHS$ when $ab = 1$

#46 Minimum value.

(1) When $x > 0$, find minimum value of $x + 2 + \dfrac{4}{x+2}$.

Since $x > 0$, $\quad x + 2 > 0$ and $\dfrac{4}{x+2} > 0$.

$$x + 2 + \dfrac{4}{x+2} \geq 2\sqrt{(x+2) \cdot \dfrac{4}{x+2}}$$

$$= 2\sqrt{4} = 4$$

$\therefore \quad x + 2 + \dfrac{4}{x+2} \geq 4 \quad$ Therefore, minimum value of $x + 2 + \dfrac{4}{x+2}$ is 4.

(2) When $x > 0$ and $y > 0$, find minimum value of $(2x + 8y)\left(\dfrac{2}{x} + \dfrac{2}{y}\right)$.

$(2x + 8y)\left(\dfrac{2}{x} + \dfrac{2}{y}\right) = 4 + 16 + \dfrac{16y}{x} + \dfrac{4x}{y} = 20 + \dfrac{16y}{x} + \dfrac{4x}{y}$

Since $\dfrac{16y}{x} > 0$ and $\dfrac{4x}{y} > 0$,

$20 + \dfrac{16y}{x} + \dfrac{4x}{y} \geq 20 + 2\sqrt{\dfrac{16y}{x} \cdot \dfrac{4x}{y}} = 20 + 2\sqrt{16 \cdot 4} = 36$

$\therefore \quad (2x + 8y)\left(\dfrac{2}{x} + \dfrac{2}{y}\right) \geq 36$

Since $20 + \dfrac{16y}{x} + \dfrac{4x}{y} = 36 \iff \dfrac{16y}{x} + \dfrac{4x}{y} = 16 \iff 4x^2 + 16y^2 = 16xy$

$\iff x^2 - 4xy + 4y^2 = 0 \iff (x - 2y)^2 = 0 \iff x - 2y = 0$

$\therefore \quad LHS = RHS$ when $x = 2y$

Therefore, minimum value of $(2x + 8y)\left(\dfrac{2}{x} + \dfrac{2}{y}\right)$ is 36.

(3) **When $x^2 + ax - b = 0$ has two real number roots α, β and $x^2 + bx - a = 0$ has two real number roots γ, δ for negative numbers a, b, find minimum value of $\dfrac{1}{\alpha} + \dfrac{1}{\beta} + \dfrac{3}{\gamma} + \dfrac{3}{\delta}$.**

$(\alpha + \beta = -a, \ \alpha\beta = -b)$ and $(\gamma + \delta = -b, \ \gamma\delta = -a)$

$\therefore \ \dfrac{1}{\alpha} + \dfrac{1}{\beta} + \dfrac{3}{\gamma} + \dfrac{3}{\delta} = \dfrac{\alpha+\beta}{\alpha\beta} + \dfrac{3(\gamma+\delta)}{\gamma\delta} = \dfrac{-a}{-b} + \dfrac{3(-b)}{-a} = \dfrac{a}{b} + \dfrac{3b}{a}$

Since $\dfrac{a}{b} > 0$ and $\dfrac{3b}{a} > 0$, $\quad \dfrac{a}{b} + \dfrac{3b}{a} \geq 2\sqrt{\dfrac{a}{b} \cdot \dfrac{3b}{a}} = 2\sqrt{3}$

$\dfrac{a}{b} + \dfrac{3b}{a} = 2\sqrt{3} \quad \Longleftrightarrow \quad a^2 + 3b^2 = 2\sqrt{3}ab \quad \Longleftrightarrow \quad (a - \sqrt{3}b)^2 = 0 \quad \Longleftrightarrow \quad a - \sqrt{3}b = 0$

$\therefore \ LHS = RHS$ when $a = \sqrt{3}b$

Therefore, minimum value of $\dfrac{1}{\alpha} + \dfrac{1}{\beta} + \dfrac{3}{\gamma} + \dfrac{3}{\delta}$ is $2\sqrt{3}$.

(4) **When $\sqrt{2x} + \sqrt{3y} = 10$ for positive numbers x and y, find minimum value of $x + y$.**

Since $\sqrt{x}$ and $\sqrt{y}$ are real numbers,

$\left\{(\sqrt{2})^2 + (\sqrt{3})^2\right\}\left\{(\sqrt{x})^2 + (\sqrt{y})^2\right\} \geq (\sqrt{2}\sqrt{x} + \sqrt{3}\sqrt{y})^2$ by Cauchy's Inequality

$\therefore \ (2 + 3)(x + y) \geq 10^2 \quad \therefore \ x + y \geq 20$

Therefore, minimum value of $x + y$ is 20.

(5) **When $x^2 + y^2 + z^2 + a \geq 6(x + y + z)$ for real numbers $x, y,$ and z, find minimum value of a (a is a real number).**

$x^2 + y^2 + z^2 + a - 6(x + y + z) = x^2 + y^2 + z^2 - 6x - 6y - 6z + a$

$\qquad\qquad\qquad\qquad\qquad = (x - 3)^2 + (y - 3)^2 + (z - 3)^2 - 27 + a \geq 0$

Since $x, y,$ and z are real numbers,

$(x - 3)^2 \geq 0, \ (y - 3)^2 \geq 0,$ and $(z - 3)^2 \geq 0$

$\therefore \ -27 + a \geq 0$

$\therefore \ a \geq 27$

Therefore, minimum value of the real number a is 27.

(6) **When $ax^2 + 3xy + by^2 \geq 0$ for any real numbers x and y, find minimum value of ab.**

Since $ax^2 + 3xy + by^2 \geq 0$ for any real numbers x and y,

i) $a > 0$

ii) The discriminant $D \leq 0$

$\quad \therefore \ D = (3y)^2 - 4 \cdot a \cdot by^2 = 9y^2 - 4aby^2 = (9 - 4ab)y^2 \leq 0$

Since $y^2 \geq 0$, $9 - 4ab \leq 0$ $\therefore$ $ab \geq \dfrac{9}{4}$

Therefore, minimum value of ab is $\dfrac{9}{4}$.

(7) When a quadratic equation $x^2 - 3ax + a - 1 = 0$ has two different positive real number solutions for real number a, find minimum value of $2a - 3 + \dfrac{2}{a-1}$.

Since $x^2 - 3ax + a - 1 = 0$ has two different positive real number solutions,

i) $D > 0$

ii) (The sum of roots) > 0

iii) (The product of roots) > 0

By i) $D = (-3a)^2 - 4(a - 1) = 9a^2 - 4a + 4 = 9\left(a^2 - \dfrac{4}{9}a\right) + 4$

$$= 9\left(\left(a - \dfrac{2}{9}\right)^2 - \left(\dfrac{2}{9}\right)^2\right) + 4 = 9\left(a - \dfrac{2}{9}\right)^2 - \dfrac{4}{9} + 4 = 9\left(a - \dfrac{2}{9}\right)^2 + \dfrac{32}{9} > 0$$

ii) $3a > 0$ $\therefore$ $a > 0$

iii) $a - 1 > 0$ $\therefore$ $a > 1$

By ii) and iii) $a > 1$

$2a - 3 + \dfrac{2}{a-1} = 2(a - 1) + \dfrac{2}{a-1} - 1$

Since $a > 1$ and $a - 1 > 0$, $2(a - 1) > 0$ and $\dfrac{2}{a-1} > 0$

$\therefore$ $2(a - 1) + \dfrac{2}{a-1} \geq 2\sqrt{2(a - 1) \cdot \dfrac{2}{a-1}} = 4$

$\therefore$ $2(a - 1) + \dfrac{2}{a-1} - 1 \geq 4 - 1 = 3$

$\therefore$ $2a - 3 + \dfrac{2}{a-1} \geq 3$

Therefore, minimum value of $2a - 3 + \dfrac{2}{a-1}$ is 3.

When $2(a - 1) = \dfrac{2}{a-1}$ (i.e., $(a - 1)^2 = 1$; $a^2 - 2a = 0$; $a(a - 2) = 0$; $a = 0$ or $a = 2$

Since $a > 1$, $a = 2$), $LHS = RHS$

(8) When $a, b,$ and c are positive real numbers,

find minimum value of $\dfrac{a+b+c}{a} + \dfrac{a+b+c}{b} + \dfrac{a+b+c}{c}$.

$\dfrac{a+b+c}{a} + \dfrac{a+b+c}{b} + \dfrac{a+b+c}{c} = \left(1 + \dfrac{b}{a} + \dfrac{c}{a}\right) + \left(1 + \dfrac{a}{b} + \dfrac{c}{b}\right) + \left(1 + \dfrac{a}{c} + \dfrac{b}{c}\right)$

$$= \left(\dfrac{b}{a} + \dfrac{a}{b}\right) + \left(\dfrac{c}{b} + \dfrac{b}{c}\right) + \left(\dfrac{a}{c} + \dfrac{c}{a}\right) + 3$$

Since $a > 0$, $b > 0$, and $c > 0$,

$$\frac{b}{a} + \frac{a}{b} \geq 2\sqrt{\frac{b}{a} \cdot \frac{a}{b}}, \quad \frac{c}{b} + \frac{b}{c} \geq 2\sqrt{\frac{c}{b} \cdot \frac{b}{c}}, \text{ and } \frac{a}{c} + \frac{c}{a} \geq 2\sqrt{\frac{a}{c} \cdot \frac{c}{a}}$$

$$\therefore \frac{b}{a} + \frac{a}{b} \geq 2, \quad \frac{c}{b} + \frac{b}{c} \geq 2, \text{ and } \frac{a}{c} + \frac{c}{a} \geq 2$$

$$\therefore \left(\frac{b}{a} + \frac{a}{b}\right) + \left(\frac{c}{b} + \frac{b}{c}\right) + \left(\frac{a}{c} + \frac{c}{a}\right) \geq 6 \quad \text{By adding all inequalities}$$

$$\therefore \frac{a+b+c}{a} + \frac{a+b+c}{b} + \frac{a+b+c}{c} \geq 9$$

When $a = b = c$, $LHS = RHS$

Therefore, minimum value of $\frac{a+b+c}{a} + \frac{a+b+c}{b} + \frac{a+b+c}{c}$ is 9.

(9) When a and b are positive numbers,

find minimum value of $a^2 - 4a + \frac{a}{b} + \frac{4b}{a}$.

$$a^2 - 4a + \frac{a}{b} + \frac{4b}{a} = (a - 2)^2 - 4 + \frac{a}{b} + \frac{4b}{a}$$

Since $a > 0$ and $b > 0$, $\frac{a}{b} > 0$ and $\frac{4b}{a} > 0$

$$\therefore \frac{a}{b} + \frac{4b}{a} \geq 2\sqrt{\frac{a}{b} \cdot \frac{4b}{a}} = 4$$

When $\frac{a}{b} = \frac{4b}{a}$ (i.e., $a = 2b$), $LHS = RHS$

Since $(a - 2)^2 \geq 0$, $(a - 2)^2 - 4 \geq -4$

$$\therefore (a - 2)^2 - 4 + \frac{a}{b} + \frac{4b}{a} \geq -4 + \frac{a}{b} + \frac{4b}{a} \geq -4 + 4 = 0$$

Therefore, $a^2 - 4a + \frac{a}{b} + \frac{4b}{a}$ has minimum value of 0.

#47 Maximum value.

(1) When $x > 0$, find maximum value of $\frac{x}{x^2+x+4}$.

Since $x > 0$, divide both numerator and denominator of $\frac{x}{x^2+x+4}$ by x.

Then, we have $\frac{1}{x+1+\frac{4}{x}}$.

Since $x > 0$ and $\frac{4}{x} > 0$, $x + \frac{4}{x} \geq 2\sqrt{x \cdot \frac{4}{x}} = 4$

$$x + \frac{4}{x} = 4 \iff x^2 - 4x + 4 = 0 \iff (x - 2)^2 = 0$$

When $x = 2$, $LHS = RHS$

$$\therefore x + \frac{4}{x} \text{ has minimum value of 4.}$$

Since $x + 1 + \frac{4}{x} \geq 5$, $\quad \frac{1}{x+1+\frac{4}{x}} \leq \frac{1}{5}$

$\therefore \frac{1}{x+1+\frac{4}{x}}$ has maximum value of $\frac{1}{5}$.

Therefore, $\frac{x}{x^2+x+4}$ has maximum value of $\frac{1}{5}$.

(2) When $ab > 0$ and $a + 2b = 2$ for any real numbers a and b,

find maximum value of c such that $\frac{2}{a} + \frac{1}{b} \geq c$.

$(a + 2b)\left(\frac{2}{a} + \frac{1}{b}\right) = 2 + \frac{4b}{a} + \frac{a}{b} + 2 = 4 + \frac{4b}{a} + \frac{a}{b}$

Since $ab > 0$, $\frac{4b}{a} > 0$ and $\frac{a}{b} > 0$

$\therefore \frac{4b}{a} + \frac{a}{b} \geq 2\sqrt{\frac{4b}{a} \cdot \frac{a}{b}} = 4$

$\therefore 4 + \frac{4b}{a} + \frac{a}{b} \geq 8 \quad \therefore (a + 2b)\left(\frac{2}{a} + \frac{1}{b}\right) \geq 8$

When $\frac{4b}{a} = \frac{a}{b}$ (i.e., $a = 2b$), $LHS = RHS$

$\therefore (a + 2b)\left(\frac{2}{a} + \frac{1}{b}\right)$ has minimum value of 8.

Since $a + 2b = 2$, $\frac{2}{a} + \frac{1}{b} \geq 4$

Therefore, maximum value of c is 4.

(3) When $\frac{x^2}{2} + \frac{y^2}{4} + \frac{z^2}{9} = 1$ for real numbers $x, y,$ and z,

find maximum value of $x + y + z$.

Since $x, y,$ and z are real numbers,

$\left\{\left(\sqrt{2}\right)^2 + 2^2 + 3^2\right\}\left\{\left(\frac{x}{\sqrt{2}}\right)^2 + \left(\frac{y}{2}\right)^2 + \left(\frac{z}{3}\right)^2\right\} \geq (\sqrt{2}\frac{x}{\sqrt{2}} + 2\frac{y}{2} + 3\frac{z}{3})^2$ by Cauchy's Inequality

$\therefore (2 + 4 + 9)\left(\frac{x^2}{2} + \frac{y^2}{4} + \frac{z^2}{9}\right) \geq (x + y + z)^2$

$\therefore 15 \cdot 1 \geq (x + y + z)^2$

$\therefore -\sqrt{15} \leq x + y + z \leq \sqrt{15}$

Therefore, maximum value of $x + y + z$ is $\sqrt{15}$.

(4) When $a + 2\sqrt{2}b = 3 - x$ and $a^2 + 4b^2 = 9 - x^2$ for real numbers $a, b,$ and x, find maximum value of x.

Since $a, b,$ and x are real numbers,

$\{1^2 + (\sqrt{2})^2\}\{a^2 + (2b)^2\} \geq (1 \cdot a + \sqrt{2} \cdot 2b)^2$ by Cauchy's Inequality

$\therefore \ 3(a^2 + 4b^2) \geq (a + 2\sqrt{2}b)^2$

Since $a + 2\sqrt{2}b = 3 - x$ and $a^2 + 4b^2 = 9 - x^2$, $\ 3(9 - x^2) \geq (3 - x)^2$

$\therefore \ 27 - 3x^2 \geq 9 - 6x + x^2$

$\therefore \ 4x^2 - 6x - 18 \leq 0 \quad \therefore \ 2x^2 - 3x - 9 \leq 0 \quad \therefore \ (2x + 3)(x - 3) \leq 0$

$\therefore \ -\dfrac{3}{2} \leq x \leq 3$

Therefore, maximum value of x is 3.

Chapter 4. Elements of Coordinate Geometry and Transformations

#1 Find the distance between two points A and B on a number line or a coordinate plane.

(1) $A\left(-2 + \sqrt{3}\right)$, $B(-6)$

$d(A, B) = \left|-6 - (-2 + \sqrt{3})\right| = \left|-4 - \sqrt{3}\right| = 4 + \sqrt{3}$

(2) $A(2, 3)$, $B(-5, -2)$

$d(A, B) = \sqrt{(-5 - 2)^2 + (-2 - 3)^2} = \sqrt{49 + 25} = \sqrt{74}$

(3) $A(-4, -1)$, $B(-2, -7)$

$d(A, B) = \sqrt{(-2 + 4)^2 + (-7 + 1)^2} = \sqrt{4 + 36} = \sqrt{40} = 2\sqrt{10}$

(4) $A(0, 0)$, $B(3, 4)$

$d(A, B) = \sqrt{(3 - 0)^2 + (4 - 0)^2} = \sqrt{9 + 16} = \sqrt{25} = 5$

#2 Find the coordinates of the point P.

(1) For two points $A(2, 2)$ and $B(-2, 4)$, $\overline{AP} = \overline{BP}$ and P is on the x-axis.

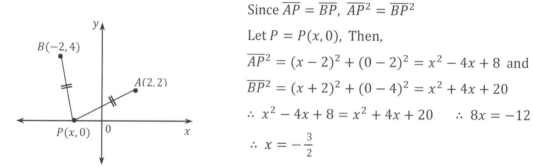

Since $\overline{AP} = \overline{BP}$, $\overline{AP}^2 = \overline{BP}^2$

Let $P = P(x, 0)$, Then,

$\overline{AP}^2 = (x - 2)^2 + (0 - 2)^2 = x^2 - 4x + 8$ and

$\overline{BP}^2 = (x + 2)^2 + (0 - 4)^2 = x^2 + 4x + 20$

$\therefore x^2 - 4x + 8 = x^2 + 4x + 20 \qquad \therefore 8x = -12$

$\therefore x = -\dfrac{3}{2}$

Therefore, $P = P\left(-\dfrac{3}{2}, 0\right)$

(2) For two points $A(1, 1)$ and $B(2, 3)$, $\overline{AP} = \overline{BP}$ and P is on the line $y = x + 2$.

Since $\overline{AP} = \overline{BP}$, $\overline{AP}^2 = \overline{BP}^2$

Let $P = P(x, x + 2)$, Then,

$\overline{AP}^2 = (x - 1)^2 + (x + 2 - 1)^2 = 2x^2 + 2$ and

$\overline{BP}^2 = (x - 2)^2 + (x + 2 - 3)^2 = 2x^2 - 6x + 5$

$\therefore 2x^2 + 2 = 2x^2 - 6x + 5 \qquad \therefore 6x = 3 \qquad \therefore x = \dfrac{1}{2}$

Therefore, $P = P\left(\dfrac{1}{2}, \dfrac{5}{2}\right)$

(3) For a triangle $\triangle ABC$, the midpoints of $\overline{AB}, \overline{BC}$, and $\overline{CA}$ are $D\left(2, \frac{7}{2}\right)$, $E\left(\frac{3}{2}, 4\right)$, and $F\left(\frac{5}{2}, \frac{9}{2}\right)$, respectively. Find the centroid P.

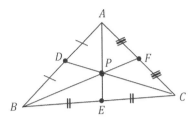

Let $A(x_1, y_1)$, $B(x_2, y_2)$, and $C(x_3, y_3)$ are the vertexes of the triangle $\triangle ABC$.

Then, $\dfrac{x_1+x_2}{2} = 2$, $\dfrac{y_1+y_2}{2} = \dfrac{7}{2}$

$\dfrac{x_2+x_3}{2} = \dfrac{3}{2}$, $\dfrac{y_2+y_3}{2} = 4$

$\dfrac{x_1+x_3}{2} = \dfrac{5}{2}$, $\dfrac{y_1+y_3}{2} = \dfrac{9}{2}$

$\therefore \dfrac{x_1+x_2}{2} + \dfrac{x_2+x_3}{2} + \dfrac{x_1+x_3}{2} = \dfrac{12}{2}$ and $\dfrac{y_1+y_2}{2} + \dfrac{y_2+y_3}{2} + \dfrac{y_1+y_3}{2} = \dfrac{24}{2}$

$\therefore 2(x_1 + x_2 + x_3) = 12$ and $2(y_1 + y_2 + y_3) = 24$

$\therefore x_1 + x_2 + x_3 = 6$ and $y_1 + y_2 + y_3 = 12$

Therefore, $P = \left(\dfrac{x_1+x_2+x_3}{3}, \dfrac{y_1+y_2+y_3}{3}\right) = (2, 4)$

(4) For two points $A(-2, a)$ and $B(b, 3)$, the midpoint of the segment $\overline{AB}$ is $P = (-2, -1)$. Find the point which divides $\overline{AB}$ externally in the ratio $\overline{AP} : \overline{PB} = 2 : 1$

Since $\dfrac{-2+b}{2} = -2$ and $\dfrac{a+3}{2} = -1$, $a = -5$ and $b = -2$

$\therefore P = \left(\dfrac{2 \cdot (-2) - 1 \cdot (-2)}{2-1}, \dfrac{2 \cdot 3 - 1 \cdot (-5)}{2-1}\right) = (-2, 11)$

(5) The line $(1 + k)x - (1 + 3k)y - 2 - 3k = 0$ passes through a point P for any k. Find the coordinates of P.

$(1 + k)x - (1 + 3k)y - 2 - 3k = 0 \Rightarrow (x - y - 2) + k(x - 3y - 3) = 0$

Since this equation of a line is always true for any k, $x - y - 2 = 0$ and $x - 3y - 3 = 0$.

$ x - y - 2 = 0$

$\underline{-)\ x - 3y - 3 = 0}$

$ 2y + 1 = 0 \quad \therefore y = -\dfrac{1}{2}, \ x = \dfrac{3}{2}$

Therefore, $P = \left(\dfrac{3}{2}, -\dfrac{1}{2}\right)$

#3 For two points $A(-3, 4)$ and $B(2, -1)$ on a coordinate plane,

find the coordinates of the point P.

(1) P is the point which divides the segment $\overline{AB}$ internally in the ratio $\overline{AP} : \overline{PB} = 2 : 3$

$$P = \left(\frac{2 \cdot 2 + 3 \cdot (-3)}{2 + 3}, \frac{2 \cdot (-1) + 3 \cdot 4}{2 + 3} \right) = \left(\frac{-5}{5}, \frac{10}{5} \right) = (-1, 2)$$

(2) P is the point which divides the segment $\overline{AB}$ externally in the ratio $\overline{AP} : \overline{PB} = 2 : 1$

$$P = \left(\frac{2 \cdot 2 - 1 \cdot (-3)}{2 - 1}, \frac{2 \cdot (-1) - 1 \cdot 4}{2 - 1} \right) = (7, -6)$$

(3) P is the midpoint of segment $\overline{AB}$

$$P = \left(\frac{-3 + 2}{2}, \frac{4 - 1}{2} \right) = \left(\frac{-1}{2}, \frac{3}{2} \right)$$

#4 Find the value of real number a.

(1) The distance between two points $A(-1, 3)$ and $B(a, -3)$ is $4\sqrt{3}$. Find the value of a.

$$\overline{AB} = \sqrt{(a + 1)^2 + (-3 - 3)^2} = \sqrt{(a + 1)^2 + 36} = 4\sqrt{3} = \sqrt{48}$$

$$\therefore \ (a + 1)^2 + 36 = 48 \quad \therefore \ (a + 1)^2 = 12 \quad \therefore a = \pm 2\sqrt{3} - 1$$

(2) For three points $A(-1, 2)$, $B(7, 1)$, and $P(x, 0)$, find the value of $a = \alpha + \beta$ where α

is minimum value of $\overline{AP^2} + \overline{BP^2}$ and β is minimum value of $\overline{AP} + \overline{BP}$.

$$\overline{AP^2} + \overline{BP^2} = (x + 1)^2 + (0 - 2)^2 + (x - 7)^2 + (0 - 1)^2$$

$$= x^2 + 2x + 1 + 4 + x^2 - 14x + 49 + 1$$

$$= 2x^2 - 12x + 55$$

$$= 2(x^2 - 6x) + 55 = 2(x - 3)^2 - 18 + 55$$

$$= 2(x - 3)^2 + 37$$

$\therefore$ When $x = 3$, minimum value of $\overline{AP^2} + \overline{BP^2}$ is 37. Therefore, $\alpha = 37$

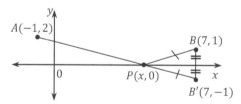

Let $B' = B'(7, 1)$ be the reflection of point $B(7, 1)$ in the x-axis. Then, $\overline{BP} = \overline{B'P}$

$\therefore \ \overline{AP} + \overline{BP} = \overline{AP} + \overline{B'P} \geq \overline{AB'} = \sqrt{(7 - (-1))^2 + (-1 - 2)^2} = \sqrt{64 + 9} = \sqrt{73}$

$\therefore$ Minimum value of $\overline{AP} + \overline{BP}$ is $\sqrt{73}$. Therefore, $\beta = \sqrt{73}$

Hence $a = \alpha + \beta = 37 + \sqrt{73}$

(3) For two points $A(-6, 4)$ and $B(-3, -5)$, a line $y = 2x + a$ passes through a point
which divides the segment $\overline{AB}$ internally in the ratio $\overline{AP} : \overline{PB} = 2 : 1$
Find the value of a.

$P = \left(\frac{2 \cdot (-3) + 1 \cdot (-6)}{2+1}, \frac{2 \cdot (-5) + 1 \cdot 4}{2+1} \right) = \left(\frac{-12}{3}, \frac{-6}{3} \right) = (-4, -2)$

Since $y = 2x + a$ passes through $(-4, -2)$, $\quad -2 = 2 \cdot (-4) + a = -8 + a$

$\therefore \ a = 6$

(4) For three points $A(-2, 1)$, $B(-3, -2)$, and $C(2, 7)$,

let $M_1, M_2,$ and M_3 be the midpoints of $\overline{AB}, \overline{BC},$ and $\overline{AC}$, respectively.

When $P(\alpha, \beta)$ is the centroid of the triangle $\Delta M_1 M_2 M_3$, find the value of $a = \alpha + \beta$.

Since the centroids of ΔABC and $\Delta M_1 M_2 M_3$ are the same,

$P(\alpha, \beta) = \left(\frac{-2-3+2}{3}, \frac{1-2+7}{3} \right) = (-1, 2)$

Therefore, $a = \alpha + \beta = -1 + 2 = 1$

(5)

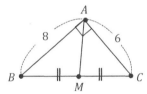

For a triangle with $m(\angle A) = 90^\circ$,
$\overline{AB} = 8$, $\overline{AC} = 6$, and M is the midpoint of the
segment $\overline{BC}$. Find the length a of $\overline{AM}$.

By Pythagorean Theorem, $\overline{BC} = \sqrt{8^2 + 6^2} = \sqrt{100} = 10$

$\therefore \ \overline{BM} = \frac{1}{2} \cdot 10 = 5$

Since $\overline{AB}^2 + \overline{AC}^2 = 2(\overline{AM}^2 + \overline{BM}^2)$ by the median theorem, $64 + 36 = 2(\overline{AM}^2 + 25)$.

$\therefore \ \overline{AM}^2 + 25 = 50 \quad \therefore \ \overline{AM}^2 = 25 \quad \therefore \ \overline{AM} = 5$

Thereore, the length a of $\overline{AM}$ is 5.

(6)

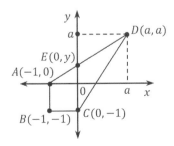

For a quadrilateral $\square ABCD$ with vertexes
$A(-1, 0), B(-1, -1), C(0, -1),$ and $D(a, a)$, y-axis
divides the area of $\square ABCD$ into 2 halves.
Find the value of a.

Let $E(0, y)$ be an intersection point of the segment $\overline{AD}$ and y-axis such that $\overline{AE} : \overline{ED} = m : n$

Then, $(0, y) = \left(\frac{m \cdot a + n \cdot (-1)}{m+n}, \frac{m \cdot a + n \cdot 0}{m+n} \right) = \left(\frac{ma-n}{m+n}, \frac{ma}{m+n} \right)$

$$\therefore \ \frac{ma-n}{m+n} = 0 \ \text{ and } \ \frac{ma}{m+n} = y$$

$$\therefore \ n = ma \ \text{ and } \ \frac{ma}{m+n} = \frac{ma}{m+ma} = \frac{ma}{m(1+a)} = \frac{a}{1+a} = y$$

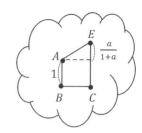

The area of the trapezoid $\square ABCE$ is $\frac{1}{2} \cdot \left\{ 1 + \left(\frac{a}{1+a} + 1 \right) \right\} \cdot 1 = \frac{1}{2} \left(\frac{a}{1+a} + 2 \right)$

The area of the triangle $\triangle ECD$ is $\frac{1}{2} \cdot \left(\frac{a}{1+a} + 1 \right) \cdot a = \frac{a}{2} \left(\frac{a}{1+a} + 1 \right)$

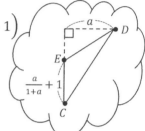

Since the two areas are the same, $\frac{1}{2} \left(\frac{a}{1+a} + 2 \right) = \frac{a}{2} \left(\frac{a}{1+a} + 1 \right)$

$$\therefore \ \frac{a}{1+a} + 2 = a \left(\frac{a}{1+a} + 1 \right)$$

$$\therefore \ a + 2(1 + a) = a(a + (1 + a))$$

$$\therefore \ 2a^2 - 2a - 2 = 0 \ ; \ a^2 - a - 1 = 0$$

By the quadratic formula, $a = \dfrac{-(-1) \pm \sqrt{(-1)^2 - 4 \cdot 1 \cdot (-1)}}{2 \cdot 1} = \dfrac{1 \pm \sqrt{5}}{2}$

Since $a > 0$, $a = \dfrac{1+\sqrt{5}}{2}$

(7) For four points $A(2,5), B(-1,1), C(\alpha,-2),$ and $D(\beta,2),$ the quadrilateral $\square ABCD$ is a rhombus. (where $\alpha > 2$) Find the value of the real number $a = \alpha\beta$.

Since the midpoints of segments $\overline{AC}$ and $\overline{BD}$ are the same,

$$\left(\frac{2+\alpha}{2} + \frac{5-2}{2} \right) = \left(\frac{-1+\beta}{2} + \frac{1+2}{2} \right)$$

$$\therefore \ \frac{2+\alpha}{2} = \frac{-1+\beta}{2} \quad \therefore \ \alpha - \beta = -3$$

Since $\overline{AB}^2 = \overline{BC}^2$,

$$(2+1)^2 + (5-1)^2 = (\alpha+1)^2 + (-2-1)^2$$

$$\therefore \ 9 + 16 = \alpha^2 + 2\alpha + 1 + 9 \quad \therefore \ \alpha^2 + 2\alpha - 15 = 0$$

$$\therefore \ (\alpha+5)(\alpha-3) = 0 \quad \therefore \alpha = -5 \text{ or } \alpha = 3$$

Since $\alpha > 2$, $\alpha = 3$ and $\beta = 6$

Therefore, $a = \alpha\beta = 3 \cdot 6 = 18$

(8) The area of a triangle $\triangle ABC$ which is surrounded by a line $ax + 3y = 3a$, x-axis, and y-axis is 24. Find the value of positive number a.

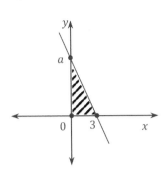

Since $a > 0$, divide both sides of the equation of the line

$ax + 3y = 3a$ by $3a$.

Then, $\frac{x}{3} + \frac{y}{a} = 1$

∴ x-intercept of the line is 3 and y-intercept of the line is a.

∴ The shaded area is $\frac{1}{2} \cdot 3 \cdot a = 24$

Therefore, $a = 16$

#5 For a regular triangle $\triangle ABC$, the coordinate of the point A is $A(-4, 6)$.

When the centroid of $\triangle ABC$ is $P(-1, 2)$, find the area of the $\triangle ABC$.

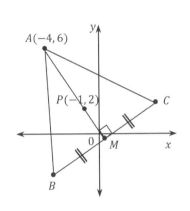

Let M be the midpoint of $\overline{BC}$.

Since $\overline{AP} : \overline{PM} = 2 : 1$, $\overline{AM} : \overline{AP} = 3 : 2$

∴ $\overline{AM} = \frac{3}{2}\overline{AP}$

Since $\overline{AP} = \sqrt{(-1+4)^2 + (2-6)^2} = \sqrt{9+16} = \sqrt{25} = 5$,

$\overline{AM} = \frac{15}{2}$

Since $\overline{AM} : \overline{BM} = \sqrt{3} : 1$

by the Pythagorean Theorem, $\overline{AM} = \sqrt{3}\,\overline{BM}$

∴ $\overline{BM} = \frac{1}{\sqrt{3}}\overline{AM} = \frac{1}{\sqrt{3}} \cdot \frac{15}{2} = \frac{15 \cdot \sqrt{3}}{2 \cdot 3} = \frac{5\sqrt{3}}{2}$

Therefore, the area of $\triangle ABC$ is $\frac{1}{2} \cdot \overline{BC} \cdot \overline{AM} = \frac{1}{2} \cdot \left(2 \cdot \frac{5\sqrt{3}}{2}\right) \cdot \frac{15}{2} = \frac{75\sqrt{3}}{4}$

#6 Find an equation in the standard form for each line.

(1) with y-intercept -3 and slope 2

$y = 2x - 3$ ∴ $2x - y - 3 = 0$

(2) with y-intercept 5 and slope 0

$y = 0 \cdot x + 5$ ∴ $y - 5 = 0$

(3) with x-intercept 5 and slope $-\frac{2}{3}$

$y = -\frac{2}{3}x + b$; $0 = -\frac{2}{3} \cdot 5 + b$; $b = \frac{10}{3}$ ∴ $y = -\frac{2}{3}x + \frac{10}{3}$

Therefore, $2x + 3y - 10 = 0$

(4) with x-intercept -3 and slope -2

$y = -2x + b$; $0 = -2 \cdot (-3) + b$; $b = -6$ ∴ $y = -2x - 6$

Therefore, $2x + y + 6 = 0$

(5) through $(1, 2)$ with slope 3

$y = 3x + b$; $2 = 3 \cdot 1 + b$; $b = -1$ ∴ $y = 3x - 1$ ∴ $3x - y - 1 = 0$

(OR, $y - 2 = 3(x - 1)$; $y = 3x - 1$ ∴ $3x - y - 1 = 0$)

(6) through $(3, -4)$ with slope -2

$y = -2x + b$; $-4 = -2 \cdot 3 + b$; $b = 2$ ∴ $y = -2x + 2$ ∴ $2x + y - 2 = 0$

(OR, $y + 4 = -2(x - 3)$; $y = -2x + 2$ ∴ $2x + y - 2 = 0$)

(7) through $(2, 3)$ with undefined slope

No slope $\Rightarrow$ No change in x ∴ $x = 2$; $x - 2 = 0$

(8) through $(-2, 3)$ with y-intercept -1

$y = \dfrac{-4}{2} x - 1 = -2x - 1$ ∴ $2x + y + 1 = 0$

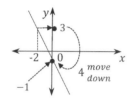

(9) through $(2, 4)$ with x-intercept -5

$y = \dfrac{4}{7} x + b$; $0 = \dfrac{4}{7} \cdot (-5) + b$; $b = \dfrac{20}{7}$ ∴ $y = \dfrac{4}{7} x + \dfrac{20}{7}$ ∴ $4x - 7y + 20 = 0$

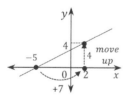

(10) through $(3, 1)$ and $(-2, 4)$

$m = \dfrac{4 - 1}{-2 - 3} = -\dfrac{3}{5}$ ∴ $y = -\dfrac{3}{5} x + b$

Thus, $1 = -\dfrac{3}{5} \cdot 3 + b$; $b = \dfrac{14}{5}$ ∴ $y = -\dfrac{3}{5} x + \dfrac{14}{5}$

Therefore, $3x + 5y - 14 = 0$

(OR, Using Point-Slope form, $y - 1 = -\dfrac{3}{5}(x - 3)$; $5y - 5 = -3x + 9$

∴ $3x + 5y - 14 = 0$)

(11) through $(-2, -3)$ and $(-1, 5)$

$m = \dfrac{5 + 3}{-1 + 2} = 8$ ∴ $y = 8x + b$

Thus, $5 = 8 \cdot (-1) + b$; $b = 13$ ∴ $y = 8x + 13$

Therefore, $8x - y + 13 = 0$

(OR, Using Point-Slope form, $y + 3 = 8(x + 2)$ $\therefore 8x - y + 13 = 0$)

(12) with x-intercept -3 and y-intercept 3

$m = \dfrac{0-3}{-3-0} = 1$ $\therefore y = x + 3$ $\therefore x - y + 3 = 0$

(13) with x-intercept $\dfrac{3}{2}$ and y-intercept -4

$m = \dfrac{-4-0}{0-\frac{3}{2}} = \dfrac{-4}{-\frac{3}{2}} = \dfrac{8}{3}$ $\therefore y = \dfrac{8}{3}x - 4$ $\therefore 8x - 3y - 12 = 0$

(14) Vertical line through $(-1, 2)$

$x = -1$ $\therefore x + 1 = 0$

(15) Horizontal line through $(3, -4)$

$y = -4$ $\therefore y + 4 = 0$

#7 Find an equation for the line through $(2, 3)$ which is:

(1) parallel to the line $y = 2x - 5$

$y = 2x + b$; $3 = 2 \cdot 2 + b$; $b = -1$ $\therefore y = 2x - 1$ $\therefore 2x - y - 1 = 0$

(2) parallel to the line $y = -3x + 1$

$y = -3x + b$; $3 = -3 \cdot 2 + b$; $b = 9$ $\therefore y = -3x + 9$ $\therefore 3x + y - 9 = 0$

(3) parallel to the line $x = 4$

$x = 2$ $\therefore x - 2 = 0$

(4) parallel to the line $y = -2$

$y = 3$ $\therefore y - 3 = 0$

(5) parallel to the line $3x + 4y = 5$

$y = -\dfrac{3}{4}x + \dfrac{5}{4}$ $\therefore y = -\dfrac{3}{4}x + b$

$3 = -\dfrac{3}{4} \cdot 2 + b$; $b = \dfrac{9}{2}$ $\therefore y = -\dfrac{3}{4}x + \dfrac{9}{2}$

Therefore, $3x + 4y - 18 = 0$

(6) perpendicular to the line $y = \frac{2}{3}x - 1$

$$y = -\frac{3}{2}x + b$$

$$3 = -\frac{3}{2} \cdot 2 + b \;\; ; \;\; b = 6 \quad \therefore \; y = -\frac{3}{2}x + 6$$

Therefore, $3x + 2y - 12 = 0$

(7) perpendicular to the line $x + 3y = -3$

$$x + 3y = -3 \;\; \Rightarrow \;\; y = -\frac{1}{3}x - 1$$

$$\therefore \; y = 3x + b$$

$$3 = 3 \cdot 2 + b \;\; ; \;\; b = -3 \quad \therefore \; y = 3x - 3$$

Therefore, $3x - y - 3 = 0$

(8) perpendicular to the line $x = 5$

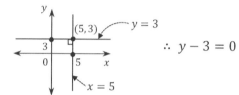

$\therefore \; y - 3 = 0$

(9) perpendicular to the line $y = -2$

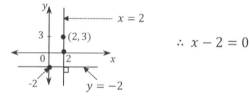

$\therefore \; x - 2 = 0$

#8 Find the value of a for each line:

(1) through $(2, 3)$ and $(1, -a)$ with slope 2

$$2 = \frac{-a-3}{1-2} = \frac{-a-3}{-1} = a + 3 \quad \therefore \; a = -1$$

(2) through $(2a - 1, -2)$ and $(-1, 1)$ with slope -2

$$-2 = \frac{1+2}{-1-2a+1} = \frac{3}{-2a} \;\; ; \;\; 4a = 3 \quad \therefore \; a = \frac{3}{4}$$

(3) through $(1, -2)$, $(-3, 2)$, and $(-a + 1, -5)$

$$m = \frac{2+2}{-3-1} = \frac{4}{-4} = -1 \quad \therefore \; y = -x + b$$

Thus, $-2 = -1 + b \;\; ; \;\; b = -1 \quad \therefore \; y = -x - 1$

$$-5 = a - 1 - 1 = a - 2 \quad \therefore \; a = -3$$

(4) through $(2a + 1, -4)$, $(2, 5)$, and $(2, -3)$

$(2, 5)$ and $(2, -3) \;\; \Rightarrow \;\;$ No change in x ; No slope ; Vertical line

$$\therefore \; 2a + 1 = 2 \quad \therefore \; a = \frac{1}{2}$$

(5) through $(-3, 3)$, $(3, a - 1)$, **and** $(0, 3)$

$(-3, 3)$ and $(0, 3) \Rightarrow$ No change in y ; Slope is 0 ; Horizontal line

$\therefore \quad a - 1 = 3 \quad \therefore \ a = 4$

(6) through $(a, 2a - 3)$ **and** $(-a - 1, 3 + 4a)$ **and parallel to the** x**-axis**

Parallel to x-axis $\Rightarrow$ Horizontal line

$\therefore \ 2a - 3 = 3 + 4a \ ; \ 2a = -6 \quad \therefore \ a = -3$

(7) through $(-3a + 1, -5)$ **and** $(2a - 1, a + 3)$ **and perpendicular to the** x**-axis**

Perpendicular to x-axis $\Rightarrow$ vertical line

$\therefore \ -3a + 1 = 2a - 1 \ ; \ 5a = 2 \quad \therefore \ a = \dfrac{2}{5}$

(8) through $(3, -2a)$ **and** $(2a - 1, -3a + 2)$ **and parallel to the** y**-axis**

Parallel to y-axis $\Rightarrow$ Vertical line

$\therefore \ 3 = 2a - 1 \ ; \ 2a = 4 \quad \therefore \ a = 2$

(9) through $(-1, 5)$ **and** $(2, -4)$ **and parallel to the line** $ax + 3y + 5 = 0$

$m = \dfrac{-4 - 5}{2 + 1} = \dfrac{-9}{3} = -3 \quad \therefore \ y = -3x + b$

$ax + 3y + 5 = 0 \ \Rightarrow \ 3y = -ax - 5 \ \Rightarrow \ y = -\dfrac{a}{3}x - \dfrac{5}{3}$

$\therefore \ -\dfrac{a}{3} = -3 \quad \therefore \ a = 9$

#9 Find the value of a **such that the line** $ax + 2y = 5$:

(1) is parallel to the line $2x + 3y = -2$.

The two lines have the same slope.

$3y = -2x - 2 \ ; \ y = -\dfrac{2}{3}x - \dfrac{2}{3} \quad \therefore$ Slope is $-\dfrac{2}{3}$

$ax + 2y = 5 \ \Rightarrow \ y = -\dfrac{1}{2}ax + \dfrac{5}{2} \quad \therefore$ Slope is $-\dfrac{1}{2}a$

Therefore, $-\dfrac{2}{3} = -\dfrac{1}{2}a \quad \therefore \ a = \dfrac{4}{3}$

(2) is perpendicular to the line $y = -2x + 3$.

Negative reciprocals of each other

Since $-2 \cdot \left(-\dfrac{1}{2}a\right) = -1, \quad a = -1$

(3) coincides with the line $6y = -4x + 15$.

$6y = -4x + 15 \ \Rightarrow \ 4x + 6y = 15 \ \Rightarrow \ \dfrac{4}{3}x + 2y = 5$

$\therefore \ a = \dfrac{4}{3}$

#10 Find the value of ab for which :

(1) the system $\begin{cases} x - 3y = a \\ 2x + by = 3 \end{cases}$ **has the intersection point $(2, 3)$.**

Substitute $(2, 3)$ into each equation.

Then, $2 - 9 = a$; $a = -7$ and $4 + 3b = 3$; $b = -\frac{1}{3}$

$\therefore$ $ab = \frac{7}{3}$

(2) the system $\begin{cases} -ax + by = 4 \\ 2ax + 3by = 2 \end{cases}$ **has the intersection point $(-1, 2)$.**

Substitute $(-1, 2)$ into each equation.

Then, $\begin{cases} a + 2b = 4 \\ -2a + 6b = 2 \end{cases}$

$ 2a + 4b = 8$

$\underline{+)\; -2a + 6b = 2}$

$ 10b = 10$; $b = 1$

$\therefore$ $a = 4 - 2b = 4 - 2 = 2$

$\therefore$ $ab = 2$

(3) the system $\begin{cases} px + y = 3 \\ 2x - 3y = q \end{cases}$ **has no intersection when $p = a$, $q \neq b$.**

Parallel $\therefore$ $\frac{p}{2} = -\frac{1}{3} \neq \frac{3}{q}$ $\therefore$ $p = -\frac{2}{3}$ and $q \neq -9$

$\therefore$ $a = -\frac{2}{3}$, $b = -9$ $\therefore$ $ab = 6$

(4) a system $\begin{cases} 2ax + 4y = -3 \\ 3x + 6y = 2b \end{cases}$ **has unlimited numbers of intersections.**

$\frac{2a}{3} = \frac{4}{6} = \frac{-3}{2b}$ $\therefore$ $2a = 2$; $a = 1$ and $4b = -9$; $b = -\frac{9}{4}$

$\therefore$ $ab = -\frac{9}{4}$

#11 Find the value of a such that :

(1) the system $\begin{cases} ax + y = -2 \\ -3x + 2y = 4 \end{cases}$ **has no solution.**

Parallel

$-\frac{a}{3} = \frac{1}{2} \neq -\frac{2}{4}$ $\therefore$ $a = -\frac{3}{2}$

(2) the system $\begin{cases} 2x - ay + 3 = 0 \\ x + 3y - 2 = 0 \\ 2x + y + 1 = 0 \end{cases}$ has one solution.

$\begin{cases} 2x - ay + 3 = 0 & \cdots\cdots ① \\ x + 3y - 2 = 0 & \cdots\cdots ② \\ 2x + y + 1 = 0 & \cdots\cdots ③ \end{cases}$

$\Rightarrow ① - ③ \; ; \; (-a - 1)y + 2 = 0$

$2 \cdot ② - ③ \; ; \; 5y - 5 = 0 \; ; \; y = 1$

$\therefore (-a - 1) + 2 = 0 \quad \therefore a = 1$

(3) the system $\begin{cases} x - 3y = 2 \\ 2x + y = -3 \end{cases}$ has a solution $(2a, -1)$.

$2x - 6y = 4$

$-) \; \underline{2x + y = -3}$

$\qquad -7y = 7 \; ; y = -1$

$\therefore x = 3y + 2 = -3 + 2 = -1 \quad \therefore 2a = -1 \quad \therefore a = -\dfrac{1}{2}$

(4) the line $2ax + 3y - 1 = 0$ passes through the intersection of the system

$\begin{cases} x - 2y = 3 \\ 2x + 2y = 1 \end{cases}.$

$2x - 4y = 6 \qquad\qquad\qquad x - 2y = 3$

$-) \; \underline{2x + 2y = 1} \qquad\qquad +) \; \underline{2x + 2y = 1}$

$\qquad -6y = 5 \; ; \; y = -\dfrac{5}{6} \qquad 3x \quad\;\; = 4 \; ; \; x = \dfrac{4}{3}$

$\therefore 2a \cdot \dfrac{4}{3} + 3 \cdot \left(-\dfrac{5}{6}\right) - 1 = 0 \quad ; \; \dfrac{8}{3}a = \dfrac{7}{2} \quad \therefore a = \dfrac{21}{16}$

#12 Find the equation of each line such that:

(1) the line passes through the intersection of the system $\begin{cases} x + 2y = 3 \\ 3x + y = -2 \end{cases}$

and runs parallel to the y-axis.

$x + 2y = 3$

$-) \; \underline{6x + 2y = -4}$

$\quad -5x \qquad = 7 \; ; \; x = -\dfrac{7}{5} \qquad \therefore 2y = -x + 3 = \dfrac{7}{5} + 3 = \dfrac{22}{5} \qquad \therefore y = \dfrac{11}{5}$

$\therefore$ The intersection is $\left(-\dfrac{7}{5}, \dfrac{11}{5}\right)$.

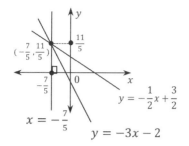

Parallel to the y-axis

$$\Rightarrow \ x = -\frac{7}{5} \qquad \therefore \ 5x + 7 = 0$$

(2) the line passes through the intersection of the system $\begin{cases} -x + y + 2 = 0 \\ 2x + y - 3 = 0 \end{cases}$

and runs perpendicular to the x-axis.

$$-x + y + 2 = 0$$
$$-) \ \underline{2x + y - 3 = 0}$$
$$-3x \quad + 5 = 0 \ ; \ x = \frac{5}{3} \quad \therefore \ y = x - 2 = \frac{5}{3} - 2 = -\frac{1}{3}$$

$\therefore$ The intersection is $\left(\frac{5}{3} , -\frac{1}{3} \right)$.

$\therefore$ Perpendicular to the x-axis $\Rightarrow x = \frac{5}{3}$

$\therefore \ 3x - 5 = 0$

(3) the line passes through the intersection of the system $\begin{cases} 2x - y + 3 = 0 \\ x + 2y + 4 = 0 \end{cases}$

and runs parallel to the line $3x + 2y = 5$.

$$2x - y + 3 = 0$$
$$-) \ \underline{2x + 4y + 8 = 0}$$
$$-5y - 5 = 0 \ ; \ y = -1$$

$\therefore \ x = -2y - 4 = -2$

$\therefore$ The intersection is $(-2 , -1)$.

$$3x + 2y = 5 \Rightarrow y = -\frac{3}{2}x + \frac{5}{2} \qquad \therefore \ m = -\frac{3}{2}$$

$\therefore \ y = -\frac{3}{2}x + b \ ; \ -1 = -\frac{3}{2} \cdot (-2) + b \ ; \ b = -4$

$\therefore \ y = -\frac{3}{2}x - 4$

$\therefore \ 3x + 2y + 8 = 0$

#13 For a line $\frac{x}{a} + \frac{y}{b} = 1 \ (a > 0, \ b > 0)$ which passes through a point $A(4,3)$,

let B be the intersection point of the line and y-axis, and C be the intersection point of the line and x-axis. Find minimum value of the area of triangle $\Delta \ OBC$ (where O is the origin $(0,0)$).

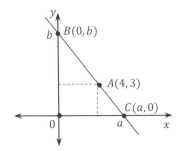

The area of $\triangle OBC$ is $\dfrac{1}{2} \cdot a \cdot b = \dfrac{ab}{2}$

Since the line $\dfrac{x}{a} + \dfrac{y}{b} = 1$ passes through the point $A(4,3)$,

$\dfrac{4}{a} + \dfrac{3}{b} = 1.$

Since $a > 0$ and $b > 0$, $\dfrac{4}{a} > 0$ and $\dfrac{3}{b} > 0$

By the relationship between arithmetic and geometric means,

$\dfrac{4}{a} + \dfrac{3}{b} \geq 2\sqrt{\dfrac{4}{a} \cdot \dfrac{3}{b}}$ (when $\dfrac{4}{a} = \dfrac{3}{b}$; i.e., $3a = 4b$, LHS = RHS)

$\therefore\ 1 \geq 4\sqrt{\dfrac{3}{ab}}$; $\dfrac{1}{4} \geq \sqrt{\dfrac{3}{ab}}$; $\dfrac{1}{16} \geq \dfrac{3}{ab}$; $ab \geq 48$

$\therefore\ $ The area of $\triangle OBC$ is $\dfrac{ab}{2} \geq \dfrac{48}{2} = 24$

Therefore, minimum value of the area of $\triangle OBC$ is 24.

#14 Find the value of the real number a.

(1) For the two points $A(1,3)$ and $B(4,-3)$, a line $y = 2x + a$ passes through the intersection point P such that $\overline{AP} : \overline{PB} = 2 : 3$

$P(x,y) = \left(\dfrac{2 \cdot 4 + 3 \cdot 1}{2+3},\ \dfrac{2 \cdot (-3) + 3 \cdot 3}{2+3} \right) = \left(\dfrac{11}{5},\ \dfrac{3}{5} \right)$

$\therefore\ y - \dfrac{3}{5} = 2\left(x - \dfrac{11}{5} \right)$

$\therefore\ y = 2x - \dfrac{22}{5} + \dfrac{3}{5}$ $\therefore\ y = 2x - \dfrac{19}{5}$

Therefore, $a = -\dfrac{19}{5}$

(2) The three points $A(-2,6), B(4,-3)$, and $C(a,3)$ are on the same line.

Since the slope of the line which passes through the points A and B is the same as the slope of the line which passes through the points B and C, $\dfrac{-3-6}{4-(-2)} = \dfrac{3-(-3)}{a-4}$

$\therefore\ \dfrac{-9}{6} = \dfrac{6}{a-4}$ $\therefore\ -9(a-4) = 36$ $\therefore\ a - 4 = -4$

Therefore, $a = 0$

#15 For two lines $l_1: ax + 2y - 2 = 0$ and $l_2: x + (a-1)y + 2 = 0$,

find the value of the constant a when $l_1 /\!/ l_2$ (parallel) and $l_1 \perp l_2$ (perpendicular).

When $l_1 /\!/ l_2$, $\dfrac{a}{1} = \dfrac{2}{a-1} \neq \dfrac{-2}{2}$

$\therefore\ a(a-1) = 2$; $a^2 - a - 2 = 0$; $(a-2)(a+1) = 0$ $\therefore\ a = 2$ or $a = -1$

Since $a \neq -1$, $a = 2$

When $l_1 \perp l_2$, $a \cdot 1 + 2 \cdot (a-1) = 0$

$\therefore\ 3a = 2 \qquad \therefore\ a = \dfrac{2}{3}$

Therefore, $l_1 \,/\!/\, l_2 \ \Rightarrow\ a = 2$

$\qquad\qquad l_1 \perp l_2 \ \Rightarrow\ a = \dfrac{2}{3}$

#16 **For a line l_1 which passes through the intersection point of two lines l_2: $2x + y + 8 = 0$ and l_3: $x - 2y - 6 = 0$, l_1 is parallel to the line $l_4 : x - 3y = 0$ and passes through a point $(a, -2)$. Find the real number a.**

l_1: $2x + y + 8 + k(x - 2y - 6) = 0$

$\therefore\ (2+k)x + (1-2k)y + 8 - 6k = 0 \ \cdots\cdots\ ①$

Since $① \,/\!/\, l_4$, $\dfrac{2+k}{1} = \dfrac{1-2k}{-3} \qquad \therefore\ 1 - 2k = -6 - 3k \qquad \therefore\ k = -7$

Substitute $k = -7$ into $①$, $-5x + 15y + 50 = 0$

$\therefore\ x - 3y - 10 = 0 \cdots\cdots ②$

Since $②$ passes through $(a, -2)$, $a + 6 - 10 = 0 \quad \therefore\ a = 4$

Alternative approach:

$\begin{cases} 2x + y + 8 = 0 \\ x - 2y - 6 = 0 \end{cases} \Rightarrow \quad - \begin{array}{|l} 2x + y + 8 = 0 \\ 2x - 4y - 12 = 0 \end{array}$

$\qquad\qquad\qquad\qquad\qquad\qquad \overline{}$

$\qquad\qquad\qquad\qquad\qquad\qquad 5y + 20 = 0 \ ; \ y = -4, \ x = -2$

$\therefore\ (-2, -4)$ is the intersection point of l_2 and l_3

Since $l_1 \,/\!/\, l_4$, l_1 and l_4 have the same slope.

$\therefore\ l_1$ is the line with slope $\dfrac{1}{3}$ and passes through $(-2, -4)$.

$\therefore\ l_1$: $y - (-4) = \dfrac{1}{3}(x - (-2)) \qquad \therefore\ l_1$: $y = \dfrac{1}{3}x - \dfrac{10}{3}$

Since $y = \dfrac{1}{3}x - \dfrac{10}{3}$ passes through $(a, -2)$, $\ -2 = \dfrac{a}{3} - \dfrac{10}{3}$

$\therefore\ \dfrac{a}{3} = \dfrac{4}{3} \qquad \therefore\ a = 4$

#17 Determine the distance between the two parallel lines.

(1) $2x + 3y - 4 = 0$, $2x + 3y + 6 = 0$

Consider a point $(2, 0)$ on the line $2x + 3y - 4 = 0$.

Then, the distance between the point $(2, 0)$ and the line $2x + 3y + 6 = 0$ is

$\dfrac{|2 \cdot 2 + 3 \cdot 0 + 6|}{\sqrt{2^2 + 3^2}} = \dfrac{|10|}{\sqrt{13}} = \dfrac{10\sqrt{13}}{13}$

Therefore, the distance between the two parallel lines is $\dfrac{10\sqrt{13}}{13}$.

(2) $x + 3y - 4 = 0$, $-x - 3y + 2 = 0$

Consider a point $(1, 1)$ on the line $x + 3y - 4 = 0$.

Then, the distance between the point $(1,1)$ and the line $-x - 3y + 2 = 0$ is

$$\frac{|(-1)\cdot 1 + (-3)\cdot 1 + 2|}{\sqrt{(-1)^2 + (-3)^2}} = \frac{|-2|}{\sqrt{10}} = \frac{2\sqrt{10}}{10} = \frac{\sqrt{10}}{5}$$

Therefore, the distance between the two parallel lines is $\dfrac{\sqrt{10}}{5}$.

#18 For two parallel lines $12x - 5y - 2 = 0$ and $12x - 5y - 3k = 0$,

find the real number k so that the distance between the lines is 3.

Consider a point $(1, 2)$ on the line $12x - 5y - 2 = 0$.

Since the distance between the point $(1, 2)$ and the line $12x - 5y - 3k = 0$ is 3,

$$\frac{|12\cdot 1 + (-5)\cdot 2 - 3k|}{\sqrt{12^2 + (-5)^2}} = 3$$

$\therefore$ $|2 - 3k| = 39$ $\quad \therefore$ $2 - 3k = \pm 39$ $\quad \therefore$ $3k = 2 \mp 39$

$\therefore$ $3k = -37$ or $3k = 41$

$\therefore$ $k = -\dfrac{37}{3}$ or $k = \dfrac{41}{3}$

#19 Find an equation of a line so that the line is perpendicular to the line $2x + 3y = 0$ and

the distance from the origin is 2.

$2x + 3y = 0 \ \Rightarrow \ y = -\dfrac{2}{3}x$ (Slope is $-\dfrac{2}{3}$)

$\therefore$ $y = \dfrac{3}{2}x + k$ ($\because$ Perpendicular to $y = -\dfrac{2}{3}x$)

$\therefore$ $3x - 2y + 2k = 0$ $\cdots\cdots$ ①

Since the distance between ① and the origin is 2, $\dfrac{|3\cdot 0 + (-2)\cdot 0 + 2k|}{\sqrt{3^2 + (-2)^2}} = 2$

$\therefore$ $\dfrac{|2k|}{\sqrt{13}} = 2$ $\quad \therefore |2k| = 2\sqrt{13}$ $\quad \therefore 2k = \pm 2\sqrt{13}$

Therefore, $3x - 2y + 2\sqrt{13} = 0$ or $3x - 2y - 2\sqrt{13} = 0$

#20 For a triangle $\Delta\, OAB$ with vertices $O(0, 0)$, $A(3, 6)$, and $B(5, -4)$,

find the area of the triangle .

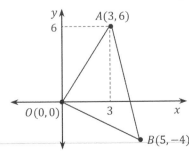

By the distance formula,

$\overline{OA}^2 = (3 - 0)^2 + (6 - 0)^2 = 9 + 36 = 45$

$\therefore \overline{OA} = \sqrt{45} = 3\sqrt{5}$

The equation of the line $\overleftrightarrow{OA}$ is $y = 2x$

; i.e., $2x - y = 0$

Since the altitude(height) h of the triangle $\triangle OAB$ is the distance between the point $B(5, -4)$

and the line $2x - y = 0$, $h = \dfrac{|2\cdot 5 + (-1)\cdot(-4)|}{\sqrt{2^2 + (-1)^2}} = \dfrac{14}{\sqrt{5}}$

Therefore, the area of triangle $\triangle OAB$ is $\dfrac{1}{2}\cdot \overline{OA}\cdot h = \dfrac{1}{2}\cdot 3\sqrt{5}\cdot \dfrac{14}{\sqrt{5}} = \dfrac{42}{2} = 21$ (unit^2)

#21 Find maximum value of the distance between the origin and the line

$2x - 3y - 2 - k(x - y) = 0$ **(where k is a real number).**

$2x - 3y - 2 - k(x - y) = 0 \quad\Rightarrow\quad (2 - k)x - (3 - k)y - 2 = 0 \cdots\cdots①$

The distance from the origin to $①$ is $\dfrac{|(2-k)\cdot 0 + (3-k)\cdot 0 - 2|}{\sqrt{(2-k)^2 + (3-k)^2}} = \dfrac{2}{\sqrt{2k^2 - 10k + 13}}$

When $\sqrt{2k^2 - 10k + 13}$ has minimum value, $\dfrac{2}{\sqrt{2k^2 - 10k + 13}}$ has maximum value.

$2k^2 - 10k + 13 = 2(k^2 - 5k) + 13 = 2\left(k - \dfrac{5}{2}\right)^2 - \dfrac{25}{2} + 13 = 2\left(k - \dfrac{5}{2}\right)^2 + \dfrac{1}{2} \geq \dfrac{1}{2}$

$\therefore\ \sqrt{2k^2 - 10k + 13} \geq \sqrt{\dfrac{1}{2}} = \dfrac{\sqrt{2}}{2}$

$\therefore\ \dfrac{2}{\sqrt{2k^2 - 10k + 13}} \leq \dfrac{4}{\sqrt{2}} = 2\sqrt{2}$

Therefore, maximum value of the distance is $2\sqrt{2}$.

#22 For any real numbers a and b, a quadratic equation $x^2 - 2(a - b)x - 2(a - b) - 1 = 0$

has a double root. Find minimum value of the distance between the two points $A(a, b)$

and $B(1, -1)$.

Let D be the discriminant of the equation.

Since $\dfrac{D}{4} = 0$, $(a - b)^2 + 2(a - b) + 1 = 0$

$\therefore\ (a - b + 1)^2 = 0 \qquad \therefore\ a - b + 1 = 0 \qquad \therefore\ b = a + 1 \cdots\cdots①$

$d(A, B) = \sqrt{(a - 1)^2 + (b + 1)^2} = \sqrt{(a - 1)^2 + (a + 2)^2}$ by $①$

$\qquad = \sqrt{2a^2 + 2a + 5} = \sqrt{2(a^2 + a) + 5} = \sqrt{2\left(a + \dfrac{1}{2}\right)^2 - \dfrac{1}{2} + 5} = \sqrt{2\left(a + \dfrac{1}{2}\right)^2 + \dfrac{9}{2}}$

Since $2\left(a + \dfrac{1}{2}\right)^2 + \dfrac{9}{2} \geq \dfrac{9}{2}$, $d(A, B) \geq \sqrt{\dfrac{9}{2}} = \dfrac{3}{\sqrt{2}} = \dfrac{3\sqrt{2}}{2}$

Therefore, minimum value of the distance is $\dfrac{3\sqrt{2}}{2}$ when $a = -\dfrac{1}{2}$.

#23

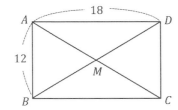

For a rectangle $\square ABCD$ with $\overline{AB} = 12$ and $\overline{AD} = 18$, let the intersection point of the two diagonals be M. When the centroid of the triangle $\triangle ABC$ is P_1 and the centroid of the triangle $\triangle CDM$ is P_2, find the distance between P_1 and P_2.

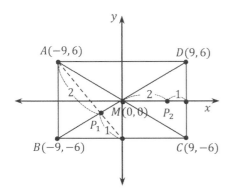

Place the rectangle $\square ABCD$ in a coordinate plane so that the intersection point M is the origin $(0,0)$. Then,

$$P_1 = P_1\left(\frac{-9-9+9}{3}, \frac{6-6-6}{3}\right) = P_1(-3,-2)$$

$$P_2 = P_2\left(\frac{9+9+0}{3}, \frac{6-6+0}{3}\right) = P_2(6,0)$$

Therefore, $\overline{P_1P_2} = \sqrt{\left(6-(-3)\right)^2 + \left(0-(-2)\right)^2}$

$$= \sqrt{81+4} = \sqrt{85}$$

#24 Determine the equation of a circle.

(1) The diameter is the segment joining the points $A(-2,2)$ and $B(4,4)$.

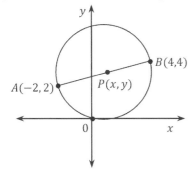

Let the center of the circle be $P(x,y)$.

Then, $x = \frac{-2+4}{2} = 1$ and $y = \frac{2+4}{2} = 3$

The length of the radius is

$$\overline{AP} = \sqrt{(1+2)^2 + (3-2)^2} = \sqrt{9+1} = \sqrt{10}$$

Therefore, the equation of the circle is

$$(x-1)^2 + (y-3)^2 = 10.$$

(2) The equation is $x^2 + y^2 - 2x + 6y - 6 = 0$.

$x^2 + y^2 - 2x + 6y - 6 = 0 \Rightarrow (x-1)^2 + (y+3)^2 - 1 - 9 - 6 = 0$

$\Rightarrow (x-1)^2 + (y+3)^2 = 4^2$

Therefore, the equation of the circle is $(x-1)^2 + (y+3)^2 = 16$.

(3) The circle passes through the three points $(-1,4), (1,2),$ and $(-1,0)$.

The equation of the circle is $x^2 + y^2 + Ax + By + C = 0$

$(-1,4) \Rightarrow 1 + 16 - A + 4B + C = 0$; $-A + 4B + C = -17 \cdots\cdots ①$

$(1,2) \Rightarrow 1 + 4 + A + 2B + C = 0$; $A + 2B + C = -5 \cdots\cdots ②$

$(-1,0) \Rightarrow 1 - A + C = 0$; $A = C + 1 \cdots\cdots ③$

Substituting ③ into ①, $B = -4 \cdots\cdots ④$

Substituting ③ and ④ into ②, $C + 1 - 8 + C = -5$ $\therefore C = 1$ and $A = 2$

$$\therefore \ x^2 + y^2 + 2x - 4y + 1 = 0 \qquad \therefore \ (x+1)^2 + (y-2)^2 = 4$$

(4) The center of the circle is on the y-axis and the circle passes through two points $(2, 2)$ and $(1, -1)$.

Since the center of the circle is on y-axis, let $(0, b)$ be the center.

Let the length of the radius be r.

Then, the equation of the circle is $(x-0)^2 + (y-b)^2 = r^2$

$(2, 2) \quad \Rightarrow \quad 4 + (2-b)^2 = r^2$

$(1, -1) \quad \Rightarrow \quad 1 + (-1-b)^2 = r^2$

$\therefore \ b^2 - 4b + 8 = b^2 + 2b + 2 \qquad \therefore \ 6b = 6 \qquad \therefore \ b = 1$ and $r^2 = 5$

Therefore, the equation of the circle is $x^2 + (y-1)^2 = 5$.

#25 The equation of a circle is $x^2 + y^2 - 4ax + 2ay + 20a - 28 = 0$. Find the coordinates of the center of the circle when the area of the circle has minimum value.

$x^2 + y^2 - 4ax + 2ay + 20a - 28 = 0$

$\Rightarrow \ (x-2a)^2 + (y+a)^2 - 4a^2 - a^2 + 20a - 28 = 0$

$\Rightarrow \ (x-2a)^2 + (y+a)^2 = 5a^2 - 20a + 28 = 5(a^2 - 4a) + 28$

$$= 5(a-2)^2 - 20 + 28 = 5(a-2)^2 + 8$$

$\therefore$ When $a = 2$, the length of the radius has minimum value $\sqrt{8}$.

$\therefore$ The area of the circle has minimum value when $a = 2$.

Therefore, the coordinates of the center of the circle are $x = 2a = 4$ and $y = -a = -2$.

#26 When two circles pass through a point $A(-3, 5)$ and intersect at only one point on both the x-axis and the y-axis, find the distance between their centers.

Since the circle is in second quadrant, the circle with radius r has the center $(-r, r)$.

$\therefore \ (x+r)^2 + (y-r)^2 = r^2 \ \cdots\cdots \ ①$

Since ① passes through $(-3, 5)$, $(-3+r)^2 + (5-r)^2 = r^2$

$\therefore \ 2r^2 - 16r + 34 = r^2 \qquad \therefore \ r^2 - 16r + 34 = 0$

Let r_1 and r_2 be the radii of the two circles.

Then, $r_1 + r_2 = 16$ and $r_1 r_2 = 34$ by the roots-coefficients relationship

Since $(r_1 - r_2)^2 = (r_1 + r_2)^2 - 4r_1 r_2 = 16^2 - 4 \cdot 34 = 120$, $|r_1 - r_2| = \sqrt{120} = 2\sqrt{30}$

Therefore, the distance between their centers $(-r_1, \ r_1)$ and $(-r_2, \ r_2)$ is

$$\sqrt{(-r_2 + r_1)^2 + (r_2 - r_1)^2} = \sqrt{2(r_1 - r_2)^2} = \sqrt{2}|r_1 - r_2| = \sqrt{2} \cdot 2\sqrt{30} = 2\sqrt{60} = 4\sqrt{15}$$

#27 For two real numbers x and y such that $(x-4)^2 + (y-3)^2 = 4$, find maximum and minimum values of $\sqrt{x^2 + y^2}$.

Note that $\sqrt{x^2 + y^2}$ is the same as the distance between a point $P(x, y)$ and the origin $(0, 0)$.

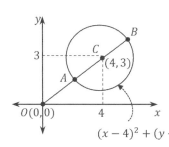

If a point P lands on a point A, then $\overline{OP}$ will have minimum value.

If a point P lands on a point B, then $\overline{OP}$ will have maximum value.

Since $\overline{OC} = \sqrt{4^2 + 3^2} = 5$, $\overline{OA} = 5 - 2 = 3$ and $\overline{OB} = 5 + 2 = 7$

Therefore, minimum value of $\sqrt{x^2 + y^2}$ is 3 and

maximum value of $\sqrt{x^2 + y^2}$ is 7.

#28 For a line which passes through the intersection points of two circles $x^2 + y^2 - 8 = 0$ and $x^2 + y^2 + 6x + ay = 0$, the line is perpendicular to $y = 2x + 1$.

Find the value of the constant a.

The equation of the line is $x^2 + y^2 - 8 + (-1) \cdot (x^2 + y^2 + 6x + ay) = 0$

That is, $-6x - ay - 8 = 0$ $\therefore$ $6x + ay + 8 = 0$ ······ ①

Since ① $\perp (y = 2x + 1)$, $-\dfrac{6}{a} \cdot 2 = -1$

Therefore, $a = 12$

#29 State how the relations between a circle $x^2 + y^2 = 1$ and a line $y = -2x + k$ are different depending on the value of k.

Substituting $y = -2x + k$ into $x^2 + y^2 = 1$, $x^2 + (-2x + k)^2 = 1$

$\therefore$ $5x^2 - 4kx + k^2 - 1 = 0$

The discriminant D of the quadratic equation is

$D = 16k^2 - 20(k^2 - 1) = -4k^2 + 20$

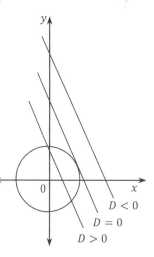

If $D > 0$; i.e., $-4k^2 + 20 > 0$; $k^2 < 5$; $-\sqrt{5} < k < \sqrt{5}$,

then the circle and the line intersect at 2 different points.

If $D = 0$; i.e., $-4k^2 + 20 = 0$; $k^2 = 5$; $k = \sqrt{5}$ or $k = -\sqrt{5}$,

then the circle and the line intersect at only one point.

If $D < 0$; i.e., $-4k^2 + 20 < 0$; $k^2 > 5$; $k > \sqrt{5}$ or $k < -\sqrt{5}$,

then the circle and the line do not intersect at any point.

#30 **For a circle with center $(2, 2)$ and radius 2, a line $y = ax + 1$ intersects the circle at 2 different points. Find the range of a.**

The equation of the circle is $(x - 2)^2 + (y - 2)^2 = 2^2$ ······ ①

Substituting $y = ax + 1$ into ①, $(x - 2)^2 + (ax - 1)^2 = 4$

$\therefore (1 + a^2)x^2 - 2(2 + a)x + 1 = 0$

Let D be the discriminant of the equation.

Since $D > 0$, $4(2 + a)^2 - 4(1 + a^2) > 0$ $\quad \therefore 16a + 12 > 0$

Therefore, $a > -\dfrac{3}{4}$

#31 **For real numbers a and b, a line $\dfrac{x}{a} + \dfrac{y}{b} = 1$ intersects a circle $x^2 + y^2 = 3$ at only one point in the first quadrant. Find minimum value of ab.**

Since the line and the circle meet at only one point, the distance between the center $(0, 0)$ of the circle and the line $\dfrac{x}{a} + \dfrac{y}{b} - 1 = 0$ is the same as the length of the radius of the circle.

$\therefore \dfrac{\left|\frac{1}{a}\cdot 0 + \frac{1}{b}\cdot 0 - 1\right|}{\sqrt{\left(\frac{1}{a}\right)^2 + \left(\frac{1}{b}\right)^2}} = \sqrt{3}$ $\quad \therefore \dfrac{1}{\left(\frac{1}{a}\right)^2 + \left(\frac{1}{b}\right)^2} = 3$ $\quad \therefore \dfrac{1}{a^2} + \dfrac{1}{b^2} = \dfrac{1}{3}$

Since $a^2 > 0$ and $b^2 > 0$, $\dfrac{1}{a^2} > 0$ and $\dfrac{1}{b^2} > 0$

By the relationship between arithmetic and geometric means, $\dfrac{1}{a^2} + \dfrac{1}{b^2} \geq 2\sqrt{\dfrac{1}{a^2}\cdot\dfrac{1}{b^2}}$

$\therefore \dfrac{1}{3} \geq 2\dfrac{1}{ab}$ (Note that a is x-intercept and b is y-intercept.

$\qquad\qquad$ Considering quadrant I, $a > 0$ and $b > 0$)

$\therefore 3 \leq \dfrac{ab}{2}$ $\quad \therefore ab \geq 6$

Therefore, minimum value of ab is 6.

#32 **Find x and y intercepts of a tangent line at $(a, 2)$ of a circle $x^2 + y^2 = 9$ $(a > 0)$.**

Since the point of tangency is on the circle $x^2 + y^2 = 9$, $a^2 + 2^2 = 9$

$\therefore a^2 = 5$ $\quad \therefore a = \sqrt{5}$ or $a = -\sqrt{5}$

Since $a > 0$, $a = \sqrt{5}$

$\therefore$ The tangent line at $(\sqrt{5}, 2)$ of the circle $x^2 + y^2 = 9$ is $\sqrt{5}x + 2y = 9$.

Therefore, x-intercept (when $y = 0$) is $x = \dfrac{9}{\sqrt{5}} = \dfrac{9\sqrt{5}}{5}$ and y-intercept (when $x = 0$) is $y = \dfrac{9}{2}$.

#33 **For a point P such that $\overline{AP} : \overline{PB} = 1 : 2$ where $A = A(3, 0)$ and $B = B(0, 3)$, find**
   maximum value of the area of $\triangle ABP$.

Let $P = P(x, y)$

Since $\overline{AP} : \overline{PB} = 1 : 2$, $\sqrt{(x-3)^2 + y^2} : \sqrt{x^2 + (3-y)^2} = 1 : 2$

$\therefore\ 2\sqrt{(x-3)^2 + y^2} = \sqrt{x^2 + (3-y)^2}$

$\therefore\ 4(x-3)^2 + 4y^2 = x^2 + (3-y)^2$

$\therefore\ 4(x^2 - 6x + 9) + 4y^2 = x^2 + (y^2 - 6y + 9)$; $3x^2 - 24x + 3y^2 + 6y + 27 = 0$

$\therefore\ x^2 - 8x + y^2 + 2y + 9 = 0$

$\therefore\ (x-4)^2 + (y+1)^2 - 16 - 1 + 9 = 0$; $(x-4)^2 + (y+1)^2 = 8$

$\therefore\ P$ is on the circle with center $C(4, -1)$ and radius $2\sqrt{2}$.

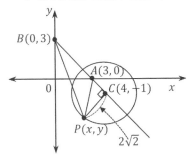

When P is on the circle such that $\overline{CP} = 2\sqrt{2}$, the area of
a triangle $\triangle ABP$ has maximum value.

Since $\overline{AB} = 3\sqrt{2}$ (Base) and altitude (height) is $2\sqrt{2}$, the

area of $\triangle ABP$ is $\dfrac{1}{2} \cdot 3\sqrt{2} \cdot 2\sqrt{2} = 6$

#34 **When a circle $x^2 + y^2 = 9$ and a line $ax - y + a + 1 = 0$ intersect at two points A and B,**
   find maximum and minimum values of $\overline{AB}$.

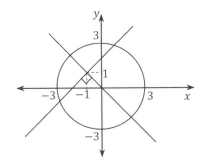

When the line passes through the center of the circle
; i.e., $\overline{AB}$ is the diameter of the circle,

$\overline{AB}$ has maximum value.

$ax - y + a + 1 = 0 \ \Rightarrow\ y = a(x + 1) + 1 \ \cdots\cdots\ ①$

$\therefore$ The line ① passes through point $(-1, 1)$.

$\therefore$ Minimum value of $\overline{AB}$ is the distance between two
points $(-3, 0)$ and $(0, 3)$.

$$\therefore\ \sqrt{(0+3)^2 + (3-0)^2} = \sqrt{9+9} = \sqrt{18} = 3\sqrt{2}$$

Therefore, maximum value of $\overline{AB}$ is 6 ($= 2 \times$ radius) and minimum value of $\overline{AB}$ is $3\sqrt{2}$.

#35 **For a point $P(x, y)$ on a circle $x^2 + y^2 - 8x + 6y + 21 = 0$, find maximum distance between the point P and a line $3x + 4y - 24 = 0$.**

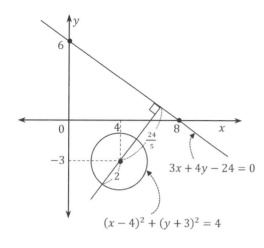

$$x^2 + y^2 - 8x + 6y + 21 = 0$$

$$\Rightarrow (x - 4)^2 + (y + 3)^2 - 16 - 9 + 21 = 0$$

$$\Rightarrow (x - 4)^2 + (y + 3)^2 = 4$$

The distance between the center $(4, -3)$ and a line $3x + 4y - 24 = 0$ is $\dfrac{|3 \cdot 4 + 4 \cdot (-3) - 24|}{\sqrt{3^2 + 4^2}} = \dfrac{24}{5}$

Therefore, maximum distance between the point on a circle and the line is $\dfrac{24}{5} + 2 = 6\dfrac{4}{5}$.

#36 **When two circles $x^2 + (y - 1)^2 = 20$ and $(x - a)^2 + (y - 1)^2 = 4$ intersect each other at two different points, find the positive number a so that the common chord has maximum value.**

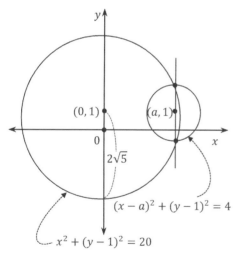

The equation of a line which passes through the intersection points of the two circles is

$$x^2 + (y - 1)^2 - 20 + (-1) \cdot \{(x - a)^2 + (y - 1)^2 - 4\} = 0$$

$$\therefore 2ax - a^2 - 16 = 0 \quad \cdots\cdots ①$$

When the line ① passes through the center of $(x - a)^2 + (y - 1)^2 = 4$,

the common chord will have maximum value.

Substituting $(a, 1)$ into ①, $2a^2 - a^2 - 16 = 0 \quad \therefore$

$$a^2 = 16 \qquad \therefore a = 4 \text{ or } a = -4$$

Since $a > 0$ (Positive number), $a = 4$

#37

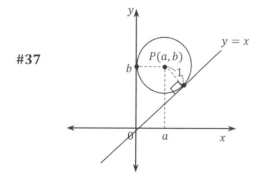

For a circle with radius 1, the center P of the circle is in quadrant I and the circle intersects y-axis at one point and has tangent line $y = x$.

Find the coordinates of P.

Let the equation of the circle be $(x - a)^2 + (y - b)^2 = 1$

Since $(0, b)$ is on the circle, $a^2 = 1$

$\therefore\ a = 1 \quad (\because\ a > 0\ ;\ \text{quadrant I})$

$\therefore$ The center P of the circle is $P = P(1, b)$

Since the distance between $P(1, b)$ and the tangent line $x - y = 0$ is 1, $\dfrac{|1 \cdot 1 + (-1) \cdot b|}{\sqrt{1^2 + (-1)^2}} = 1$

$\therefore\ |1 - b| = \sqrt{2} \qquad \therefore\ 1 - b = \pm\sqrt{2} \qquad \therefore\ b = 1 - \sqrt{2}\ \text{ or }\ b = 1 + \sqrt{2}$

Since $a < b,\ b > 1$

$\therefore\ b = 1 + \sqrt{2}$

$\therefore\ P = P(1,\ 1 + \sqrt{2})$

Therefore, the coordinates of P are $x = 1$ and $y = 1 + \sqrt{2}$.

#38 When a circle $x^2 + y^2 + 4ax - 2ay + 8a - 4 = 0$ passes through two points A and B,

not depending on the value of real number a, find the length of the segment $\overline{AB}$.

$x^2 + y^2 + 4ax - 2ay + 8a - 4 = 0 \quad \Rightarrow \quad x^2 + y^2 - 4 + 2a(2x - y + 4) = 0 \ \cdots\cdots\ \textcircled{1}$

$\therefore\ \textcircled{1}$ passes through two intersection points of a circle $x^2 + y^2 = 4$ and a line $2x - y + 4 = 0$

Substituting $y = 2x + 4$ into $x^2 + y^2 = 4$, $x^2 + (2x + 4)^2 = 4$

$\therefore\ 5x^2 + 16x + 12 = 0 \qquad \therefore\ x = \dfrac{-8 \pm \sqrt{64 - 60}}{5} = \dfrac{-8 \pm 2}{5} \qquad \therefore\ x = -\dfrac{6}{5}\ \text{ or }\ x = -2$

If $x = -\dfrac{6}{5}$, then $y = 2 \cdot \left(-\dfrac{6}{5}\right) + 4 = \dfrac{8}{5}$

If $x = -2$, then $y = 2 \cdot (-2) + 4 = 0$

$\therefore\ (x, y) = \left(-\dfrac{6}{5}, \dfrac{8}{5}\right)\ \text{ or }\ (x, y) = (-2, 0)$

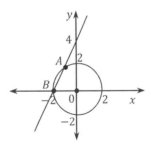

Therefore, for two points $A\left(-\dfrac{6}{5}, \dfrac{8}{5}\right)$ and $B(-2, 0)$,

$$\overline{AB} = \sqrt{\left(-2 + \dfrac{6}{5}\right)^2 + \left(0 - \dfrac{8}{5}\right)^2} = \sqrt{\dfrac{16}{25} + \dfrac{64}{25}} = \sqrt{\dfrac{80}{25}} = \dfrac{4\sqrt{5}}{5}$$

#39 Find a tangent line at $(2, 1)$ of a circle $(x - 1)^2 + (y + 2)^2 = 10$.

The line which passes through $(2, 1)$ and the center $(1, -2)$ of the circle has slope $\frac{-2-1}{1-2} = 3$.

$\therefore$ The tangent line has slope $-\frac{1}{3}$ and passes through $(2, 1)$.

$\therefore$ The equation of the tangent line is $y - 1 = -\frac{1}{3}(x - 2)$.

Therefore, $y = -\frac{1}{3}x + \frac{5}{3}$

#40 For points of tangency A and B of a circle, find the length of the segment $\overline{AB}$.

(1) When two tangent lines of a circle $x^2 + y^2 - 6x - 2y + 6 = 0$ are drawn from a point $P(3, 5)$.

$x^2 + y^2 - 6x - 2y + 6 = 0 \quad \Rightarrow \quad (x - 3)^2 + (y - 1)^2 - 9 - 1 + 6 = 0$

$\Rightarrow \quad (x - 3)^2 + (y - 1)^2 = 4$; A circle with center $C(3, 1)$ and radius 2

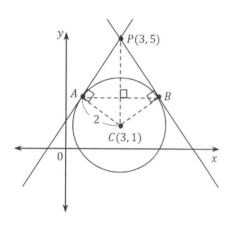

The distance between the center $C(3, 1)$ of the circle and a point $P(3, 5)$ is

$\overline{CP} = \sqrt{(3 - 3)^2 + (5 - 1)^2} = \sqrt{4^2} = 4$

Since $\overline{AC}$ is the length of radius, $\overline{AC} = 2$

By the Pythagorean Theorem,

$\overline{AP} = \sqrt{4^2 - 2^2} = \sqrt{12} = 2\sqrt{3}$

The area of $\triangle PAC$ is $\frac{1}{2} \cdot \overline{AC} \cdot \overline{AP} = \frac{1}{2} \cdot \overline{CP} \cdot \left(\frac{1}{2}\overline{AB}\right)$

$\therefore \frac{1}{2} \cdot 2 \cdot 2\sqrt{3} = \frac{1}{2} \cdot 4 \cdot \left(\frac{1}{2}\overline{AB}\right)$

Therefore, $\overline{AB} = 2\sqrt{3}$

(2) When two tangent lines of a circle $x^2 + y^2 = 5$ are drawn from a point $P(4, 2)$.

$\overline{OP} = \sqrt{4^2 + 2^2} = \sqrt{20} = 2\sqrt{5}$

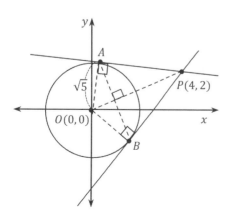

Since $\overline{OA} = \sqrt{5}$,

$\overline{AP} = \sqrt{(2\sqrt{5})^2 - (\sqrt{5})^2} = \sqrt{20 - 5} = \sqrt{15}$

The area of $\triangle OAP$ is

$\frac{1}{2} \cdot \sqrt{5} \cdot \sqrt{15} = \frac{1}{2} \cdot \overline{OP} \cdot \left(\frac{1}{2}\overline{AB}\right)$

$\therefore \frac{1}{2} \cdot 5\sqrt{3} = \frac{1}{2} \cdot 2\sqrt{5} \cdot \left(\frac{1}{2}\overline{AB}\right)$; $\frac{5\sqrt{3}}{2} = \frac{\sqrt{5}}{2}\overline{AB}$

Therefore, $\overline{AB} = \sqrt{15}$

#41 Two tangent lines of a circle $(x-3)^2 + y^2 = 6$ are drawn from a point $P(a,b)$ which lies on the x-axis. When the lines are perpendicular, find the coordinates of $P(a,b)$. (where $a > 0$)

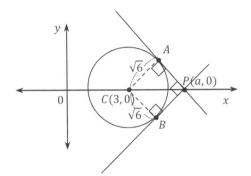

Let $P(a,b) = P(a,0)$

Let two points A and B be the points of tangency and C be the center of the circle.

Then, $\square ACBP$ is a square.

By the Pythagorean Theorem, $\overline{CP} = \sqrt{12} = 2\sqrt{3}$

Since $a - 3 = 2\sqrt{3}$, $a = 3 + 2\sqrt{3}$

$\therefore P(a,b) = P(3 + 2\sqrt{3}, 0)$

Therefore, the coordinates of $P(a,b)$ are $a = 3 + 2\sqrt{3}$ and $b = 0$.

#42 Find the value of $a + b$ for which:

(1) The graph of $y = ax + 2$ is translated by b along the y-axis from a graph of $y = 3x - 5$.

$y = 3x - 5 + b = ax + 2$ $\therefore a = 3$ and $-5 + b = 2$; $b = 7$

$\therefore a + b = 3 + 7 = 10$

(2) The graph is translated by a along the y-axis from a graph of $y = 2x + 4$ and passes through both points $(a + 1, -2)$ and $\left(-\frac{1}{3}, b\right)$.

Since the translated graph is $y = 2x + 4 + a$ and

this graph passes through a point $(a + 1, -2)$, $-2 = 2(a + 1) + 4 + a$

Thus, $3a = -8$; $a = -\frac{8}{3}$

$\therefore$ The translated graph is $y = 2x + 4 - \frac{8}{3} = 2x + \frac{4}{3}$.

Since this graph passes through the point $\left(-\frac{1}{3}, b\right)$, $b = 2\left(-\frac{1}{3}\right) + \frac{4}{3} = \frac{2}{3}$

Therefore, $a + b = -\frac{8}{3} + \frac{2}{3} = -2$

(3) A point $(-1, 1)$ is on the graph of $y = -2x + a$. If the graph is translated by b along the y-axis, then the translated line will pass through the point $(3, -4)$.

$1 = -2(-1) + a$; $a = -1$ $\therefore y = -2x - 1$

The translated graph is $y = -2x - 1 + b$.

Thus, $-4 = -2 \cdot 3 - 1 + b$; $b = 3$

$\therefore a + b = -1 + 3 = 2$

#43 Determine the equation whose graph is translated from the graph of

$x^2 + y^2 - 2x + 4y = 0$ **by -1 unit in x-axis and 2 units in y-axis.**

$$x^2 + y^2 - 2x + 4y = 0 \quad \Rightarrow \quad (x-1)^2 + (y+2)^2 - 1 - 4 = 0$$
$$\Rightarrow \quad (x-1)^2 + (y+2)^2 = 5$$

$\begin{array}{l} x \longrightarrow x+1 \\ y \longrightarrow y-2 \end{array}$ $\quad \Rightarrow \quad (x+1-1)^2 + (y-2+2)^2 = 5 \qquad \therefore \quad x^2 + y^2 = 5$

#44 Find the value of the real number a.

(1) When a line $2x + 3y = 5$ is translated by -3 units in the x-axis and $+2$ units in the y-axis, the translated line will divide the area of a circle $(x+a)^2 + (y-3)^2 = 6$ in half.

The translated line is $2(x+3) + 3(y-2) = 5$; i.e., $2x + 3y = 5$

To split the area of a circle in half, the line should pass through the center $(-a, 3)$ of the circle. $\quad \therefore \quad 2(-a) + 3(3) = 5 \quad \therefore \quad -2a = -4 \quad \therefore \quad a = 2$

(2) When a line $2x - y + 1 = 0$ is translated by a units in the x-axis, the translated line will be a tangent line of a circle $(x-2)^2 + (y-3)^2 = 12$.

The translated line is $2(x-a) - y + 1 = 0$; i.e., $2x - y - 2a + 1 = 0 \quad \cdots\cdots ①$

Since ① is the tangent line of $(x-2)^2 + (y-3)^2 = 12$,

the distance between the center $(2, 3)$ of the circle and the line ① is $\sqrt{12} = 2\sqrt{3}$.

$\therefore \quad \dfrac{|2 \cdot 2 + (-1) \cdot 3 - 2a + 1|}{\sqrt{2^2 + (-1)^2}} = 2\sqrt{3} \qquad \therefore \quad \dfrac{|2 - 2a|}{\sqrt{5}} = 2\sqrt{3} \qquad \therefore \quad |2 - 2a| = 2\sqrt{15}$

$\therefore \quad 2 - 2a = \pm 2\sqrt{15} \qquad \therefore \quad a = 1 - \sqrt{15} \text{ or } a = 1 + \sqrt{15}$

(3) If a circle $x^2 + y^2 = 4$ is translated by $a \ (a > 0)$ units in the x-axis, then it will meet a line $2x - 4y - 6 = 0$ at one point.

The translated circle is $(x-a)^2 + y^2 = 4$

Since the distance between the center $(a, 0)$ of the circle and the line $2x - 4y - 6 = 0$ is the length of the radius, $\dfrac{|2 \cdot a + (-4) \cdot 0 - 6|}{\sqrt{2^2 + (-4)^2}} = 2$

$\therefore \quad |2a - 6| = 2\sqrt{20} = 4\sqrt{5}$

$\therefore \quad 2a - 6 = \pm 4\sqrt{5} \qquad \therefore \quad a = 3 - 2\sqrt{5} \text{ or } a = 3 + 2\sqrt{5}$

Since $a > 0$, $a = 3 + 2\sqrt{5}$

(4) If a circle $x^2 + y^2 + 2ax - 4y + a^2 = 0$ is translated by 1 unit in the x-axis and -2 units in the y-axis, then it will be divided in half by a line $2x - y + 4 = 0$.

$$x^2 + y^2 + 2ax - 4y + a^2 = 0 \quad \Rightarrow \quad (x+a)^2 + (y-2)^2 - a^2 - 4 + a^2 = 0$$
$$\Rightarrow \quad (x+a)^2 + (y-2)^2 = 4$$

The translated circle is $(x-1+a)^2 + (y+2-2)^2 = 4$. ; i.e., $(x-1+a)^2 + y^2 = 4$

To split the area of a circle in half, the line should pass through the center $(1-a, 0)$ of the

circle.　∴ $2(1-a) + 4 = 0$　∴ $a - 1 = 2$　∴ $a = 3$

#45 **If a line $y = ax + b$ is translated by -2 units in the x-axis and $+1$ unit in the y-axis, then the translated line and a line $x - 2y - 4 = 0$ will be perpendicular to each other on the y-axis. Find the values of the real numbers a and b.**

The translated line is $y - 1 = a(x+2) + b$; i.e., $y = ax + 2a + b + 1$

$x - 2y - 4 = 0 \Rightarrow y = \frac{1}{2}x - 2$

Thus, the line which perpendicular to $y = \frac{1}{2}x - 2$ on the y-axis is $y = -2x - 2$

∴ $ax + 2a + b + 1 = -2x - 2$

Therefore, $a = -2$ and $b = 1$

#46 Find the value of the real number a.

(1) The reflection of a circle $(x-3)^2 + (y+2)^2 = 4$ in line $y = x$ has the center which lies on $y = ax + 1$.

The reflection is $(y-3)^2 + (x+2)^2 = 4$.

Since the center $(-2, 3)$ lies on $y = ax + 1$, $3 = -2a + 1$

∴ $a = -1$

(2) The reflection of a line $2x - 3y + a = 0$ in line $y = x$ is the tangent line of a circle $x^2 + (y-2)^2 = 4$.

The reflection is $2y - 3x + a = 0$; $-3x + 2y + a = 0$ ⋯⋯ ①

Since the distance between the center $(0, 2)$ of the circle and the line ① is 2 (radius),

$\frac{|-3\cdot0 + 2\cdot2 + a|}{\sqrt{(-3)^2 + 2^2}} = 2$　∴ $\frac{|4+a|}{\sqrt{13}} = 2$

∴ $|4 + a| = 2\sqrt{13}$　∴ $4 + a = \pm 2\sqrt{13}$

Therefore, $a = -4 + 2\sqrt{13}$ or $a = -4 - 2\sqrt{13}$

(3) The reflection of a line $y = ax + 1$ in the y-axis divides the area of a circle $(x+2)^2 + (y-3)^2 = 4$ in half.

The reflection of a line $y = ax + 1$ in the y-axis is $y = -ax + 1$

Since $y = -ax + 1$ should pass through the center $(-2, 3)$ of the circle to split the area of the circle in half, $3 = -a \cdot (-2) + 1$

Therefore, $a = 1$

#47 For three points $A(-2, 2)$, $B(4, 3)$, and $C(a, 0)$, find minimum value of $\overline{AC} + \overline{CB}$.

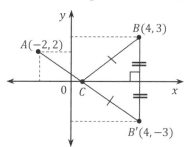

Let B' be the reflection of point B in the x-axis.

Then, $\overline{CB} = \overline{CB'}$

$\therefore \ \overline{AC} + \overline{CB} = \overline{AC} + \overline{CB'} \geq \overline{AB'}$

Since $\overline{AB'} = \sqrt{(4+2)^2 + (-3-2)^2} = \sqrt{36 + 25} = \sqrt{61}$,

minimum value of $\overline{AC} + \overline{CB}$ is $\sqrt{61}$.

#48 Find a reflection point of a point $A(2, 3)$ in line $y = 2x + 3$.

Let the reflection point be $A'(a, b)$.

Then, the midpoint of $\overline{AA'}$ lies on the line $y = 2x + 3$.

$\therefore$ The line $y = 2x + 3$ passes through the midpoint $\left(\frac{a+2}{2}, \frac{b+3}{2}\right)$.

$\therefore \ \frac{b+3}{2} = 2 \cdot \frac{a+2}{2} + 3$; $b + 3 = 2(a+2) + 6$; $b = 2a + 7$ $\cdots\cdots$ ①

Since the line $\overleftrightarrow{AA'}$ and the line $y = 2x + 3$ are perpendicular each other,

the line $\overleftrightarrow{AA'}$ has slope $-\frac{1}{2}$.

$\therefore \ \frac{b-3}{a-2} = -\frac{1}{2}$ $\therefore \ a - 2 = -2b + 6$

By ①, $a - 2 = -2(2a+7) + 6$ $\therefore \ 5a = -6$ $\therefore \ a = -\frac{6}{5}$ and $b = \frac{23}{5}$

Therefore, the reflection point of A is $\left(-\frac{6}{5}, \frac{23}{5}\right)$.

#49 For two points $A(2, 2)$ and $B(3, 2)$, a point $P(x, y)$ lies on a line $y = -2x + 1$.

Find the coordinates of $P(x, y)$ so that $\overline{PA} + \overline{PB}$ has minimum value.

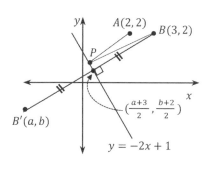

Let the reflection point of $B(3, 2)$ in a line $y = -2x + 1$

be $B'(a, b)$.

Then, the midpoint $\left(\frac{a+3}{2}, \frac{b+2}{2}\right)$ of $\overline{BB'}$ lies on $y = -2x + 1$.

$\therefore \ \frac{b+2}{2} = -2 \cdot \frac{a+3}{2} + 1$ $\therefore \ b + 2 = -2(a+3) + 2$ $\therefore$

$b = -2a - 6$ $\cdots\cdots$ ①

Since $\overleftrightarrow{BB'} \perp$ (line $y = -2x + 1$), slope of $\overleftrightarrow{BB'}$ is $\frac{1}{2}$.

$\therefore \ \frac{b-2}{a-3} = \frac{1}{2}$ $\therefore \ a - 3 = 2b - 4$

By ①, $a - 3 = 2(-2a - 6) - 4 = -4a - 16$ $\therefore \ 5a = -13$ $\therefore \ a = -\frac{13}{5}$ and $b = -\frac{4}{5}$

$\therefore \ B'(a, b) = B'\left(-\frac{13}{5}, -\frac{4}{5}\right)$

Since minimum value of $\overline{PA} + \overline{PB}$ is $\overline{AB'}$,

P lies on the intersection point of $\overleftrightarrow{AB'}$ and $y = -2x + 1$.

The line which passes through $B'\left(-\frac{13}{5}, -\frac{4}{5}\right)$ and $A(2, 2)$ is

$$y - 2 = \frac{2 + \frac{4}{5}}{2 + \frac{13}{5}}(x - 2) = \frac{14}{23}(x - 2)$$

$$\therefore \ y = \frac{14}{23}x + \frac{18}{23}$$

Now, solve the system of two equations of lines $\begin{cases} y = \frac{14}{23}x + \frac{18}{23} \\ y = -2x + 1 \end{cases}$

$$\Rightarrow \quad \frac{14}{23}x + \frac{18}{23} = -2x + 1 \quad \Rightarrow \quad 14x + 18 = -46x + 23 \quad \Rightarrow \quad 60x = 5$$

$$\therefore \ x = \frac{1}{12} \ \text{ and } \ y = \frac{5}{6}$$

Therefore, $P(x, y) = P\left(\frac{1}{12}, \frac{5}{6}\right)$

#50 For any positive real numbers a and b, two lines $y = ax$ and $y = bx$ are translations of each other in line $y = x$. For a point $P(a, b)$, find minimum value of the distance between the origin O and the point P.

The translation of $y = ax$ in line $y = x$ is $x = ay$; i.e., $y = \frac{1}{a}x$

Since $y = \frac{1}{a}x$ is equal to $y = bx$, $\quad \frac{1}{a} = b \quad \therefore \ ab = 1$

Since $a > 0$ and $b > 0$, $a^2 > 0$ and $b^2 > 0$

By the relationship between arithmetic and geometric means,

$a^2 + b^2 \geq 2\sqrt{a^2 b^2}$ (When $a = b$, $LHS = RHS$)

$\therefore \ a^2 + b^2 \geq 2ab = 2$

$\therefore \ \overline{OP} = \sqrt{a^2 + b^2} \geq \sqrt{2}$

Therefore, minimum value of the distance between the origin O and the point P is $\sqrt{2}$.

#51 Find an equation of a circle C which is translated by a circle $x^2 + y^2 = 4$ in line $y = 2x + 1$.

Let the center of a circle C be $P(a, b)$.

Then, the point $P(a, b)$ is a translation of the center $O(0, 0)$ of a circle $x^2 + y^2 = 4$

in line $y = 2x + 1$.

$\therefore$ The midpoint of $\overline{OP}$ lies on the line $y = 2x + 1$.

$\therefore$ $y = 2x + 1$ passes through the midpoint $\left(\frac{a}{2}, \frac{b}{2}\right)$.

$\therefore$ $\frac{b}{2} = 2 \cdot \frac{a}{2} + 1$ $\therefore$ $b = 2a + 2$ $\cdots\cdots$ ①

Since $\overleftrightarrow{OP} \perp$ (line $y = 2x + 1$), the slope of $\overleftrightarrow{OP}$ is $-\frac{1}{2}$

$\therefore$ $\frac{b-0}{a-0} = -\frac{1}{2}$ $\therefore$ $a = -2b$

By ①, $b = 2(-2b) + 2 = -4b + 2$; $5b = 2$ $\therefore$ $b = \frac{2}{5}$ and $a = -\frac{4}{5}$

$\therefore$ The center of a circle C is $P\left(-\frac{4}{5}, \frac{2}{5}\right)$

Therefore, the equation of the circle C is $\left(x + \frac{4}{5}\right)^2 + \left(y - \frac{2}{5}\right)^2 = 4$

#52 Graph the region of the following inequalities on a coordinate plane.

(1) $\begin{cases} x^2 + y^2 - 9 < 0 \\ x + y - 2 > 0 \end{cases}$

$x^2 + y^2 - 9 < 0 \Rightarrow x^2 + y^2 < 3^2$ $x + y - 2 > 0 \Rightarrow y > -x + 2$

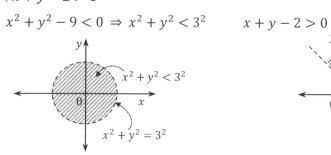

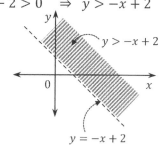

$\therefore$ The common region is the solution of the system.

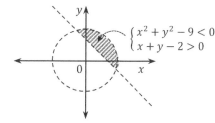

(2) $|x - y| \le 2$

$\Rightarrow -2 \le x - y \le 2$

$\Rightarrow -2 - x \le -y \le 2 - x$

$\Rightarrow -(-2 - x) \ge y \ge -(2 - x)$

$\Rightarrow x - 2 \le y \le x + 2$

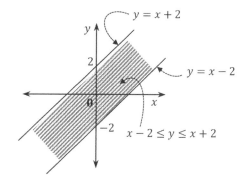

(3) $(x - y + 2)(x^2 + y^2 - 4) < 0$

Consider the union of regions of two systems:

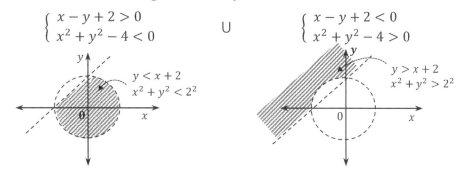

$$\begin{cases} x - y + 2 > 0 \\ x^2 + y^2 - 4 < 0 \end{cases} \quad \cup \quad \begin{cases} x - y + 2 < 0 \\ x^2 + y^2 - 4 > 0 \end{cases}$$

∴ The union of the regions is the solution for the product of polynomials.

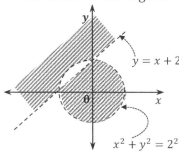

(4) $xy(x - y + 2) \geq 0$

Consider the union of regions of two systems:

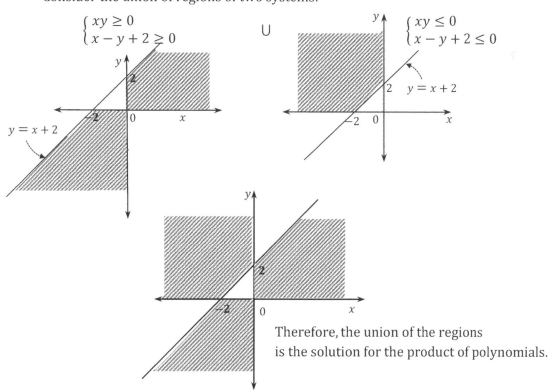

$$\begin{cases} xy \geq 0 \\ x - y + 2 \geq 0 \end{cases} \quad \cup \quad \begin{cases} xy \leq 0 \\ x - y + 2 \leq 0 \end{cases}$$

Therefore, the union of the regions
is the solution for the product of polynomials.

#53 When a point $(1, a)$ is in the region of a circle $(x - 2)^2 + y^2 < 13$ and a point $(-1, a)$ is not in the region of the circle, determine the range of the real number a.

Since a point $(1, a)$ satisfies the inequality $(x - 2)^2 + y^2 < 13$, $(1 - 2)^2 + a^2 < 13$

$\therefore\ a^2 < 12 \qquad \therefore\ -2\sqrt{3} < a < 2\sqrt{3}$

Since a point $(-1, a)$ does not satisfy the inequality $(x - 2)^2 + y^2 < 13$,

$(-1, a)$ satisfies the inequality $(x - 2)^2 + y^2 \geq 13$.

$\therefore\ (-1 - 2)^2 + a^2 \geq 13 \qquad \therefore\ a^2 \geq 4 \qquad \therefore\ a \geq 2$ or $a \leq -2$

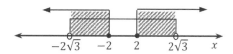

Therefore $-2\sqrt{3} < a \leq -2$ or $2 \leq a < 2\sqrt{3}$

#54 When a positive number a satisfies the inequality $x^2 + y^2 \leq a$, the inequality $3x + 4y - 20 \leq 0$ is always true. Find maximum value of a.

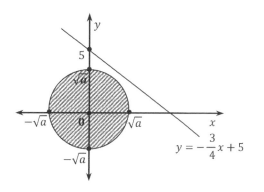

$3x + 4y - 20 \leq 0 \ \Rightarrow\ y \leq -\dfrac{3}{4}x + 5$

To satisfy the condition, the line should be a tangent line of the circle or the circle should be located below the line.

Since the distance between the center $(0, 0)$ of the circle and the line $3x + 4y - 20 = 0$ is greater than or equal to the radius of the circles,

$$\dfrac{|3 \cdot 0 + 4 \cdot 0 - 20|}{\sqrt{3^2 + 4^2}} \geq \sqrt{a}$$

$\therefore\ \dfrac{20}{5} \geq \sqrt{a} \qquad \therefore\ \sqrt{a} \leq 4 \qquad \therefore\ a \leq 16$

Since $a > 0$, $\quad 0 < a \leq 16 \qquad$ Therefore, maximum value of a is 16.

#55 Find the range of a constant a.

(1) By a line $ax + y - 2 = 0$, two points $A(1, 1)$ and $B(3, -4)$ lie in different regions.

Let $f(x, y) = ax + y - 2$.

To satisfy the condition, $f(1, 1)f(3, -4) < 0$

$\therefore\ (a + 1 - 2)(3a - 4 - 2) < 0$

$\therefore\ (a - 1)(3a - 6) < 0$

Therefore, $1 < a < 2$

(2) A line $x + ay + 2 + a = 0$ runs through between two points $A(1, 1)$ and $B(-1, 1)$.

Two points should lie in two different regions which are divided by the line.

Let $f(x, y) = x + ay + 2 + a$.

To satisfy the condition, $f(1, 1)f(-1, 1) < 0$

$\therefore \ (1 + a + 2 + a)(-1 + a + 2 + a) < 0$

$\therefore \ (2a + 3)(2a + 1) < 0$

Therefore, $\ -\dfrac{3}{2} < a < -\dfrac{1}{2}$

Alternative approach:

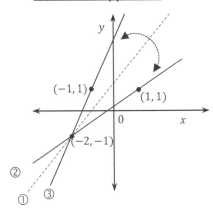

$x + ay + 2 + a = 0$

$\Rightarrow \ y = -\dfrac{1}{a}(x + 2) - 1 \ \cdots\cdots \ ①$

$\therefore \ $ Slope of the line $①$ is $-\dfrac{1}{a}$ and

$①$ passes through a point $(-2, -1)$.

The slope of the line $②$ is $\dfrac{1+1}{1+2} = \dfrac{2}{3}$ and

the slope of the line $③$ is $\dfrac{1+1}{-1+2} = 2$

$\therefore \ \dfrac{2}{3} < -\dfrac{1}{a} < 2 \qquad \therefore \ -\dfrac{2}{3} > \dfrac{1}{a} > -2 \qquad \therefore \ -\dfrac{3}{2} < a < -\dfrac{1}{2}$

#56 Find maximum value.

(1) When a point $P(x, y)$ satisfies all inequalities:

$x \geq 0, \ y \geq 0, \ 2x + y \leq 5,$ and $3x + 4y \leq 10,$ find maximum value of $x + y$.

$\begin{cases} 2x + y = 5 \\ 3x + 4y = 10 \end{cases} \Rightarrow \ \begin{array}{r} 8x + 4y = 20 \\ -\underline{\ \ 3x + 4y = 10\ \ } \\ 5x \qquad\ \ = 10 \end{array} \qquad \therefore \ x = 2, \ y = 1$

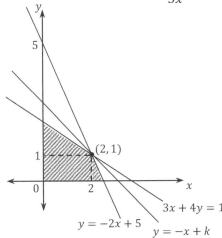

$3x + 4y = 10$

$y = -2x + 5$

$y = -x + k$

The region which satisfies all inequalities is

the shaded part.

Let $x + y = k$ (k is a constant.)

When the line $y = -x + k$ passes through the

intersection point $(2, 1)$ of two lines $2x + y = 5$ and

$3x + 4y = 10$, the constant k has maximum value.

Substituting $(2, 1)$ into $y = -x + k$, $\ 1 = -2 + k$

$\therefore \ k = 3$

Therefore, maximum value of $x + y$ is 3.

(2) When a point $P(x, y)$ satisfies all inequalities: $x - y \geq 0$, $2x - 3y \leq 0$, and $x + 3y \leq 4$, find maximum value of $x + y$.

The region which satisfies all inequalities is the shaded part.

Let $x + y = k$ (k is a constant.)

Then, $y = -x + k$

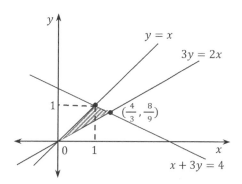

Now, find the intersection point of $x + 3y = 4$ and $2x - 3y = 0$

Then, $x + (2x) = 4$ $\quad \therefore x = \dfrac{4}{3}$ and $y = \dfrac{8}{9}$

Since the line $y = -x + k$ should pass through the intersection point $(\dfrac{4}{3}, \dfrac{8}{9})$,

we have $\dfrac{8}{9} = -\dfrac{4}{3} + k$ $\quad \therefore k = \dfrac{20}{9}$

Therefore, maximum value of $x + y$ is $\dfrac{20}{9}$.

(3) When a point $P(x, y)$ satisfies the inequality: $x^2 + y^2 - 2x - 4y \leq 0$, find maximum value of $2x + y$.

$x^2 + y^2 - 2x - 4y \leq 0 \quad \Rightarrow \quad (x - 1)^2 + (y - 2)^2 \leq 5$

Let $2x + y = k$

When the line $y = -2x + k$ is a tangent line of the circle, k has maximum value.

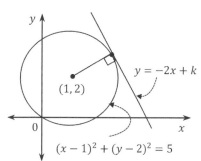

The distance between the center $(1, 2)$ of the circle and the line $2x + y - k = 0$ is $\dfrac{|2 \cdot 1 + 1 \cdot 2 - k|}{\sqrt{2^2 + 1^2}} = \sqrt{5}$

$\therefore \dfrac{|4 - k|}{\sqrt{5}} = \sqrt{5}$ $\quad \therefore |4 - k| = 5$ $\quad \therefore 4 - k = \pm 5$

$\therefore k = -1$ or $k = 9$

Since $k > 0$, $k = 9$

Therefore, maximum value of $2x + y$ is 9.

(4) When a point $P(x, y)$ satisfies the system of inequalities: $\begin{cases} y \geq x^2 \\ y \leq -x + 2 \end{cases}$, find maximum value of $y - 2x$.

$x^2 = -x + 2 \quad \Rightarrow \quad x^2 + x - 2 = 0 \quad \Rightarrow \quad (x + 2)(x - 1) = 0 \quad \therefore x = -2$ or $x = 1$

$\therefore$ The intersection points are $(-2, 4)$ and $(1, 1)$

Let $y - 2x = k$

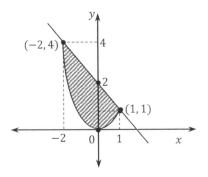

When the line passes through the intersection point

$(-2, 4)$, k has maximum value.

$\therefore \quad 4 - 2(-2) = k \qquad \therefore \quad k = 8$

Therefore, maximum value of $y - 2x$ is 8.

#57 When a point $P(x, y)$ satisfies the inequality $x^2 - 4y^2 - 2x + 1 \leq 0$,

find minimum value of $x^2 + y^2$.

$x^2 - 4y^2 - 2x + 1 \leq 0 \quad \Rightarrow \quad (x - 1)^2 - (2y)^2 \leq 0$

$\Rightarrow \quad (x - 1 + 2y)(x - 1 - 2y) \leq 0$

$\Rightarrow \quad \begin{cases} x - 1 + 2y \geq 0 \\ x - 1 - 2y \leq 0 \end{cases} \ \cup \ \begin{cases} x - 1 + 2y \leq 0 \\ x - 1 - 2y \geq 0 \end{cases}$

Therefore, the union of the regions
is the solution for the systems.

Let $x^2 + y^2 = k^2$, $(k \geq 0)$

Then, k is the radius of a circle with center $(0, 0)$.

$\therefore$ The radius k has minimum value when the lines $y = \dfrac{1}{2}x - \dfrac{1}{2}$ and $y = -\dfrac{1}{2}x + \dfrac{1}{2}$ are

tangents of the circle $x^2 + y^2 = k^2$.

∴ The distance between $(0,0)$ and a line $y = -\dfrac{1}{2}x + \dfrac{1}{2}$; i.e., $x + 2y - 1 = 0$ is

$\dfrac{|1\cdot 0 + 2\cdot 0 - 1|}{\sqrt{1^2 + 2^2}} = k$ ∴ $k = \dfrac{1}{\sqrt{5}}$ ∴ $k^2 = \dfrac{1}{5}$

Therefore, minimum value of $x^2 + y^2$ is $\dfrac{1}{5}$.

#58 Two points $A(a, b)$ and $B(c, d)$ satisfy a system of inequalities $\begin{cases} y \geq x \\ x + y \leq 1 \\ y - 2x \leq 0 \end{cases}$

Find maximum and minimum values of $\dfrac{b+d}{a+c}, (a + c \neq 0)$.

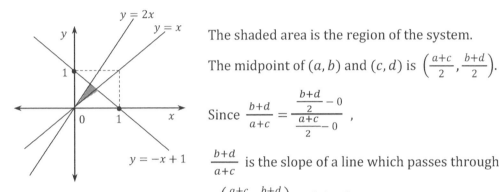

The shaded area is the region of the system.

The midpoint of (a, b) and (c, d) is $\left(\dfrac{a+c}{2}, \dfrac{b+d}{2}\right)$.

Since $\dfrac{b+d}{a+c} = \dfrac{\dfrac{b+d}{2} - 0}{\dfrac{a+c}{2} - 0}$,

$\dfrac{b+d}{a+c}$ is the slope of a line which passes through the points

$\left(\dfrac{a+c}{2}, \dfrac{b+d}{2}\right)$ and $(0, 0)$.

∴ $\dfrac{b+d}{a+c}$ has maximum value when the two points (a, b) and (c, d) lie on $y = 2x$ and

has minimum value when the two points (a, b) and (c, d) lie on $y = x$.

Therefore, maximum value is 2 (slope of $y = 2x$) and minimum value is 1 (slope of $y = x$).

Alternative approach:

Since the shaded area is in the first quadrant, $a \geq 0$, $b \geq 0$, $c \geq 0$, and $d \geq 0$.

Since $y \geq x$ and $y - 2x \leq 0$, $b \geq a$ and $b - 2a \leq 0$

∴ $a \leq b \leq 2a$

Similarly, $d \geq c$ and $d - 2c \leq 0$ ∴ $c \leq d \leq 2c$

Thus, $a + c \leq b + d \leq 2a + 2c$

∴ $1 \leq \dfrac{b+d}{a+c} \leq 2$ ($\because a + c \neq 0$)

Therefore, maximum value of $\dfrac{b+d}{a+c}$ is 2 and minimum value of $\dfrac{b+d}{a+c}$ is 1.

Chapter 5. Functions

#1 For a function $f(x) = -\dfrac{1}{2}x + 4$, find each value.

(1) $f(0) = -\dfrac{1}{2} \cdot 0 + 4 = 4$

(2) $f(2) = -\dfrac{1}{2} \cdot 2 + 4 = 3$

(3) $f(-4) = -\dfrac{1}{2} \cdot (-4) + 4 = 6$

(4) $f\left(\dfrac{1}{2}\right) = -\dfrac{1}{2} \cdot \left(\dfrac{1}{2}\right) + 4 = \dfrac{15}{4}$

#2 The domain of a function $f(x) = -2x + 3$ is $\{0,\ 1,\ 2,\ 3\}$. Find the range of $f(x)$.

$x = 0 \ \Rightarrow\ f(0) = -2 \cdot 0 + 3 = 3$

$x = 1 \ \Rightarrow\ f(1) = -2 \cdot 1 + 3 = 1$

$x = 2 \ \Rightarrow\ f(2) = -2 \cdot 2 + 3 = -1$

$x = 3 \ \Rightarrow\ f(3) = -2 \cdot 3 + 3 = -3$

$\therefore$ The range of $f(x)$ is $\{-3, -1,\ 1,\ 3\}$.

#3 Find the domain.

(1) The range of a function $f(x) = 2x$ is $\{-8,\ 0,\ 4,\ 8\}$. Find the domain of $f(x)$.

$2x = -8 \Rightarrow\ x = -4$

$2x = 0 \Rightarrow\ x = 0$

$2x = 4 \Rightarrow\ x = 2$

$2x = 8 \Rightarrow\ x = 4$

$\therefore$ The domain of $f(x)$ is $\{-4,\ 0,\ 2,\ 4\}$.

(2) The range of a function $g(x) = ax$ is $\{-2,\ 0,\ 2\}$ when $g(2) = -1$.

Find the domain of the function.

Since $g(2) = 2a = -1$, $a = -\dfrac{1}{2}$ $\qquad \therefore g(x) = -\dfrac{1}{2}x$

Since $-\dfrac{1}{2}x = -2 \Rightarrow x = 4$, $\ -\dfrac{1}{2}x = 0 \Rightarrow x = 0$, and $\ -\dfrac{1}{2}x = 2 \Rightarrow x = -4$,

the domain of the function is $\{-4,\ 0,\ 4\}$.

(3) Find the domain of $f(x) = \frac{1}{x-3}$

The domain of f is the set of x's in the real numbers such that x is not equal to 3.

We exclude 3 to avoid division by 0.

(4) Find the domain of $g(x) = \sqrt{9 - x^2}$.

We must restrict x so that $9 - x^2 \geq 0$ in order to avoid non-real values for $\sqrt{9 - x^2}$.

The domain is $\{x \mid x$ is the real number such that $|x| \leq 3\} = [-3, 3]$

#4 Find the value of $f(4)$.

(1) For any real number x, a function $f(x)$ satisfies $f\left(\frac{3x-1}{2}\right) = 3x - 4$.

$\frac{3x-1}{2} = 4 \quad \Rightarrow \quad 3x - 1 = 8 \quad \Rightarrow \quad 3x = 9 \quad \therefore \quad x = 3$

Therefore, $f(4) = 3 \cdot 3 - 4 = 5$

Alternative Approach:

Let $\frac{3x-1}{2} = a$.

Then, $3x - 1 = 2a \quad \therefore \quad x = \frac{2a+1}{3}$

Since $f\left(\frac{3x-1}{2}\right) = 3x - 4$, $f(a) = 3 \cdot \frac{2a+1}{3} - 4 = 2a + 1 - 4 = 2a - 3$

Therefore, $f(4) = 2 \cdot 4 - 3 = 5$

(2) For any real number x, a function $f(x)$ satisfies $f(x) + 3f\left(\frac{1}{x}\right) = 2$.

$x = 4 \Rightarrow \quad f(4) + 3f\left(\frac{1}{4}\right) = 2 \quad \cdots\cdots ①$

$x = \frac{1}{4} \Rightarrow \quad f\left(\frac{1}{4}\right) + 3f(4) = 2 \quad \cdots\cdots ②$

$① + ② \Rightarrow \quad 4f(4) + 4f\left(\frac{1}{4}\right) = 4 \quad \therefore \quad f(4) + f\left(\frac{1}{4}\right) = 1 \quad \cdots\cdots ③$

$② - ③ \Rightarrow 2f(4) = 1 \quad$ Therefore, $f(4) = \frac{1}{2}$

#5 A function $f(x)$ satisfies $f(x) + 2f\left(\frac{1}{x}\right) = 6x$ for any non-zero real number x.
Find the value of real number x such that $f(x) = f(-x)$.

Given $f(x) + 2f\left(\frac{1}{x}\right) = 6x$,

substituting $x = a$ gives $f(a) + 2f\left(\frac{1}{a}\right) = 6a \quad \cdots\cdots ①$

substituting $x = \frac{1}{a}$ gives $f\left(\frac{1}{a}\right) + 2f(a) = \frac{6}{a} \quad \cdots\cdots ②$

①+② ⇒ $3f(a) + 3f\left(\frac{1}{a}\right) = 6a + \frac{6}{a} = \frac{6a^2+6}{a}$ ∴ $f(a) + f\left(\frac{1}{a}\right) = \frac{2a^2+2}{a}$ ······③

②−③ ⇒ $f(a) = \frac{6-2a^2-2}{a} = \frac{-2a^2+4}{a} = -2a + \frac{4}{a}$

∴ $f(x) = -2x + \frac{4}{x}$

Since $f(x) = f(-x)$ and $f(-x) = 2x - \frac{4}{x}$, $-2x + \frac{4}{x} = 2x - \frac{4}{x}$

∴ $4x = \frac{8}{x}$ ∴ $x = \frac{2}{x}$ ∴ $x^2 = 2$

Therefore, $x = \sqrt{2}$ or $x = -\sqrt{2}$

#6 Is $f(x) = \dfrac{x^3 + 3x}{x^4 - 2x^2 + 3}$ even, odd, or neither?

Since $f(-x) = \dfrac{(-x)^3 + 3(-x)}{(-x)^4 - 2(-x)^2 + 3} = \dfrac{-x^3 - 3x}{x^4 - 2x^2 + 3} = -\dfrac{x^3 + 3x}{x^4 - 2x^2 + 3} = -f(x)$, f is an odd function.

#7 Specify whether the given function is even, odd, or neither, and then sketch its graph.

(1) $f(x) = -2$

(2) $f(x) = 2x + 1$

(3) $f(x) = 3x^2 + 2x - 1$

(4) $f(x) = \sqrt{x - 1}$

(5) $f(x) = |2x|$

(6) $f(x) = \left[\dfrac{x}{2}\right]$

(7) $f(x) = x^3 - 2x$

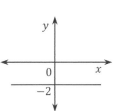

(1) Even

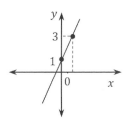

(2) Neither

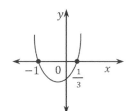

(3) Neither

$f(x) = 3x^2 + 2x - 1$
$= (3x - 1)(x + 1)$

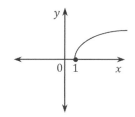

(4) Neither

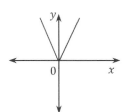

(5) Even

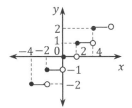

(6) Neither

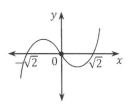

(7) Odd

$f(-x) = (-x)^3 - 2(-x)$

$= -x^3 + 2x = -(x^3 - 2x) = -f(x)$

#8 For a function $f(x)$, $f(x+3) = f(x)$ for any real number x

and $f(x) = 3 - |x|$ when $-\dfrac{3}{2} \leq x \leq \dfrac{3}{2}$. Find the value of $f(11)$.

Since $f(x+3) = f(x)$, $f(11) = f(8) = f(5) = f(2) = f(-1)$.

Since -1 is in the range of $-\dfrac{3}{2} \leq x \leq \dfrac{3}{2}$, $f(-1) = 3 - |-1| = 3 - 1 = 2$

Therefore, $f(11) = 2$

#9 Find the range of a.

(1) For a function $f: X = \{x \mid 1 \leq x \leq 2\} \rightarrow Y = \{y \mid 0 \leq y \leq 5\}$, $f(x) = ax - 1$

when a is a constant.

i) When $a > 0$,

$1 \leq x \leq 2 \Rightarrow a \leq ax \leq 2a \Rightarrow a - 1 \leq ax - 1 \leq 2a - 1$

$\therefore$ The range of $f(x) = ax - 1$ is $\{y \mid a - 1 \leq y \leq 2a - 1\}$.

Since $Y = \{y \mid 0 \leq y \leq 5\}$, $a - 1 \geq 0$ and $2a - 1 \leq 5$

$\therefore$ $1 \leq a \leq 3$

ii) When $a < 0$,

$1 \leq x \leq 2 \Rightarrow a \geq ax \geq 2a \Rightarrow a - 1 \geq ax - 1 \geq 2a - 1$

$\therefore$ The range of $f(x) = ax - 1$ is $\{y \mid 2a - 1 \leq y \leq a - 1\}$.

Since $Y = \{y \mid 0 \leq y \leq 5\}$, $2a - 1 \geq 0$ and $a - 1 \leq 5$

$\therefore$ $\dfrac{1}{2} \leq a \leq 6$ ……①

Since $a < 0$, ① is not satisfied.

iii) When $a = 0$,

Since $f(x) = ax - 1 = -1$, $f(x)$ is not included in $Y = \{y \mid 0 \leq y \leq 5\}$.

Therefore, by i), ii), and iii), the range of a is $1 \leq a \leq 3$.

Range of $f(x)$

(2) For a function $f(x) = \begin{cases} -x & , \ x \leq 0 \\ (a^2 - a)x, & x > 0 \end{cases}$, $f(x)$ is one-to-one correspondence.

To be one-to-one correspondence, the slopes of the two lines must have the same sign.

$\therefore$ $a^2 - a < 0$ $\therefore$ $a(a - 1) < 0$

Therefore, $0 < a < 1$

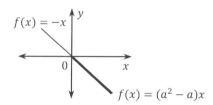

#10 Find the value.

(1) For a function $f(x) = ax$, $f(3) = -4$. Find the value of $f(9)$.

Since $f(3) = 3a = -4$, $a = -\dfrac{4}{3}$ $\quad \therefore f(x) = -\dfrac{4}{3}x$

Therefore, $f(9) = -\dfrac{4}{3} \cdot 9 = -12$

(2) Find the value of $f(3) - f(2) + f(4)$ for the function $f(x) = \dfrac{3}{x}$.

Since $f(3) = \dfrac{3}{3} = 1$, $f(2) = \dfrac{3}{2}$, and $f(4) = \dfrac{3}{4}$,

$f(3) - f(2) + f(4) = 1 - \dfrac{3}{2} + \dfrac{3}{4} = \dfrac{4-6+3}{4} = \dfrac{1}{4}$

(3) For the two functions $f(x) = ax + 2$ and $g(x) = \dfrac{b}{x} - 2$, $\quad f(1) = g(-1) = 3$.

Find the value of $a + b$.

Since $f(1) = g(-1) = 3$, $a + 2 = -b - 2 = 3$

$\therefore a = 1$, $b = -5$

Therefore, $a + b = -4$

(4) For the two functions $f(x) = \dfrac{a}{x} + 2$ and $g(x) = -\dfrac{3}{x} + 5$, $\quad 3f(-2) = 2g(-3)$.

Find the value of b which satisfies $f(b) = g(b)$.

Since $3f(-2) = 2g(-3)$,

$3\left(\dfrac{a}{-2} + 2\right) = 2\left(-\dfrac{3}{-3} + 5\right)$; $-\dfrac{3a}{2} + 6 = 12$; $-\dfrac{3a}{2} = 6$; $a = -4$

Since $f(b) = g(b)$, $\quad \dfrac{a}{b} + 2 = -\dfrac{3}{b} + 5$; $\dfrac{-4}{b} + 2 = -\dfrac{3}{b} + 5$; $\dfrac{1}{b} = -3$; $b = -\dfrac{1}{3}$

Therefore, the value of b is $-\dfrac{1}{3}$.

(5) For the function $f(3x - 2) = 2x - a$, $f(4) = 3$. Find the value of $f(1)$.

To find $f(4)$, $3x - 2 = 4$; $3x = 6$; $x = 2$

Since $f(4) = 3$, $f(4) = f(3 \cdot 2 - 2) = 2 \cdot 2 - a = 3$ $\quad \therefore a = 1$

$\therefore f(3x - 2) = 2x - 1$

Therefore, $f(1) = f(3 \cdot 1 - 2) = 2 \cdot 1 - 1 = 1$

(6) For the two functions $f(x) = 2ax$ and $g(x) = \frac{2}{x} - 1$, $g(f(2)) = 3$.

Find the value of a.

Since $f(2) = 2a \cdot 2 = 4a$, $\quad g\big(f(2)\big) = g(4a) = \frac{2}{4a} - 1 = \frac{1}{2a} - 1 = 3$

$\therefore \frac{1}{2a} = 4 \qquad$ Therefore, $a = \frac{1}{8}$

(7) Two points $P(a+2, 4-2a)$ and $Q(2-2b, 3b+1)$ are on the x-axis and y-axis, respectively. Find the value of $a + b$.

$4 - 2a = 0$ and $2 - 2b = 0$

$\therefore \ a = 2$ and $b = 1 \qquad$ Therefore, $a + b = 3$

(8) The function $f(x) = -\frac{3}{2}x$ passes through a point $(a+1, 2a-3)$. Find the value of a.

$2a - 3 = -\frac{3}{2}(a+1)$; $\ 4a - 6 = -3a - 3$; $\ 7a = 3 \qquad \therefore \ a = \frac{3}{7}$

(9) The function $y = ax$ passes through a point $(3, -15)$ and $(b, 10)$.

Find the value of $a - b$.

Since $-15 = 3a$, $a = -5$

Since $10 = ab = -5b$, $b = -2$

Therefore, $a - b = -5 - (-2) = -3$

(10) For any constants a and b, the function $f(x) = \frac{2a}{x}$ passes through the points

$(-2, 8)$ and $(4, b)$. Find the value of $a + b$.

$8 = \frac{2a}{-2}$; $a = -8 \qquad \therefore f(x) = -\frac{16}{x}$

$b = -\frac{16}{4}$; $b = -4$

Therefore, $a + b = -12$

(11) For any constants $a, b,$ and c, the function $f(x) = \frac{a}{x}$ passes through the points

$(b, 1)$, $(1, c)$, and $(3, -1)$. Find the value of $a + b + c$.

$-1 = \frac{a}{3}$; $a = -3 \qquad \therefore f(x) = -\frac{3}{x}$

Since $1 = -\frac{3}{b}$, $b = -3$

Since $c = -\frac{3}{1}$, $c = -3$

Therefore, $a + b + c = -3 + (-3) + (-3) = -9$

(12) Two functions $f(x) = ax$ and $g(x) = \dfrac{b}{x}$ intersect at the points $(3, 9)$ and $(-3, c)$.

Find the value of $a + b + c$.

Since $f(x) = ax$, $9 = 3a$; $a = 3$

Since $g(x) = \dfrac{b}{x}$, $9 = \dfrac{b}{3}$; $b = 27$

Since $(-3, c)$ is on $f(x) = 3x$, $c = 3(-3) = -9$

Therefore, $a + b + c = 3 + 27 + (-9) = 21$

(13) Two functions $y = -ax$ and $y = -\dfrac{2}{x}$ intersect at Point $A(b, 8)$.

Find the value of ab.

Since $y = -\dfrac{2}{x}$, $8 = -\dfrac{2}{b}$ $\quad \therefore b = -\dfrac{1}{4}$

Since $y = -ax$, $8 = -ab = \dfrac{1}{4}a$ $\quad \therefore a = 32$

Therefore, $ab = 32 \cdot \left(-\dfrac{1}{4}\right) = -8$

#11 Find the following values for the given linear functions with a condition:

(1) $f(1)$ for $f(x) = 2ax + 1$ with $f(-1) = 3$

$f(-1) = -2a + 1 = 3$; $2a = -2$; $a = -1$ $\quad \therefore f(x) = -2x + 1$

Therefore, $f(1) = -2 + 1 = -1$

(2) $\dfrac{a}{2}$ for $f(x) = \dfrac{1}{2}x + 5$ with $f\left(\dfrac{a}{2}\right) = -a$

$f\left(\dfrac{a}{2}\right) = \dfrac{1}{2} \cdot \dfrac{a}{2} + 5 = \dfrac{a}{4} + 5 = -a$; $\dfrac{5a}{4} = -5$; $a = -4$

Therefore, $\dfrac{a}{2} = -2$

(3) $a + b$ for $f(x) = 3ax - 2$ with $f(-1) = 4$ and $f(b) = 1$

$f(-1) = -3a - 2 = 4$; $3a = -6$; $a = -2$ $\quad \therefore f(x) = -6x - 2$

Thus, $f(b) = -6b - 2 = 1$; $b = -\dfrac{1}{2}$

Therefore, $a + b = -\dfrac{5}{2}$

(4) $a + \dfrac{1}{a}$ when $f(x) = 3x - 1$ passes through the point $(a, a + 3)$.

$a + 3 = 3a - 1$; $2a = 4$; $a = 2$ $\quad$ Therefore, $a + \dfrac{1}{a} = \dfrac{5}{2}$

(5) $a - b$ when $f(x) = ax + 2$ passes through both point $(1, 3)$ and point $(2, b)$.

Substitute $(1, 3)$ into $f(x) = ax + 2$. Then, $3 = a + 2$; $a = 1$ ∴ $f(x) = x + 2$

Substitute $(2, b)$ into $f(x) = ax + 2$. Then, $b = 2a + 2 = 2 + 2 = 4$

Therefore, $a - b = 1 - 4 = -3$

#12 Find the value of $a + b$ for which :

(1) The graph of $y = ax + 2$ is translated by b along the y-axis from a graph of $y = 3x - 5$.

$y = 3x - 5 + b = ax + 2$ ∴ $a = 3$ and $-5 + b = 2$; $b = 7$

∴ $a + b = 3 + 7 = 10$

(2) The graph is translated by a along the y-axis from a graph of $y = 2x + 4$ and passes through both point $(a + 1, -2)$ and point $\left(-\frac{1}{3}, b\right)$.

Since the translated graph is $y = 2x + 4 + a$ and

this graph passes through a point $(a + 1, -2)$, $-2 = 2(a + 1) + 4 + a$

Thus, $3a = -8$; $a = -\frac{8}{3}$

∴ The translated graph is $y = 2x + 4 - \frac{8}{3} = 2x + \frac{4}{3}$.

Since this graph passes through the point $\left(-\frac{1}{3}, b\right)$, $b = 2\left(-\frac{1}{3}\right) + \frac{4}{3} = \frac{2}{3}$

Therefore, $a + b = -\frac{8}{3} + \frac{2}{3} = -2$

(3) A point $(-1, 1)$ is on the graph of $y = -2x + a$. If the graph is translated by b along the y-axis, then it will pass through the point $(3, -4)$.

$1 = -2(-1) + a$; $a = -1$ ∴ $y = -2x - 1$

The translated graph is $y = -2x - 1 + b$.   Thus, $-4 = -2(3) - 1 + b$; $b = 3$

Therefore, $a + b = -1 + 3 = 2$

#13 Find the x-intercept and y-intercept.

(1) The linear function $y = ax + b$ passes through both point $(1, 2)$ and point $(-1, 4)$.

$\qquad 2 = a + b$

$+)\ \underline{4 = -a + b}$

$\qquad 6 = \qquad 2b$; $b = 3$ ∴ $a = -1$ ∴ $y = -x + 3$

Therefore, x-intercept (when $y = 0$) is 3 and y-intercept (when $x = 0$) is 3.

(2) The graph of $y = ax + b$ intersects the graph of $y = 2x + 3$ on the x-axis.

It also intersects the graph of $y = -5x - 6$ on the y-axis.

x-intercept is $-\frac{3}{2}$; $(-\frac{3}{2}, 0)$ and y-intercept is -6 ; $(0, -6)$

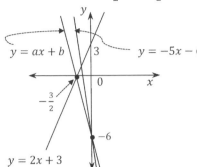

(3) The area surrounded by the graph of $y = \frac{1}{2}x + a$ $(a > 0)$, the x-axis, and

the y-axis is 36.

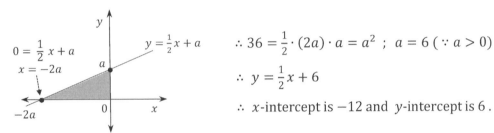

$\therefore 36 = \frac{1}{2} \cdot (2a) \cdot a = a^2$; $a = 6 (\because a > 0)$

$\therefore y = \frac{1}{2}x + 6$

$\therefore x$-intercept is -12 and y-intercept is 6.

#14 Find an equation in the standard form for each line.

(1) with y-intercept -3 and slope 2

$y = 2x - 3$ $\quad \therefore 2x - y - 3 = 0$

(2) with y-intercept 5 and slope 0

$y = 0 \cdot x + 5$ $\quad \therefore y - 5 = 0$

(3) with x-intercept 5 and slope $-\frac{2}{3}$

$y = -\frac{2}{3}x + b$; $0 = -\frac{2}{3} \cdot 5 + b$; $b = \frac{10}{3}$ $\quad \therefore y = -\frac{2}{3}x + \frac{10}{3}$

Therefore, $2x + 3y - 10 = 0$

(4) with x-intercept -3 and slope -2

$y = -2x + b$; $0 = -2 \cdot (-3) + b$; $b = -6$ $\quad \therefore y = -2x - 6$

Therefore, $2x + y + 6 = 0$

(5) through $(1, 2)$ with slope 3

$y = 3x + b$; $2 = 3 \cdot 1 + b$; $b = -1$ $\quad \therefore y = 3x - 1$ $\quad \therefore 3x - y - 1 = 0$

(OR $y - 2 = 3(x - 1)$; $y = 3x - 1$ $\quad \therefore 3x - y - 1 = 0$)

(6) through $(3, -4)$ with slope -2

$y = -2x + b$; $-4 = -2 \cdot 3 + b$; $b = 2$ $\therefore$ $y = -2x + 2$ $\therefore$ $2x + y - 2 = 0$

(OR $y + 4 = -2(x - 3)$; $y = -2x + 2$ $\therefore$ $2x + y - 2 = 0$)

(7) through $(2, 3)$ with undefined slope

No slope $\Rightarrow$ No change in x $\therefore$ $x = 2$; $x - 2 = 0$

(8) through $(-2, 3)$ with y-intercept -1

$y = \dfrac{-4}{2}x - 1 = -2x - 1$ $\therefore$ $2x + y + 1 = 0$

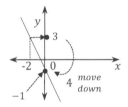

(9) through $(2, 4)$ with x-intercept -5

$y = \dfrac{4}{7}x + b$; $0 = \dfrac{4}{7} \cdot (-5) + b$; $b = \dfrac{20}{7}$ $\therefore$ $y = \dfrac{4}{7}x + \dfrac{20}{7}$ $\therefore$ $4x - 7y + 20 = 0$

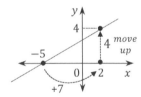

(10) through $(3, 1)$ and $(-2, 4)$

$m = \dfrac{4-1}{-2-3} = -\dfrac{3}{5}$ $\therefore$ $y = -\dfrac{3}{5}x + b$

Thus, $1 = -\dfrac{3}{5} \cdot 3 + b$; $b = \dfrac{14}{5}$ $\therefore$ $y = -\dfrac{3}{5}x + \dfrac{14}{5}$

Therefore, $3x + 5y - 14 = 0$

(OR Using Point-Slope form,

$y - 1 = -\dfrac{3}{5}(x - 3)$; $5y - 5 = -3x + 9$ $\therefore$ $3x + 5y - 14 = 0$)

(11) through $(-2, -3)$ and $(-1, 5)$

$m = \dfrac{5+3}{-1+2} = 8$ $\therefore$ $y = 8x + b$

Thus, $5 = 8 \cdot (-1) + b$; $b = 13$ $\therefore$ $y = 8x + 13$

Therefore, $8x - y + 13 = 0$

(OR Using Point-Slope form, $y + 3 = 8(x + 2)$ $\therefore$ $8x - y + 13 = 0$)

(12) with x-intercept -3 and y-intercept 3

$$m = \frac{0-3}{-3-0} = 1 \quad \therefore \; y = x + 3 \quad \therefore \; x - y + 3 = 0$$

(13) with x-intercept $\frac{3}{2}$ and y-intercept -4

$$m = \frac{-4-0}{0-\frac{3}{2}} = \frac{-4}{-\frac{3}{2}} = \frac{8}{3} \quad \therefore \; y = \frac{8}{3}x - 4 \quad \therefore \; 8x - 3y - 12 = 0$$

(14) Vertical line through $(-1, 2)$

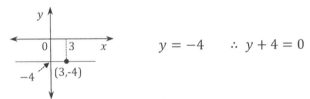

$$x = -1 \quad \therefore \; x + 1 = 0$$

(15) Horizontal line through $(3, -4)$

$$y = -4 \quad \therefore \; y + 4 = 0$$

#15 Find an equation for the line through $(2, 3)$ which is:

(1) parallel to the line $y = 2x - 5$

$$y = 2x + b \; ; \; 3 = 2 \cdot 2 + b \; ; \; b = -1 \quad \therefore \; y = 2x - 1 \quad \therefore \; 2x - y - 1 = 0$$

(2) parallel to the line $y = -3x + 1$

$$y = -3x + b \; ; \; 3 = -3 \cdot 2 + b \; ; \; b = 9 \quad \therefore \; y = -3x + 9 \quad \therefore \; 3x + y - 9 = 0$$

(3) parallel to the line $x = 4$

$$x = 2 \quad \therefore \; x - 2 = 0$$

(4) parallel to the line $y = -2$

$$y = 3 \quad \therefore \; y - 3 = 0$$

(5) parallel to the line $3x + 4y = 5$

$$y = -\frac{3}{4}x + \frac{5}{4} \quad \therefore \; y = -\frac{3}{4}x + b$$

$$3 = -\frac{3}{4} \cdot 2 + b \; ; \; b = \frac{9}{2} \quad \therefore \; y = -\frac{3}{4}x + \frac{9}{2}$$

Therefore, $3x + 4y - 18 = 0$

(6) perpendicular to the line $y = \frac{2}{3}x - 1$

$$y = -\frac{3}{2}x + b$$

$$3 = -\frac{3}{2} \cdot 2 + b \ ; \ b = 6 \quad \therefore \ y = -\frac{3}{2}x + 6$$

Therefore, $3x + 2y - 12 = 0$

(7) perpendicular to the line $x + 3y = -3$

$$x + 3y = -3 \ \Rightarrow \ y = -\frac{1}{3}x - 1$$

$$\therefore \ y = 3x + b$$

$$3 = 3 \cdot 2 + b \ ; \ b = -3 \quad \therefore \ y = 3x - 3$$

Therefore, $3x - y - 3 = 0$

(8) perpendicular to the line $x = 5$

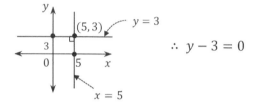

$$\therefore \ y - 3 = 0$$

(9) perpendicular to the line $y = -2$

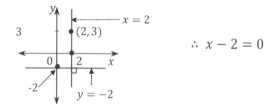

$$\therefore \ x - 2 = 0$$

#16 Find the value of a for the following lines :

(1) through $(2, 3)$ **and** $(1, -a)$ **with slope 2**

$$2 = \frac{-a-3}{1-2} = \frac{-a-3}{-1} = a + 3 \quad \therefore \ a = -1$$

(2) through $(2a - 1, -2)$ **and** $(-1, 1)$ **with slope** -2

$$-2 = \frac{1+2}{-1-2a+1} = \frac{3}{-2a} \ ; \ 4a = 3 \quad \therefore \ a = \frac{3}{4}$$

(3) through $(1, -2)$ **,** $(-3, 2)$**, and** $(-a + 1, -5)$

$$m = \frac{2+2}{-3-1} = \frac{4}{-4} = -1 \quad \therefore \ y = -x + b$$

Thus, $-2 = -1 + b \ ; \ b = -1 \quad \therefore \ y = -x - 1$

$$-5 = a - 1 - 1 = a - 2 \quad \therefore \ a = -3$$

(4) through $(2a + 1, -4)$, $(2, 5)$, and $(2, -3)$

$(2, 5)$ and $(2, -3)$ $\Rightarrow$ No change in x ; No slope ; Vertical line

$\therefore\ 2a + 1 = 2$　　$\therefore\ a = \dfrac{1}{2}$

(5) through $(-3, 3)$, $(3, a - 1)$, and $(0, 3)$

$(-3, 3)$ and $(0, 3)$ $\Rightarrow$ No change in y ; Slope is 0 ; Horizontal line

$\therefore\ a - 1 = 3$　　$\therefore\ a = 4$

(6) through $(a, 2a - 3)$ and $(-a - 1, 3 + 4a)$, and parallel to the x-axis

Parallel to x-axis $\Rightarrow$ Horizontal line

$\therefore\ 2a - 3 = 3 + 4a$; $2a = -6$　　$\therefore\ a = -3$

(7) through $(-3a + 1, -5)$ and $(2a - 1, a + 3)$, and perpendicular to the x-axis

Perpendicular to x-axis $\Rightarrow$ Vertical line

$\therefore\ -3a + 1 = 2a - 1$; $5a = 2$　　$\therefore\ a = \dfrac{2}{5}$

(8) through $(3, -2a)$ and $(2a - 1, -3a + 2)$, and parallel to the y-axis

Parallel to y-axis $\Rightarrow$ Vertical line

$\therefore\ 3 = 2a - 1$; $2a = 4$　　$\therefore\ a = 2$

(9) through $(-1, 5)$ and $(2, -4)$, and parallel to the line $ax + 3y + 5 = 0$

$m = \dfrac{-4 - 5}{2 + 1} = \dfrac{-9}{3} = -3$　　$\therefore\ y = -3x + b$

$ax + 3y + 5 = 0$ $\Rightarrow$ $3y = -ax - 5$ $\Rightarrow$ $y = -\dfrac{a}{3}x - \dfrac{5}{3}$

$\therefore\ -\dfrac{a}{3} = -3$　　$\therefore\ a = 9$

#17 Find the value of a such that the line $ax + 2y = 5$:

(1) is parallel to the line $2x + 3y = -2$.

The same slope

$3y = -2x - 2$; $y = -\dfrac{2}{3}x - \dfrac{2}{3}$　　$\therefore$ Slope $= -\dfrac{2}{3}$

$ax + 2y = 5$ $\Rightarrow$ $y = -\dfrac{1}{2}ax + \dfrac{5}{2}$　　$\therefore$ Slope $= -\dfrac{1}{2}a$

Therefore, $-\dfrac{2}{3} = -\dfrac{1}{2}a$　　$\therefore\ a = \dfrac{4}{3}$

(2) is perpendicular to the line $y = -2x + 3$.

Negative reciprocals of each other

Since $-2 \cdot \left(-\dfrac{1}{2}a\right) = -1$, $\quad a = -1$

(3) coincides with the line $6y = -4x + 15$.

$$6y = -4x + 15 \Rightarrow 4x + 6y = 15 \Rightarrow \frac{4}{3}x + 2y = 5 \qquad \therefore a = \frac{4}{3}$$

#18 Find the value of ab for which:

(1) the system $\begin{cases} x - 3y = a \\ 2x + by = 3 \end{cases}$ has the intersection point $(2, 3)$.

Substitute $(2, 3)$ into each equation.

Then, $2 - 9 = a$; $a = -7$ and $4 + 3b = 3$; $b = -\frac{1}{3}$

$\therefore ab = \frac{7}{3}$

(2) the system $\begin{cases} -ax + by = 4 \\ 2ax + 3by = 2 \end{cases}$ has the intersection point $(-1, 2)$.

Substitute $(-1, 2)$ into each equation.

Then, $\begin{cases} a + 2b = 4 \\ -2a + 6b = 2 \end{cases}$

$\quad 2a + 4b = 8$

$\underline{+)\ -2a + 6b = 2}$

$\ \ 10b = 10$; $b = 1$

$\therefore a = 4 - 2b = 4 - 2 = 2 \qquad \therefore ab = 2$

(3) the system $\begin{cases} px + y = 3 \\ 2x - 3y = q \end{cases}$ has no intersection when $p = a$, $q \neq b$.

Parallel $\quad \therefore \dfrac{p}{2} = -\dfrac{1}{3} \neq \dfrac{3}{q} \qquad \therefore p = -\dfrac{2}{3}$ and $q \neq -9$

$\therefore a = -\dfrac{2}{3}, b = -9 \qquad \therefore ab = 6$

(4) the system $\begin{cases} 2ax + 4y = -3 \\ 3x + 6y = 2b \end{cases}$ has unlimited numbers of intersections.

$\dfrac{2a}{3} = \dfrac{4}{6} = \dfrac{-3}{2b} \qquad \therefore 2a = 2$; $a = 1$ and $4b = -9$; $b = -\dfrac{9}{4}$

$\therefore ab = -\dfrac{9}{4}$

#19 Find the value of a such that :

(1) the system $\begin{cases} ax + y = -2 \\ -3x + 2y = 4 \end{cases}$ has no solution.

Parallel

$-\dfrac{a}{3} = \dfrac{1}{2} \neq -\dfrac{2}{4} \qquad \therefore a = -\dfrac{3}{2}$

(2) the system $\begin{cases} 2x - ay + 3 = 0 \\ x + 3y - 2 = 0 \\ 2x + y + 1 = 0 \end{cases}$ has one solution.

$\begin{cases} 2x - ay + 3 = 0 \cdots\cdots ① \\ x + 3y - 2 = 0 \cdots\cdots ② \\ 2x + y + 1 = 0 \cdots\cdots ③ \end{cases}$

$\Rightarrow ① - ③ ; (-a - 1)y + 2 = 0$

$\quad 2 \cdot ② - ③ ; 5y - 5 = 0 ; y = 1$

$\therefore (-a - 1) + 2 = 0 \qquad \therefore a = 1$

(3) the system $\begin{cases} x - 3y = 2 \\ 2x + y = -3 \end{cases}$ has a solution $(2a, -1)$.

$\quad\quad 2x - 6y = 4$

$-) \ \underline{2x + y = -3}$

$\quad\quad\quad -7y = 7 \ ; y = -1$

$\therefore x = 3y + 2 = -3 + 2 = -1 \qquad \therefore 2a = -1 \qquad \therefore a = -\dfrac{1}{2}$

(4) the line $2ax + 3y - 1 = 0$ passes through the intersection of the system

$\begin{cases} x - 2y = 3 \\ 2x + 2y = 1 \end{cases}.$

$\quad\quad 2x - 4y = 6 \qquad\qquad\qquad x - 2y = 3$

$-) \ \underline{2x + 2y = 1} \qquad\qquad +) \ \underline{2x + 2y = 1}$

$\quad\quad -6y = 5 \ ; y = -\dfrac{5}{6} \qquad 3x \quad = 4 \ ; \ x = \dfrac{4}{3}$

$\therefore 2a \cdot \dfrac{4}{3} + 3 \cdot \left(-\dfrac{5}{6}\right) - 1 = 0 \quad ; \quad \dfrac{8}{3}a = \dfrac{7}{2} \qquad \therefore a = \dfrac{21}{16}$

#20 Find the equation of each line such that:

(1) the line passes through the intersection of the system $\begin{cases} x + 2y = 3 \\ 3x + y = -2 \end{cases}$

and runs parallel to the y-axis.

$\quad\quad x + 2y = 3$

$-) \ \underline{6x + 2y = -4}$

$\quad -5x \quad = 7 \ ; \ x = -\dfrac{7}{5} \quad \therefore 2y = -x + 3 = \dfrac{7}{5} + 3 = \dfrac{22}{5} \qquad \therefore y = \dfrac{11}{5}$

$\therefore$ The intersection point is $\left(-\dfrac{7}{5}, \dfrac{11}{5}\right)$.

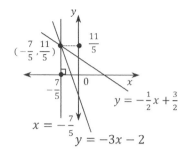

Parallel to the y-axis

$\Rightarrow x = -\frac{7}{5} \quad \therefore 5x + 7 = 0$

(2) the line passes through the intersection of the system $\begin{cases} -x + y + 2 = 0 \\ 2x + y - 3 = 0 \end{cases}$

and runs perpendicular to the x-axis.

$ -x + y + 2 = 0$

$\underline{-)\ \ 2x + y - 3 = 0}$

$ -3x + 5 = 0 \ ; \ x = \frac{5}{3} \qquad \therefore y = x - 2 = \frac{5}{3} - 2 = -\frac{1}{3}$

$\therefore$ The intersection is $\left(\frac{5}{3}, -\frac{1}{3} \right)$.

$\therefore$ Perpendicular to the x-axis $\Rightarrow x = \frac{5}{3}$

$\therefore 3x - 5 = 0$

(3) the line passes through the intersection of the system $\begin{cases} 2x - y + 3 = 0 \\ x + 2y + 4 = 0 \end{cases}$

and runs parallel to the line $3x + 2y = 5$.

$ 2x - y + 3 = 0$

$\underline{-)\ \ 2x + 4y + 8 = 0}$

$ -5y - 5 = 0 \ ; \ y = -1$

$\therefore x = -2y - 4 = -2$

$\therefore$ The intersection is $(-2, -1)$.

$3x + 2y = 5 \Rightarrow y = -\frac{3}{2}x + \frac{5}{2} \qquad \therefore m = -\frac{3}{2}$

$\therefore y = -\frac{3}{2}x + b \ ; \ -1 = -\frac{3}{2} \cdot (-2) + b \ ; \ b = -4$

$\therefore y = -\frac{3}{2}x - 4$

$\therefore 3x + 2y + 8 = 0$

#21 Find the value.

(1) Two functions $f(x) = x^2 - 1$ and $g(x) = ax + b$ have the same domain $\{-2, 1\}$.

Find the value of ab such that $f(x) = g(x)$ for any real number x.

Since $f(x) = g(x)$,

$f(-2) = g(-2) \implies 4 - 1 = -2a + b \qquad \therefore -2a + b = 3 \ \cdots\cdots \ ①$

$f(1) = g(1) \implies 1 - 1 = a + b \qquad \therefore a + b = 0 \ ; \ a = -b \ \cdots\cdots \ ②$

Substituting ② into ①, $3b = 3 \qquad \therefore b = 1, \ a = -1$

Therefore, $ab = -1$

(2) For a function $f : X = \{x \mid -1 \leq x \leq 2\} \to Y = \{y \mid -2 \leq y \leq 4\}$,

$f(x) = ax + b$ is one-to-one correspondence. Find the value of ab (where $a < 0$).

Since $f(x) = ax + b \ (a < 0)$ is one-to-one correspondence, the graph of $f(x)$ passes

through the points $(-1, 4)$ and $(2, -2)$.

$\therefore \ 4 = -a + b \cdots\cdots ① \quad$ and $\ -2 = 2a + b \cdots\cdots ②$

$① - ②: \ 6 = -3a \qquad \therefore a = -2, \ \text{and } b = 2$

Therefore, $ab = -4$

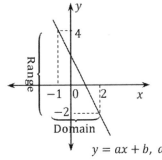
$y = ax + b, \ a < 0$

Alternative Approach:

The line through points $(-1, 4)$ and $(2, -2)$ is

$f(x) = \frac{-2-4}{2-(-1)}\left(x - (-1)\right) + 4 = \frac{-6}{3}(x + 1) + 4 = -2(x + 1) + 4 = -2x + 2$

$\therefore \ a = -2 \text{ and } b = 2 \ \text{ Therefore, } ab = -4$

(3) For a function $f(x) = ax + b$, $-1 \leq f(0) \leq 1$, $0 \leq f(1) \leq 3$, and $m \leq f(2) \leq n$.

Find the value of $m + n$.

Since $f(0) = b, \ -1 \leq b \leq 1 \cdots\cdots ①$

Since $f(1) = a + b, \ 0 \leq a + b \leq 3 \cdots\cdots ②$

Note that $f(2) = 2a + b$.

$② \times 2: \ 0 \leq 2a + 2b \leq 6$

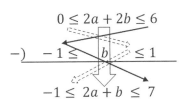

$0 \leq 2a + 2b \leq 6$

$-) \quad -1 \leq b \leq 1$

$-1 \leq 2a + b \leq 7$

$\therefore \ -1 \leq f(2) \leq 7 \qquad \therefore \ m = -1 \text{ and } n = 7 \qquad \text{Therefore, } m + n = 6$

#22 Find the number of a function f.

> For a function $f: X \to Y$,
>
> let the number of elements in a set X be $n(X) = a$ and the number of elements in a set Y be $n(Y) = b$. Then,
>
> ① The number of possible functions is b^a.
>
> ② The number of possible functions such that f is one-to-one is
>
> $\quad b \times (b-1) \times (b-2) \times \cdots\cdots \times (b-a+1), \ b \geq a$
>
> ③ The number of correspondence is $b \times (b-1) \times (b-2) \times \cdots\cdots \times 3 \times 2 \times 1, \ a = b$

(1) For a function $f: X = \{-1, \ 0, \ 1\} \to Y = \{-2, -1, \ 0, \ 1, \ 2, \ 3\}, \ xf(x) = 0$ for any x in X.

For any x in X, $xf(x) = 0$

$\Rightarrow$ i) When $x = -1$,

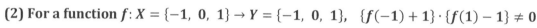

$\qquad -f(-1) = 0 \quad \therefore \ f(-1) = 0$

ii) When $x = 1$, $\qquad f(1) = 0$

iii) When $x = 0$,

$\qquad 0f(0) = 0 \qquad \therefore$ The number of $f(0)$ is one of the numbers: $-2, -1, \ 0, \ 1, \ 2, \ 3$.

Therefore, the number of the function f is 6 according to the value of $f(0)$.

(2) For a function $f: X = \{-1, \ 0, \ 1\} \to Y = \{-1, \ 0, \ 1\}, \ \{f(-1) + 1\} \cdot \{f(1) - 1\} \neq 0$

Since $f(-1) \neq -1$ and $f(1) \neq 1$,

$x = -1 \Rightarrow f(-1) = 0$ or $f(-1) = 1$ (Number of 2)

$x = 1 \ \Rightarrow f(1) = -1$ or $f(1) = 0$ (Number of 2)

$x = 0 \ \Rightarrow f(0) = -1$ or $f(0) = 0$ or $f(0) = 1$ (Number of 3)

Therefore, the number of possible functions is $2 \times 2 \times 3 = 12$.

Alternative Approach:

Total possible number of functions is $3^3 = 27$.

The number of functions such that $f(-1) = -1$ is $3 \times 3 = 9$.

The number of functions such that $f(1) = 1$ is $3 \times 3 = 9$.

The number of functions such that $f(-1) = -1$ and $f(1) = 1$ is 3.

Therefore, the number of possible functions is $27 - (9 + 9 - 3) = 12$.

(3) For a function $f: X = \{0, 1, 2\} \to Y = \{1, 2, 3, 4, 5, 6\}$, $f(1) = 3$,

and if $x_1 < x_2$, then $f(x_1) < f(x_2)$

Since $f(1) = 3$, the value of $f(2)$ is the one of the numbers: 4, 5, and 6.

And the value of $f(0)$ is the one of the numbers: 1 and 2.

Therefore, the number of function f is $3 \times 2 = 6$

(4) For a function $f: X = \{1, 2, 3, 4\} \to Y = \{1, 2, 3, 4\}$, $f(x) \geq x$ **for any** x **in** X.

$f(1) \geq 1 \Rightarrow f(1) = 1, f(1) = 2, f(1) = 3, \text{ or } f(1) = 4$ (Number of 4)

$f(2) \geq 2 \Rightarrow f(2) = 2, f(2) = 3, \text{ or } f(2) = 4,$ (Number of 3)

$f(3) \geq 3 \Rightarrow f(3) = 3 \text{ or } f(3) = 4$ (Number of 2)

$f(4) \geq 4 \Rightarrow f(4) = 4$ (Number of 1)

Therefore, the number of function f is $4 \times 3 \times 2 \times 1 = 24$

#23 A function such that $f(x) = \begin{cases} x, & x \text{ is rational} \\ 1-x, & x \text{ is irrational} \end{cases}$ **is defined in the domain**

$X = \{x \mid 0 \leq x \leq 1\}$. **Find the value of** $f(x) + f(1-x)$.

If x is rational, $1 - x$ is also rational.

$\therefore \; f(x) + f(1-x) = x + (1-x) = 1$

If x is irrational, $1 - x$ is also irrational.

$\therefore \; f(x) + f(1-x) = (1-x) + (1-(1-x)) = 1$

Therefore, $f(x) + f(1-x) = 1$

#24 For two functions $f(x) = 2x + 1$ **and** $g(x) = x - 3$, **find each composite function.**

(1) $(f \circ f)(x) = f(f(x)) = f(2x+1) = 2(2x+1) + 1 = 4x + 3$

(2) $(f \circ g)(x) = f(g(x)) = f(x-3) = 2(x-3) + 1 = 2x - 5$

(3) $(f \circ f \circ f)(x) = (f \circ f)(f(x)) = (f \circ f)(2x+1) = 4(2x+1) + 3 = 8x + 7$

(4) $(g \circ f \circ g)(x) = g \circ ((f \circ g)(x)) = g(2x-5) = 2x - 5 - 3 = 2x - 8$

#25 For two functions $f(x) = 2x - 1$ **and** $g(x) = x + 2$, **find the function** $h(x)$ **such that:**

(1) $(f \circ h)(x) = g(x)$

$(f \circ h)(x) = f(h(x)) = g(x) \quad \therefore \; 2h(x) - 1 = x + 2 \quad \therefore \; h(x) = \dfrac{x}{2} + \dfrac{3}{2}$

(2) $(g \circ h)(x) = f(x)$

$\quad (g \circ h)(x) = g\big(h(x)\big) = f(x) \qquad \therefore \ h(x) + 2 = 2x - 1 \qquad \therefore \ h(x) = 2x - 3$

#26 Find the value.

(1) For a function f, $f_1 = f$, $f_2 = f \circ f_1$, $f_3 = f \circ f_2$, $\cdots\cdots$, $f_n = f \circ f_{n-1}$

When $f(x) = \dfrac{x-1}{x}$, find the value of $f_{10}(2)$.

$$f_2(x) = (f \circ f_1)(x) = f\big(f_1(x)\big) = f\big(f(x)\big) = \frac{\frac{x-1}{x} - 1}{\frac{x-1}{x}} = \frac{\frac{x-1-x}{x}}{\frac{x-1}{x}} = \frac{\frac{-1}{x}}{\frac{x-1}{x}} = \frac{-1}{x-1}$$

$$f_3(x) = (f \circ f_2)(x) = f\big(f_2(x)\big) = \frac{\frac{-1}{x-1} - 1}{\frac{-1}{x-1}} = \frac{\frac{-1-x+1}{x-1}}{\frac{-1}{x-1}} = \frac{\frac{-x}{x-1}}{\frac{-1}{x-1}} = \frac{-x}{-1} = x$$

$$f_4(x) = (f \circ f_3)(x) = f\big(f_3(x)\big) = \frac{x-1}{x} = f(x) = f_1(x)$$

Therefore, $f_{10}(2) = f_7(2) = f_4(2) = f_1(2) = f(2) = \dfrac{2-1}{2} = \dfrac{1}{2}$

(2) For two functions $f(x) = \begin{cases} x^2, & x \geq 0 \\ \dfrac{x}{2}, & x < 0 \end{cases}$ and $g(x) = x + 2$,

a function h satisfies $h \circ g = f$. Find the value of $h(-2)$.

$(h \circ g)(x) = h\big(g(x)\big) = f(x)$

When $g(x) = -2$, $x + 2 = -2 \qquad \therefore \ x = -4$

$\therefore \ h(-2) = f(-4) = \dfrac{-4}{2} = -2$

(3) For two functions $f(x) = x - a$ and $g(x) = x^2 + 1$,

find the value of a constant a such that $(f \circ g)(x) = (g \circ f)(x)$.

$(f \circ g)(x) = \big(f(g(x)\big) = (x^2 + 1) - a$

$(g \circ f)(x) = g\big(f(x)\big) = (x - a)^2 + 1$

$\therefore \ (x^2 + 1) - a = (x - a)^2 + 1 \ ; \quad 1 - a = -2ax + a^2 + 1 \ ; \quad a^2 - 2ax + a = 0 \cdots\cdots ①$

Since ① is always true for all x, $a = 0$ $\quad (\because \ (a^2 + a) - 2ax = 0 = 0 + 0x)$

(4) For any real number x, $f\big(f(x)\big) = x$ and $f(0) = 1$. Find the value of $f(-2)$.

The function f such that $f\big(f(x)\big) = x$ has the form $f(x) = ax + b$, $a \neq 0$.

Since $f(0) = 1$, $f(x) = ax + 1$

Since $f\big(f(x)\big) = x$, $a(ax + 1) + 1 = x$; i.e., $a^2 x + a + 1 = x$

$\therefore \ (a^2 - 1)x + a + 1 = 0 \cdots\cdots ①$

Since ① is always true for all x, $a^2 - 1 = 0$ and $a + 1 = 0$

$\therefore\ a = -1$ $\quad \therefore\ f(x) = -x + 1$

Therefore, $f(-2) = 3$

#27 Sketch all possible graphs of a function $y = f(x)$ such that $(f \circ f)(x) = x$.

The function f such that $f(f(x)) = x$ has the form $f(x) = ax + b,\ a \neq 0$.

$\therefore\ (f \circ f)(x) = f(f(x)) = a(ax + b) + b = a^2x + ab + b$

$\therefore\ a^2x + ab + b = x \cdots\cdots$ ①

Since ① is always true for all x, $a^2 = 1$ and $ab + b = 0$

If $a = 1$, then $2b = 0$ $\therefore\ b = 0$ Thus, $f(x) = x$

If $a = -1$, then $-b + b = 0$ $\therefore\ b$ is any real number.

Thus, $f(x) = -x + b$ where b is a real number.

Therefore, the possible graphs of the function f are:

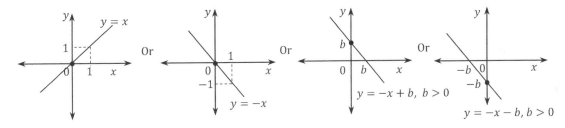

#28 Tell whether x and y show direct variation, inverse variation, or neither.

(1) $xy = 5$ $\Rightarrow$ $y = \dfrac{5}{x}$ $\quad \therefore$ Inverse variation

(2) $x = \dfrac{3}{y}$ $\Rightarrow$ $y = \dfrac{3}{x}$ $\quad \therefore$ Inverse variation

(3) $y = x + 1$ $\quad \therefore$ Neither

(4) $\dfrac{y}{2} = x$ $\Rightarrow$ $y = 2x$ $\quad \therefore$ Direct variation

(5) $2x = y$ $\Rightarrow$ $y = 2x$ $\quad \therefore$ Direct variation

(6) $x = 3y$ $\Rightarrow$ $y = \dfrac{1}{3}x$ $\quad \therefore$ Direct variation

#29 Determine which of the functions have inverse functions.

(1) $f(x) = x^2$, for all x

(2) $g(x) = x^2$, $x \geq 0$

(3) $h(x) = x^2$, $x \leq 0$

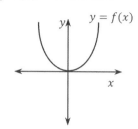 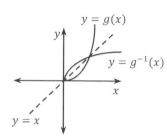 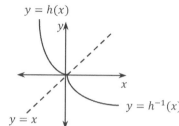

$f(x) = x^2$, for all x $g(x) = x^2$, $x \geq 0$ $h(x) = x^2$, $x \leq 0$

$g^{-1}(x) = \sqrt{x}$, $x \geq 0$ $h^{-1}(x) = -\sqrt{x}$, $x \geq 0$

By the horizontal test, f fails. But, g and h pass it.

Therefore, g and h have inverse functions.

#30 Find the inverse of each function and sketch the graphs of f and f^{-1}.

(1) $f(x) = \dfrac{3-5x}{2}$

$$y = \frac{3-5x}{2} \qquad \text{Rewrite in form } y = f(x)$$

$$\Rightarrow \quad x = \frac{3-5y}{2} \qquad \text{Switch } x \text{ and } y$$

$$\Rightarrow \quad 2x = 3 - 5y$$

$$\Rightarrow \quad y = -\frac{2x}{5} + \frac{3}{5} \qquad \text{Solve for } y$$

$$\therefore \quad f^{-1}(x) = -\frac{2x}{5} + \frac{3}{5} \qquad \text{Replace } y \text{ by } f^{-1}(x)$$

The domain and range of both f and f^{-1} are all real numbers.

(2) $f(x) = \sqrt{2x - 1}$

$$y = \sqrt{2x - 1} \qquad \text{Rewrite in form } y = f(x)$$

$$\Rightarrow \quad x = \sqrt{2y - 1} \qquad \text{Switch } x \text{ and } y$$

$$\Rightarrow \quad x^2 = 2y - 1 \qquad \text{Square both sides}$$

$$\Rightarrow \quad y = \frac{x^2}{2} + \frac{1}{2}, \ x \geq 0 \qquad \text{Solve for } y$$

$$\therefore \quad f^{-1}(x) = \frac{x^2}{2} + \frac{1}{2}, \ x \geq 0 \qquad \text{Replace } y \text{ by } f^{-1}(x)$$

The domain of f is the interval $\left[\frac{1}{2}, \infty\right)$ and the range of f is the interval $[0, \infty)$.

The domain of f^{-1} is the interval $[0, \infty)$ and the range of f^{-1} is the interval $\left[\frac{1}{2}, \infty\right)$.

(3) $f(x) = x^3$

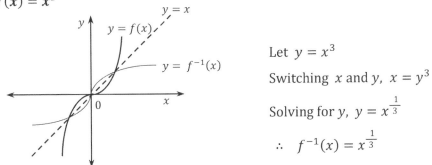

Let $y = x^3$

Switching x and y, $x = y^3$

Solving for y, $y = x^{\frac{1}{3}}$

$\therefore$ $f^{-1}(x) = x^{\frac{1}{3}}$

(4) $f(x) = \dfrac{1}{x-1}$

Let $y = \dfrac{1}{x-1}$

Switching x and y, $x = \dfrac{1}{y-1}$

Solving for y, $y - 1 = \dfrac{1}{x}$ $\therefore$ $y = \dfrac{1}{x} + 1$

$\therefore$ $f^{-1}(x) = \dfrac{1}{x} + 1$

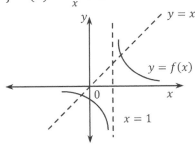

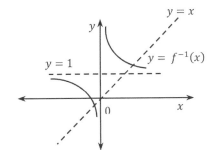

#31 Find the value.

(1) A function f with the real number domain is one-to-one correspondence.

When $f\left(\dfrac{3x-1}{3}\right) = -4x + 5$, find the value of $f^{-1}(-3)$.

$f\left(\dfrac{3x-1}{3}\right) = -4x + 5$ $\Rightarrow$ $f^{-1}(-4x + 5) = \dfrac{3x-1}{3}$

If $-4x + 5 = -3$, then $x = 2$.

$\therefore$ $f^{-1}(-3) = \dfrac{3 \cdot 2 - 1}{3} = \dfrac{5}{3}$

(2) For an inverse function $f^{-1}(x) = ax + b$ of $f(x)$, $f(-2) = 1$ and $f^{-1}(3) = -4$.

Find the value of $a + b$ (a, b are real numbers).

Since $f(-2) = 1$, $f^{-1}(1) = -2$ $\therefore a + b = -2$ ······ ①

Since $f^{-1}(3) = -4$, $3a + b = -4$ ······ ②

②−① : $2a = -2$

$\therefore a = -1$, $b = -1$ Therefore, $a + b = -2$

(3) For two functions $f(x) = x + a$ and $g(x) = bx + c$,

$(f^{-1} \circ g)(x) = 2x + 3$ and $g^{-1}(3) = 2$. Find the value of $a + b + c$.

$f(x) = x + a$

$\Rightarrow y = x + a$ Rewrite $f(x)$ with y

$\Rightarrow x = y + a$ Switch x and y

$\Rightarrow y = x - a$ Solve for y

$\therefore f^{-1}(x) = x - a$

Since $(f^{-1} \circ g)(x) = f^{-1}(g(x)) = bx + c - a = 2x + 3$, $b = 2$ and $c - a = 3$ ······ ①

Since $g^{-1}(3) = 2$, $g(2) = 3$ $\therefore 2b + c = 3$ ······ ②

Substituting $b = 2$ into ② gives $c = -1$.

Substituting $c = -1$ into ① gives $a = -4$.

Therefore, $a + b + c = -4 + 2 - 1 = -3$

(4) For two functions $f(x) = \dfrac{1}{2}x + a$ and $g(x) = bx - 2$ with the real number domains,

$(f \circ g)(x) = x + 3$. Find the value of $f^{-1}(g(-1))$.

$(f \circ g)(x) = f(g(x)) = \dfrac{1}{2}(bx - 2) + a = \dfrac{bx}{2} - 1 + a$

Since $(f \circ g)(x) = x + 3$, $\dfrac{b}{2} = 1$ and $-1 + a = 3$ $\therefore b = 2$, $a = 4$

$\therefore f(x) = \dfrac{1}{2}x + 4$ and $g(x) = 2x - 2$

Note that $f^{-1}(g(-1)) = f^{-1}(-4)$

Let $f^{-1}(-4) = k$. Then, $f(k) = -4$. $\therefore \dfrac{1}{2}k + 4 = -4$ $\therefore k = -16$

Therefore, $f^{-1}(g(-1)) = -16$

(5) For a function $f(x) = x|x| + a$, a is a real number, $f^{-1}(3) = -1$.

Find the value of $(f^{-1} \circ f^{-1})(3)$.

$f(x) = x|x| + a$

i) $x \geq 0 \implies f(x) = x^2 + a$

ii) $x < 0 \implies f(x) = -x^2 + a$

$f^{-1}(3) = -1 \implies f(-1) = 3 \quad \therefore -(-1)^2 + a = 3 \quad \therefore a = 4$

$\therefore f(x) = \begin{cases} x^2 + 4 & (x \geq 0) \\ -x^2 + 4 & (x < 0) \end{cases}$

Note that $(f^{-1} \circ f^{-1})(3) = f^{-1}(f^{-1}(3)) = f^{-1}(-1)$

Let $f^{-1}(-1) = k$. Then, $f(k) = -1$

If $k \geq 0$, then $f(k) = k^2 + 4 \geq 4$ (This is not true because $f(k) = -1 < 0$.)

Thus, $k < 0$ and $f(k) = -k^2 + 4 = -1$

$\therefore k^2 = 5 \quad \therefore k = -\sqrt{5} \quad (\because k < 0)$

Therefore, $(f^{-1} \circ f^{-1})(3) = f^{-1}(-1) = k = -\sqrt{5}$

(6) For a function $f(x) = \begin{cases} x + a & (x \geq 3) \\ 2x + 1 & (x < 3) \end{cases}$ with the real number domain, the inverse

function of f exists. Find the value of $(f^{-1} \circ f^{-1})(10)$. (Where a is a constant.)

To have an inverse function,

the values of $f(3)$ must be the same when $x \geq 3$ and when $x < 3$.

That is, $f(3) = 3 + a = 2 \cdot 3 + 1 \quad \therefore a = 4$

$\therefore f(x) = \begin{cases} x + 4 & (x \geq 3) \\ 2x + 1 & (x < 3) \end{cases}$

Rewriting $f(x)$ with y, $y = \begin{cases} x + 4 & (x \geq 3) \\ 2x + 1 & (x < 3) \end{cases}$

Switching x and y, $x = \begin{cases} y + 4 & (y \geq 7) \\ 2y + 1 & (y < 7) \end{cases}$

Solving for y, $y = \begin{cases} x - 4 \\ \dfrac{1}{2}x - \dfrac{1}{2} \end{cases} \qquad \therefore f^{-1}(x) = \begin{cases} x - 4 & (x \geq 7) \\ \dfrac{1}{2}x - \dfrac{1}{2} & (x < 7) \end{cases}$

Therefore, $(f^{-1} \circ f^{-1})(10) = f^{-1}(f^{-1}(10)) = f^{-1}(6) = \dfrac{1}{2} \cdot 6 - \dfrac{1}{2} = \dfrac{5}{2}$

(7) When two functions $f(x)$ and $g(x)$ are one-to-one correspondences,

$g(x) = f(3x - 1)$ and $f^{-1}(1) = 5$. Find the value of $g^{-1}(1)$.

$f^{-1}(1) = 5 \Rightarrow f(5) = 1$

$\therefore \ 3x - 1 = 5 \Rightarrow x = 2$

$\therefore \ g(2) = f(3 \cdot 2 - 1) = f(5) = 1 \qquad \therefore \ g^{-1}(1) = 2$

(8) For a function $f(x)$, $f^{-1}(x)$ is the inverse function of $f(x)$ such that $f^{-1}(0) = 3$.
For a function $h(x)$ such that $h(x) = f(2x - 1)$, $h^{-1}(x)$ is the inverse function of $h(x)$. Find the value of $h^{-1}(0)$.

$f^{-1}(0) = 3 \Rightarrow f(3) = 0$

Since $h(x) = f(2x - 1)$,

When $2x - 1 = 3$; i.e., $x = 2$, $h(2) = f(3) = 0$

Therefore, $h^{-1}(0) = 2$

(9) For an inverse function $g(x)$ of $f(x)$, $f\left(2g(x) - \dfrac{2x-3}{x+2}\right) = x$. Find the value of $f(1)$.

Let $h(x) = 2g(x) - \dfrac{2x-3}{x+2}$

Then, $f\big(h(x)\big) = x$

$\therefore \ h(x)$ is the inverse function of $f(x)$.

$\therefore \ h(x) = g(x)$

$\therefore \ 2g(x) - \dfrac{2x-3}{x+2} = g(x) \qquad \therefore \ g(x) = \dfrac{2x-3}{x+2}$

Let $y = f(x)$

Then, $f^{-1}(y) = x$

$\therefore \ f^{-1}(y) = g(y) = \dfrac{2y-3}{y+2} = x$

$\therefore \ xy + 2x = 2y - 3 \qquad \therefore \ (x - 2)y = -2x - 3$

$\therefore \ y = \dfrac{-2x-3}{x-2} = f(x)$

Therefore, $f(1) = \dfrac{-2(1)-3}{1-2} = \dfrac{-5}{-1} = 5$

#32 Find the range.

(1) For a function $f(x) = |2x - 1| + ax + 3$ which is defined in the real number system, find the range of a so that f has an inverse function.

To have an inverse function, f must be one-to-one correspondence.

When $x \geq \frac{1}{2}$, $f(x) = (2x - 1) + ax + 3 = (2 + a)x + 2$

When $x < \frac{1}{2}$, $f(x) = -(2x - 1) + ax + 3 = (-2 + a)x + 4$

Since the signs of the slopes must be the same, $(2 + a)(-2 + a) > 0$.

Therefore, $a > 2$ or $a < -2$

(2) When $g(x)$ is an inverse function of $f(x) = \frac{x^2}{2} + a \ (x \geq 0)$, find the range of a so that the equation $f(x) = g(x)$ has two different non-negative real number solutions.

The intersection points of the graphs of $y = f(x)$ and $y = g(x)$ are the same as the intersection points of the graph of $y = f(x)$ and a line $y = x$.

$$\therefore \ \frac{x^2}{2} + a = x \qquad \therefore \ x^2 - 2x + 2a = 0 \cdots\cdots ①$$

Since the discriminant D of the equation ① is $D > 0$, $D = 4 - 8a > 0 \qquad \therefore \ a < \frac{1}{2}$

Note that $x \geq 0$. Thus, $a \geq 0$ (to have two different non-negative real number solutions)

Therefore, $0 \leq a < \frac{1}{2}$

(3) For two functions $f(x) = x^2 - x - 12$ and $g(x) = x^2 + ax + 5$, find the range of a real number a such that $(f \circ g)(x) \geq 0$ for any real number x.

$(f \circ g)(x) = f(g(x)) \geq 0$

Let $g(x) = k$. Then, $f(k) \geq 0$

$\therefore \ k^2 - k - 12 \geq 0 \quad \therefore \ (k - 4)(k + 3) \geq 0 \quad \therefore \ k \geq 4$ or $k \leq -3$

$\therefore \ g(x) \geq 4$ or $g(x) \leq -3$

i) When $g(x) \geq 4$; i.e., $x^2 + ax + 5 \geq 4$,

$x^2 + ax + 1 \geq 0 \iff$ The discriminant D of the equation $x^2 + ax + 1 = 0$ is $D \leq 0$.

$\therefore \ a^2 - 4 \leq 0 \qquad \therefore \ -2 \leq a \leq 2$

ii) When $g(x) \leq -3$; i.e., $x^2 + ax + 5 \leq -3$,

Consider $x^2 + ax + 8 \leq 0$ for any real number x.

Since $x^2 \geq 0$, $x^2 + 8 > 0$

$\therefore$ There is no a such that $x^2 + ax + 8 \leq 0$ for any real number x.

Therefore, by i) and ii), $-2 \leq a \leq 2$

#33 Find the distance.

(1) When a function $f(x) = \frac{1}{2}(x^2 + x - 2)$, $x \geq -1$, and its inverse function intersect at two points A and B, find the length of the segment $\overline{AB}$.

Since the two points A and B are the intersection points of the graph of $f(x)$ and a line $= x$,

we have $\frac{1}{2}(x^2 + x - 2) = x$

$\therefore \ x^2 + x - 2 = 2x$ $\quad \therefore \ x^2 - x - 2 = 0$ $\quad \therefore (x - 2)(x + 1) = 0$ $\quad \therefore x = 2$ or $x = -1$

Thus, $A = A(2, 2)$ $\ B = B(-1, -1)$

Therefore, $\overline{AB} = \sqrt{(2 - (-1))^2 + (2 - (-1))^2} = \sqrt{3^2 + 3^2} = \sqrt{18} = 3\sqrt{2}$

(2) When a function $g(x)$ is an inverse of $f(x) = x^2 - 3x + 3$ $(x \geq 1)$, find the distance between the intersection points of the graph of $y = f(x)$ and a line $y = x$.

Since $g(x)$ is the inverse function of $f(x)$, the intersection points of the graphs $y = f(x)$ and $y = g(x)$ are the same as the intersection points of the graph of $y = f(x)$ and a line $y = x$.

Let $x^2 - 3x + 3 = x$. Then, $x^2 - 4x + 3 = 0$

$\therefore \ (x - 1)(x - 3) = 0$ $\quad \therefore \ x = 1$ or $x = 3$

$\therefore$ The intersection points are $(1, 1)$ and $(3, 3)$.

Therefore, the distance between the two points is

$$\sqrt{(3 - 1)^2 + (3 - 1)^2} = \sqrt{2^2 + 2^2} = \sqrt{8} = 2\sqrt{2}$$

#34 When $g(x)$ is an inverse function of $f(x)$, find the inverse function of $f(2x)$.

Let $y = f(2x)$. Then, $f^{-1}(y) = 2x$ $\quad \therefore \ x = \frac{1}{2}f^{-1}(y)$

Switching x and y, $\ y = \frac{1}{2}f^{-1}(x)$

Since $g(x) = f^{-1}(x)$, $\ y = \frac{1}{2}g(x)$

Therefore, the inverse function of $f(2x)$ is $\frac{1}{2}g(x)$.

#35 For a function $f(x) = \begin{cases} x, & x \geq 0 \\ 2x, & x \leq 0 \end{cases}$, answer the question.

(1) Find the value of $f(f(10))$.

Since $f(10) = 10$, $f(f(10)) = f(10) = 10$

(2) Find the value of $f^{-1}(-4)$.

Since $f(-2) = -4$, $f^{-1}(-4) = -2$

(3) Find the number of intersection points of the graphs of the function $y = f(x)$ and its inverse function $y = f^{-1}(x)$.

The intersection points of the graphs of $y = f(x)$ and $y = f^{-1}(x)$ are the same as the intersection points of the graph of $y = f(x)$ and a line $y = x$.

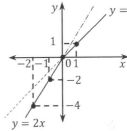

i) When $x \geq 0$,

The graph of $y = f(x)$ coincides with the line $y = x$.

∴ There are unlimited numbers of intersection points.

ii) When $x \leq 0$,

The graph of $y = 2x$ and the line $y = x$ intersect at only one point $x = 0$.

#36 Let $\mathbb{R} = \{x \mid x \text{ is a real number}\}$. For a function $f: \mathbb{R} \to \mathbb{R}$, the inverse function f^{-1} exists. When $f(a + b) = f^{-1}(a) + f^{-1}(b)$ for any real numbers a and b, determine whether the following statements are true or false.

(1) $f(f(a) + f(b)) = a + b$

Since $f(a + b) = f^{-1}(a) + f^{-1}(b)$,

rewriting a with $f(a)$ and b with $f(b)$, we have

$f(f(a) + f(b)) = f^{-1}(f(a)) + f^{-1}(f(b)) = a + b$ ∴ True

(2) If $f(-1) = 2$, then $f(-4) = -1$

$f(-1) = 2 \Rightarrow f^{-1}(2) = -1$

Since $f(a + b) = f^{-1}(a) + f^{-1}(b)$, $f^{-1}(2) + f^{-1}(2) = f(2 + 2) = f(4)$

∴ $f(4) = -2$ ∴ $f^{-1}(-2) = 4$

$f^{-1}(-2) + f^{-1}(-2) = f(-2 - 2) = f(-4) = 8$ ∴ False

(3) $f^{-1}(a + b) = f(a) + f(b)$

Let $f(a) = \alpha$, $f(b) = \beta$ Then, $f^{-1}(\alpha) = a$, $f^{-1}(\beta) = b$

∴ $f(\alpha + \beta) = f^{-1}(\alpha) + f^{-1}(\beta) = a + b$

∴ $f^{-1}(a + b) = \alpha + \beta = f(a) + f(b)$ ∴ True

#37 Minimum and Maximum values.

(1) When the domain of a function $y = (x - 1)^2 - 2(x - 1) - 2$ is $\{x| \, 0 \leq x \leq 2\}$, find minimum and maximum values for the function.

Let $x - 1 = k$.

Since $0 \leq x \leq 2$, $-1 \leq k \leq 1$

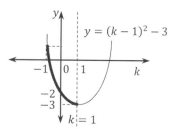

The given function is $y = k^2 - 2k - 2 = (k - 1)^2 - 3$

When $k = 1$, minimum value of the function is -3.

When k $= -1$, maximum value of the function is 1.

(2) When the domain of a function $y = -3x + a$ is $\{x| -1 \leq x \leq 3\}$, minimum value of the function is 1. Find maximum value of the function.

Since $y = -3x + a$ is a decreasing function, y has minimum value when $x = 3$.

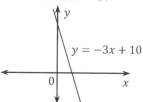

$\therefore \ 1 = -3 \cdot 3 + a \qquad \therefore \ a = 10$

$\therefore \ y = -3x + 10$

When $x = -1$, the function has maximum value.

The maximum value of the function is $-3 \cdot (-1) + 10 = 13$

(3) For a function $y = |x - 1| + |2x - 3|$, find minimum value of the function.

i)   When $x < 1$,

$y = -(x - 1) - (2x - 3) = -3x + 4$

ii) When $1 \leq x < \dfrac{3}{2}$

$y = (x - 1) - (2x - 3) = -x + 2$

iii) When $x \geq \dfrac{3}{2}$

$y = (x - 1) + (2x - 3) = 3x - 4$

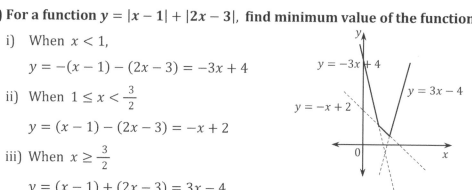

Therefore, when $x = \dfrac{3}{2}$, the minimum value of $f(x)$ is $3 \cdot \dfrac{3}{2} - 4 = \dfrac{9}{2} - 4 = \dfrac{1}{2}$

(4) For a function $f(x) = x^2 - 2x + 4$ with the domain $\{x| \, 0 \leq x \leq 3\}$, find maximum value of $f(f(x))$.

$f(x) = x^2 - 2x + 4 = (x - 1)^2 - 1 + 4 = (x - 1)^2 + 3$

Since $f(0) = 4$, $f(1) = 3$, and $f(3) = 7$, the range of the function is $3 \leq f(x) \leq 7$

Let $f(x) = k$.

Then $f(f(x)) = f(k) = (k - 1)^2 + 3$, $3 \leq k \leq 7$

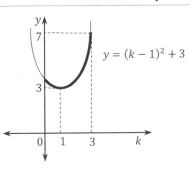

If $k = 3$, then $f(3) = 7$

If $k = 7$, then $f(7) = 39$

$\therefore$ The range is $7 \leq f(k) \leq 39$

Therefore, maximum number of $f(f(x))$ is 39.

(5) The graph of a function $ax + by + 1 = 0$ is as shown in Figure.

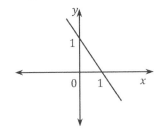

Find maximum and minimum values of $y = ax^2 + 2bx$, $0 \leq x \leq 2$ where a and b are real numbers.

From the graph, the equation of the line is $y = -x + 1$, $0 \leq x \leq 2$

$\therefore -x - y + 1 = 0$ $\qquad \therefore a = -1, b = -1$

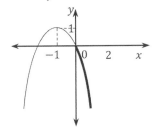

$y = ax^2 + 2bx = -x^2 - 2x = -(x^2 + 2x)$

$\qquad = -(x + 1)^2 + 1$

$\therefore$ The graph has maximum value 0 when $x = 0$

and minimum value -8 when $x = 2$.

(6) For a function $y = |x^2 - 4x + 3|$ with the domain $\{x | 1 \leq x \leq 3\}$, find maximum and minimum values of the function.

Let $x^2 - 4x + 3 = 0$. Then $(x - 3)(x - 1) = 0$

$\therefore x = 3$ or $x = 1$ are x-intercepts.

$1 \leq x \leq 3 \Rightarrow 0 \leq y \leq 1$

When $x = 2$, the function has maximum value 1.

When $x = 1$ or $x = 3$, the function has minimum value 0.

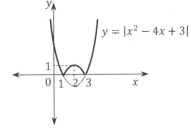

(7) For a quadratic function $y = f(x)$, the leading coefficient of the function is positive.

When $f(4 - x) - f(x) = 0$ for any real number x, find minimum value of $f(x)$.

$f(4 - x) - f(x) = 0 \Rightarrow f(4 - x) = f(x)$

Rewriting x with $x + 2$, we have $f(2 - x) = f(2 + x)$

$\therefore$ The graph of the function f is a parabola that opens upwards

with vertex $(2, f(2))$ and the axis of symmetry $x = 2$.

Therefore, minimum value of $f(x)$ is $f(2)$.

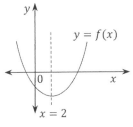

(8) When $x > 0$, find maximum value of $y = -(x + \frac{1}{x})^2 + 4\left(x + \frac{1}{x}\right) + 5$.

Let $x + \frac{1}{x} = k$.

Then $y = -k^2 + 4k + 5 = -(k^2 - 4k) + 5 = -(k - 2)^2 + 9$

Since $x > 0$,

$x + \frac{1}{x} \geq 2\sqrt{x \cdot \frac{1}{x}} = 2$ (When $x = 1$, $LHS = RHS$)

by the relationship between the arithmetic and geometric means.

$\therefore k \geq 2$

Therefore, the given function has maximum value 9 when $k = 2$.

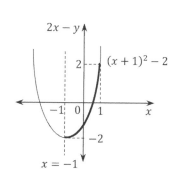

(9) For any real numbers x and y, $x^2 + y = 1$.

Find maximum and minimum values of $2x - y$, $y \geq 0$.

$x^2 + y = 1 \Rightarrow y = 1 - x^2$

Since $y \geq 0$, $1 - x^2 \geq 0$ $\therefore -1 \leq x \leq 1$

$2x - y = 2x - (1 - x^2) = x^2 + 2x - 1 = (x + 1)^2 - 2$

Therefore, maximum value is 2 (when $x = 1$) and

minimum value is -2 (when $x = -1$).

(10) For any real number x, $f(x) + 2f(1 - x) = x^2$.

Find minimum value of $f(x)$, $-1 \leq x \leq 1$.

$f(x) + 2f(1 - x) = x^2$ ······ ①

Rewriting x with $1 - x$, $f(1 - x) + 2f(x) = (1 - x)^2$ ······ ②

① $-$ ② $\times 2$: $-3f(x) = x^2 - 2(1 - x)^2 = -x^2 + 4x - 2$

$\therefore f(x) = \frac{1}{3}(x^2 - 4x) + \frac{2}{3} = \frac{1}{3}(x - 2)^2 - \frac{4}{3} + \frac{2}{3}$

$= \frac{1}{3}(x - 2)^2 - \frac{2}{3}$

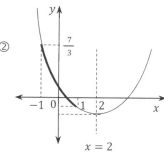

Therefore, minimum value of $f(x)$ in $[-1, 1]$ is $f(1) = -\frac{1}{3}$.

#38 Find the range of a.

(1) For a function $y = 2ax - a + 1$, find the range of a such that $y > 0$ for $1 < x < 3$.

$y = 2ax - a + 1 = 2a\left(x - \frac{1}{2}\right) + 1$

$\therefore y$ is a line that passes through a point $(\frac{1}{2}, 1)$.

Let $f(x) = 2ax - a + 1$.

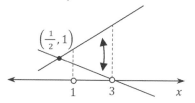

Since $y > 0$ when $1 < x < 3$, $f(3) \geq 0$

$\therefore$ $2a \cdot 3 - a + 1 = 5a + 1 \geq 0$

$\therefore$ $a \geq -\dfrac{1}{5}$

(2) For any real number x, $(a + 2)x^2 + 2(a + 2)x - 1 < 0$ is always true.

Find the range of the real number a.

i) $a + 2 < 0$ $\therefore$ $a < -2$ $\cdots\cdots$ ①

ii) Let $(a + 2)x^2 + 2(a + 2)x - 1 = 0$.

Then, the discriminant D of the equation is $D < 0$

$\therefore$ $(a + 2)^2 + (a + 2) = a^2 + 5a + 6 = (a + 2)(a + 3) < 0$

$\therefore$ $-3 < a < -2$ $\cdots\cdots$ ②

By ① and ②, $-3 < a < -2$

(3) When the graph of a function $y = x^2 + ax + b$ and the x-axis intersect at one point,

find the range of the real number a so that the graph of the function and a line $y = 4x$ don't intersect at any point.

The equation of the x-axis is $y = 0$.

Since $y = x^2 + ax + b$ and $y = 0$ intersect at one point, the discriminant D_1 of the

equation $x^2 + ax + b = 0$ is $D_1 = 0$.

That is, $a^2 - 4b = 0$ $\cdots\cdots$ ①

Since $y = x^2 + ax + b$ and $y = 4x$ don't intersect at any point, the discriminant D_2 of the

equation $x^2 + ax + b = 4x$ is $D_2 < 0$.

That is, $x^2 + (a - 4)x + b = 0$ and $D_2 = (a - 4)^2 - 4b < 0$

$\therefore$ $a^2 - 8a + 16 - 4b < 0$ $\cdots\cdots$ ②

Substituting ① into ② gives $-8a + 16 < 0$

Therefore, $a > 2$

(4) For a function $f(x) = x^2 - (2a + 1)x + 4$, find the range of a such that $f(x) > x$

for any real number x.

Since $f(x) > x$, $x^2 - (2a + 1)x + 4 > x$

$\therefore$ $x^2 - 2(a + 1)x + 4 > 0$ $\cdots\cdots$ ①

Since ① is always true for any real number x, the discriminant D of the equation $x^2 -$

$2(a + 1)x + 4 = 0$ is $D < 0$.

$\therefore$ $D = (a + 1)^2 - 4 = a^2 + 2a - 3 = (a + 3)(a - 1) < 0$

Therefore, $-3 < a < 1$

(5) When an equation $x^2 + ax - 3a + 3 = 0$ has two different roots, one is less than 2 and the other one is greater than 2. Find the range of the real number a.

Let $f(x) = x^2 + ax - 3a + 3$

Since 2 is in-between the two roots of $f(x) = 0$, $f(2) < 0$

$\therefore$ $4 + 2a - 3a + 3 = -a + 7 < 0$

Therefore, $a > 7$

(6) Find the range of the real number a so that one root of an equation $x^2 + 2x + a = 0$ is in-between the two roots of an equation $x^2 - 4x + 3 = 0$.

$x^2 - 4x + 3 = 0 \Rightarrow (x - 3)(x - 1) = 0 \quad \therefore x = 3$ or $x = 1$

Let $f(x) = x^2 + 2x + a$

Then, $f(x) = (x + 1)^2 - 1 + a$

$\therefore$ $f(1) < 0$ and $f(3) > 0$

$\therefore$ $3 + a < 0$ and $15 + a > 0 \quad \therefore a < -3$ and $a > -15$

Therefore, $-15 < a < -3$

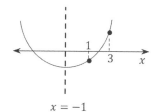

(7) Find the range of the real number a so that an equation $\left|x^2 - 1\right| = x + a$ has four different real number solutions.

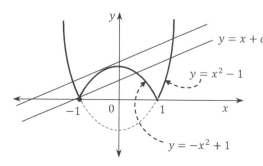

The equation $\left|x^2 - 1\right| = x + a$ has four different real number solutions.

$\Leftrightarrow$ The graph of $y = |x^2 - 1|$ and the line $y = x + a$ intersect at 4 different points.

When the line $y = x + a$ passes through a point $(-1, 0)$, $0 = -1 + a \quad \therefore a = 1 \cdots\cdots$ ①

When the line $y = x + a$ and the graph of $y = -x^2 + 1$ intersect at one point,

$-x^2 + 1 = x + a \quad \therefore x^2 + x + a - 1 = 0$

Since the discriminant D of the equation $x^2 + x + a - 1 = 0$ is $D = 0$,

$1 - 4a + 4 = 0 \quad \therefore a = \dfrac{5}{4} \cdots\cdots$ ②

Therefore, by ① and ②, $1 < a < \dfrac{5}{4}$

(8) Find the range of the real number a so that the solution of an inequality

$x^2 + ax + a^2 - 4 \leq 0$ **includes the interval** $[0, 2] = \{x | 0 \leq x \leq 2\}$.

Let $f(x) = x^2 + ax + a^2 - 4$.

Since $f(0) \leq 0$ and $f(2) \leq 0$,

$a^2 - 4 \leq 0$ and $2a + a^2 = a(2 + a) \leq 0$

$\therefore -2 \leq a \leq 2$ and $-2 \leq a \leq 0$

Therefore, $-2 \leq a \leq 0$

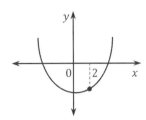

(9) Find the range of the real number a so that the vertex of the graph of a function $y =$

$x^2 + 4ax - 2a + 1$ **lies on the second quadrant.**

$y = x^2 + 4ax - 2a + 1 = (x + 2a)^2 - 4a^2 - 2a + 1$

Vertex is $(-2a, -4a^2 - 2a + 1)$

$\therefore -2a < 0$ and $-4a^2 - 2a + 1 > 0$

$\therefore a > 0$ and $4a^2 + 2a - 1 < 0 \quad \cdots\cdots ①$

Let $4a^2 + 2a - 1 = 0$. Then, $a = \dfrac{-1 \pm \sqrt{1+4}}{4} = \dfrac{-1 \pm \sqrt{5}}{4}$

$\therefore \dfrac{-1-\sqrt{5}}{4} < a < \dfrac{-1+\sqrt{5}}{4} \quad \cdots\cdots ②$

By ① and ②, $0 < a < \dfrac{-1+\sqrt{5}}{4}$

(10) Find the range of the real number a so that an equation $\left|x^2 - 1\right| = ax - 3a + 2$ **has**

four different real number solutions.

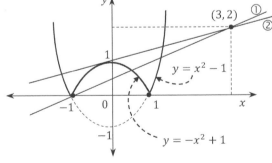

The equation $\left|x^2 - 1\right| = ax - 3a + 2$ has

four different real number solutions.

$\Leftrightarrow$ The graph of $y = \left|x^2 - 1\right|$ and the line

$y = ax - 3a + 2$ intersect at 4 different

points.

For the line $y = ax - 3a + 2$, $y = a(x - 3) + 2$

$\therefore$ The line y passes through a point $(3, 2)$.

The line $y = ax - 3a + 2$ should be in-between the lines ① and ②.

i) For the line ①, ① passes through a point $(-1, 0)$. $\quad \therefore 0 = -4a + 2 \quad \therefore a = \dfrac{1}{2}$

ii) For the line ②, the graph $y = -x^2 + 1$ and the line ② have only one intersection point.

Let $-x^2 + 1 = ax - 3a + 2$; i.e., $x^2 + ax - 3a + 1 = 0$

Then, the discriminant D of the equation is $D = 0$

$\therefore\ a^2 - 4(-3a + 1) = a^2 + 12a - 4 = 0$

By the quadratic formula, $a = \dfrac{-6 \pm \sqrt{36+4}}{1} = \dfrac{-6 \pm \sqrt{40}}{1} = -6 \pm 2\sqrt{10}$

Since $a > 0$, $a = -6 + 2\sqrt{10}$

By i) and ii), $-6 + 2\sqrt{10} < a < \dfrac{1}{2}$

#39 Let $[x]$ be the greatest integer less than or equal to x. Find the range of f.

(1) f is a function such that $f(x) = [2x] - 2[x]$ with the domain $\{x | x$ is a real number$\}$.

For any integer n,

i) $n \le x < n + \dfrac{1}{2}\ \Rightarrow\ 2n \le 2x < 2n + 1$

$\therefore\ [x] = n,\ [2x] = 2n$

$\therefore\ f(x) = [2x] - 2[x] = 2n - 2n = 0$

ii) $n + \dfrac{1}{2} \le x < n + 1\ \Rightarrow\ 2n + 1 \le 2x < 2n + 2$

$\therefore\ [x] = n,\ [2x] = 2n + 1$

$\therefore\ f(x) = [2x] - 2[x] = 2n + 1 - 2n = 1$

Therefore, the range of f is $\{0, 1\}$.

(2) f is a function such that $f(x) = \left[\dfrac{[x]}{x}\right]$ for any positive real number x.

i) When x is not an integer,

$0 \le [x] < x$ for any $x > 0$

$\therefore\ 0 \le \dfrac{[x]}{x} < 1$ $\therefore\ f(x) = \left[\dfrac{[x]}{x}\right] = 0$

ii) When x is an integer,

$[x] = x$

$\therefore\ f(x) = \left[\dfrac{[x]}{x}\right] = [1] = 1$

Therefore, the range of f is $\{0, 1\}$.

(3) f is a function such that $f(x) = [x] + [-x]$ for any positive real number x.

For any integer n, $n \le x < n + 1\ \Rightarrow\ [x] = n$ and $-n - 1 < -x \le -n$

i) $-n - 1 < -x < -n\ \Rightarrow\ [-x] = -n - 1$

$\therefore\ f(x) = [x] + [-x] = n + (-n - 1) = -1$

ii) $-x = -n \Rightarrow [-x] = -n$

$\therefore f(x) = [x] + [-x] = n + (-n) = 0$

Therefore, the range of f is $\{0, -1\}$.

#40 **For a function $f(x) = |x - 1| + |x - a|$, find all possible numbers for an integer k so that $f(x)$ has minimum value 3 when $x = k$.**

i) When $a \geq 1$,

① $x < 1 \Rightarrow f(x) = -(x - 1) - (x - a) = -2x + 1 + a$

② $1 \leq x < a \Rightarrow f(x) = (x - 1) - (x - a) = a - 1$

③ $x \geq a \Rightarrow f(x) = (x - 1) + (x - a) = 2x - 1 - a$

Since the minimum value of $f(x)$ is 3, $a - 1 = 3$. $\therefore a = 4$

Thus, $f(x)$ has minimum value 3 when $1 \leq x \leq 4$.

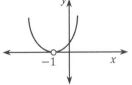

ii) When $a < 1$,

① $x < a \Rightarrow f(x) = -(x - 1) - (x - a) = -2x + 1 + a$

② $a \leq x < 1 \Rightarrow f(x) = -(x - 1) + (x - a) = 1 - a$

③ $x \geq 1 \Rightarrow f(x) = (x - 1) + (x - a) = 2x - 1 - a$

Since the minimum value of $f(x)$ is 3, $1 - a = 3$. $\therefore a = -2$

Thus, $f(x)$ has minimum value 3 when $-2 \leq x \leq 1$.

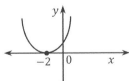

Therefore, the range of k so that $f(x)$ has minimum value 3 is $-2 \leq x \leq 4$.

$\therefore k = -2, -1, 0, 1, 2, 3, 4$

Hence, 7 integers

#41 Solve the following inequalities by using graphs.

(1) $x^2 + 2x + 1 > 0$

$y = x^2 + 2x + 1 = (x + 1)^2 > 0$

$\therefore x < -1$ or $x > -1$

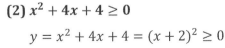

(2) $x^2 + 4x + 4 \geq 0$

$y = x^2 + 4x + 4 = (x + 2)^2 \geq 0$

Since $y \geq 0$ for all values of x, the solution is all real numbers.

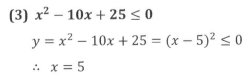

(3) $x^2 - 10x + 25 \leq 0$

$y = x^2 - 10x + 25 = (x - 5)^2 \leq 0$

$\therefore x = 5$

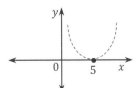

(4) $4x^2 < 4x - 1$

$\Rightarrow\ 4x^2 - 4x + 1 < 0 \quad \Rightarrow\quad (2x - 1)^2 < 0$

Since $(2x - 1)^2 \geq 0$, there is no value of x such that $(2x - 1)^2 < 0$.

Therefore, there is no solution.

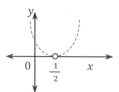

#42 **Determine the values of the coefficients a and b so that the following inequalities have the given solutions.**

(1) $x^2 + ax + b > 0$, **the solution is** $x < -1,\ x > 2$

From the solution, $(x + 1)(x - 2) > 0 \quad \therefore\ x^2 - x - 2 > 0$

Therefore, $a = -1,\ b = -2$

(2) $x^2 + ax + b \geq 0$, **the solution is** $x \geq -2,\ x \leq -4$

From the solution, $(x + 2)(x + 4) \geq 0 \quad \therefore\ x^2 + 6x + 8 \geq 0$

Therefore, $a = 6,\ b = 8$

(3) $ax^2 + bx - 24 \leq 0,\ (a > 0)$, **the solution is** $-2 \leq x \leq 3$

$a(x + 2)(x - 3) \leq 0 \quad \therefore\ a(x^2 - x - 6) \leq 0 \quad \therefore\ ax^2 - ax - 6a \leq 0$

$\therefore\ -a = b,\ -6a = -24$

Therefore, $a = 4,\ b = -4$

(4) $ax^2 - bx + 6 > 0,\ (a < 0)$, **the solution is** $-1 < x < 3$

Form the solution, $(x + 1)(x - 3) < 0$

Since $a < 0,\ a(x + 1)(x - 3) > 0$

$\therefore\ ax^2 - 2ax - 3a > 0$

$\therefore\ -2a = -b,\ -3a = 6$

Therefore, $a = -2,\ b = -4$

Another approach:

Since $a < 0,\ -ax^2 + bx - 6 < 0$

From the solution, $-a(x + 1)(x - 3) < 0 \quad \therefore\ -a(x^2 - 2x - 3) < 0$

$\therefore\ -ax^2 + 2ax + 3a < 0 \quad \therefore\ ax^2 - 2ax - 3a > 0\ ;\ -ax^2 + 2ax + 3a < 0$

$\therefore\ 2a = b,\ 3a = -6$

Therefore, $a = -2,\ b = -4$

#43 Determine the range of values of the constant a so that the solution of each inequality is all real numbers.

(1) $x^2 - (a+1)x + a + 2 > 0$

The discriminant D of the equation $x^2 - (a+1)x + a + 2 = 0$ is $D < 0$.

$\therefore$ $D = (a+1)^2 - 4a - 8 = a^2 - 2a - 7 < 0$

Since $a = \frac{1 \pm \sqrt{1+7}}{1} = 1 \pm \sqrt{8} = 1 \pm 2\sqrt{2}$, $\quad 1 - 2\sqrt{2} < a < 1 + 2\sqrt{2}$

(2) $-x^2 + (a+1)x - a^2 < 0$

$-x^2 + (a+1)x - a^2 < 0 \Rightarrow x^2 - (a+1)x + a^2 > 0$

The discriminant D of the equation $x^2 - (a+1)x + a^2 = 0$ is $D < 0$.

$\therefore$ $D = (a+1)^2 - 4a^2 = -3a^2 + 2a + 1 = -(3a^2 - 2a - 1) = -(3a+1)(a-1) < 0$

$\therefore$ $(3a+1)(a-1) > 0$ $\qquad \therefore$ $a > 1$ or $a < -\frac{1}{3}$

(3) $ax^2 - 2ax + 3 \geq 0$, $a > 0$

The discriminant D of the equation $ax^2 - 2ax + 3 = 0$ is $D \leq 0$.

$\therefore$ $D = a^2 - 3a = a(a-3) \leq 0$ $\qquad \therefore$ $0 \leq a \leq 3$

Since $a > 0$, $\quad 0 < a \leq 3$

#44 Find the range of a constant a.

(1) The graph of $y = x^2 + ax + 4$ and a line $y = x - 5$ do not intersect at any points.

Let $x^2 + ax + 4 = x - 5$.

Then $x^2 + (a-1)x + 9 = 0$ ······ ①

Since ① has no solution, the discriminant D of the equation ① is $D < 0$

$\therefore$ $D = (a-1)^2 - 36 = a^2 - 2a - 35 = (a-7)(a+5) < 0$

$\therefore$ $-5 < a < 7$

(2) The graph of $y = x^2 - ax + 3$ and the x-axis do not intersect at any points.

The equation of the x-axis is $y = 0$.

Since the graph of $y = x^2 - ax + 3$ and the line $y = 0$ have no intersection point,

the equation $x^2 - ax + 3 = 0$ has no solution.

$\therefore$ The discriminant D of the equation $x^2 - ax + 3 = 0$ is $D < 0$.

$\therefore$ $D = a^2 - 12 < 0$ $\qquad \therefore$ $-2\sqrt{3} < a < 2\sqrt{3}$

(3) The two graphs of $f(x) = 2x^2 + ax + 3$ and $g(x) = x^2 - 2x - a$ intersect at two different points.

Since the equation $2x^2 + ax + 3 = x^2 - 2x - a$ has two different real number solutions,

the discriminant D of the equation $x^2 + (a + 2)x + 3 + a = 0$ is $D > 0$.

$\therefore$ $D = (a + 2)^2 - 4(3 + a) = a^2 - 8 = (a + 2\sqrt{2})(a - 2\sqrt{2}) > 0$

$\therefore$ $a > 2\sqrt{2}$ or $a < -2\sqrt{2}$

(4) The two roots of the equation $x^2 - 2ax + a + 12 = 0$ are less than 1.

i) Since the equation $x^2 - 2ax + a + 12 = 0$ has two roots, the discriminant D of the

equation is $D \geq 0$.

$\therefore$ $D = a^2 - (a + 12) = a^2 - a - 12 = (a - 4)(a + 3) \geq 0$

$\therefore$ $a \geq 4$ or $a \leq -3$

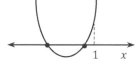

ii) Let $f(x) = x^2 - 2ax + a + 12$. Then, $f(1) > 0$

$\therefore$ $1 - 2a + a + 12 = -a + 13 > 0$ $\therefore a < 13$

iii) Since $f(x) = x^2 - 2ax + a + 12 = (x - a)^2 - a^2 + a + 12$, the axis of symmetry is

$x = a$ and $a < 1$

By i), ii), and iii), $a \leq -3$

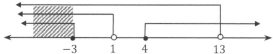

$$-3 \quad 1 \quad 4 \quad\quad 13$$

(5) One root of the equation $2x^2 - ax - 1 = 0$ is in-between -1 and 0, and the other root of the equation is in-between 0 and 1.

i) Since the equation $2x^2 - ax - 1 = 0$ has two different real number solutions,

the discriminant D of the equation is $D > 0$.

$\therefore$ $D = (-a)^2 - 4 \cdot 2 \cdot (-1) = a^2 + 8 > 0$

$\therefore$ a is all real number.

ii) Let $f(x) = 2x^2 - ax - 1$.

Then, $f(-1) > 0$, $f(0) < 0$, and $f(1) > 0$

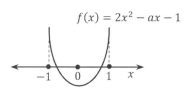

$f(x) = 2x^2 - ax - 1$

$\therefore$ $2 + a - 1 > 0$, $-1 < 0$, and $2 - a - 1 > 0$

$\therefore$ $a > -1$ and $a < 1$; i.e., $-1 < a < 1$

Therefore, by i) and ii), $-1 < a < 1$

(6) For any real number x, $(a - 2)x^2 - (a - 2)x + 2 > 0$ is always true.

i) $a - 2 = 0 \Rightarrow 2 > 0$ $\therefore$ It is always true. $\therefore a = 2$

ii) $a - 2 \neq 0$

$\Rightarrow a - 2 > 0$ and

the discriminant D of the equation $(a - 2)x^2 - (a - 2)x + 2 = 0$ is $D < 0$.

$\therefore \ D = (a - 2)^2 - 8(a - 2) = (a - 2)(a - 2 - 8) = (a - 2)(a - 10) < 0$

$\therefore \ 2 < a < 10$

Therefore, by i) and ii), $2 \leq a < 10$

(7) The equation $x^2 - ax + 2a - 3 = 0$ has at least one real number solution in the range $-2 \leq x \leq 4$.

i) When the equation has two real number solutions in the region $-2 \leq x \leq 4$,

let D be the discriminant of the equation. Then, $D \geq 0$

$\therefore \ D = a^2 - 4(2a - 3) = a^2 - 8a + 12 = (a - 2)(a - 6) \geq 0$

$\therefore \ a \geq 6$ or $a \leq 2 \cdots\cdots$ ①

Let $f(x) = x^2 - ax + 2a - 3$

Then, $f(-2) \geq 0$ and $f(4) \geq 0$

$\therefore \ 4 + 2a + 2a - 3 = 4a + 1 \geq 0 \ ; \ a \geq -\dfrac{1}{4} \ \cdots\cdots$ ②

$16 - 4a + 2a - 3 = 13 - 2a \geq 0 \ ; \ a \leq \dfrac{13}{2} \ \cdots\cdots$ ③

Since $f(x) = x^2 - ax + 2a - 3 = (x - \dfrac{a}{2})^2 - \dfrac{a^2}{4} + 2a - 3$,

the axis of symmetry is $x = \dfrac{a}{2}$.

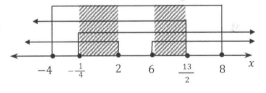

$\therefore \ -2 \leq \dfrac{a}{2} \leq 4 \qquad \therefore \ -4 \leq a \leq 8 \cdots\cdots$ ④

Thus, by ①, ②, ③, and ④,

$-\dfrac{1}{4} \leq a \leq 2$ or $6 \leq a \leq \dfrac{13}{2} \ \cdots\cdots$ ❶

ii) When the equation has one real number solution in the region $-2 \leq x \leq 4$,

Since $f(-2)f(4) \leq 0$, $(4a + 1)(13 - 2a) \leq 0$ $\therefore (4a + 1)(2a - 13) \geq 0$

$\therefore \ a \geq \dfrac{13}{2}, \ a \leq -\dfrac{1}{4} \ \cdots\cdots$ ❷

Therefore, $a \geq 6$ or $a \leq 2$ by ❶ and ❷

(8) For a function $f(x) = x^2 - 2ax + 3$, $f(x) \geq a$ (where $-1 \leq x \leq 1$).

$f(x) \geq a \Rightarrow x^2 - 2ax + 3 \geq a \qquad \therefore x^2 - 2ax + 3 - a \geq 0$

Let $g(x) = x^2 - 2ax + 3 - a = (x - a)^2 - a^2 + 3 - a$

i) When the axis of symmetry $x = a$ is in-between the range $-1 \leq x \leq 1$,

the minimum value of $g(x)$ is $g(a)$ when $x = a$.

$\therefore \; g(a) = -a^2 + 3 - a \geq 0 \qquad \therefore \; a^2 + a - 3 \leq 0$

By the quadratic formula, $a = \dfrac{-1 \pm \sqrt{1+12}}{2} = \dfrac{-1 \pm \sqrt{13}}{2}$

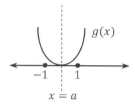

$\therefore \; \dfrac{-1-\sqrt{13}}{2} \leq a \leq \dfrac{-1+\sqrt{13}}{2}$

ii) When the axis of symmetry $x = a$ is not in-between the range $-1 \leq x \leq 1$,

Case 1. The minimum value of $g(x)$ is $g(-1)$ and $g(-1) \geq 0$

$\qquad \therefore \; 1 + 2a + 3 - a = a + 4 \geq 0 \qquad \therefore \; a \geq -4$

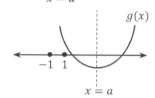

Since $a \leq -1$, $\; -4 \leq a \leq -1$

Case 2. The minimum value of $g(x)$ is $g(1)$ and $g(1) \geq 0$

$\qquad \therefore \; 1 - 2a + 3 - a = -3a + 4 \geq 0 \qquad \therefore \; a \leq \dfrac{4}{3}$

Since $1 \leq a$, $\; 1 \leq a \leq \dfrac{4}{3}$

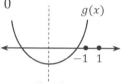

By Case 1 and Case 2, $\; -4 \leq a \leq -1$ or $1 \leq a \leq \dfrac{4}{3}$

Therefore, by i) and ii),

$-4 \leq a \leq \dfrac{4}{3}$

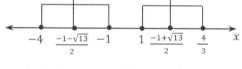

(9) For a function $f(x) = x^2 - 2ax + a$, $\; f(x) > 0$ in the range $0 < x < 1$.

$x^2 - 2ax + a > 0 \quad \Rightarrow \quad x^2 > 2ax - a$

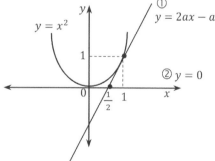

Consider a parabola $y = x^2$ and a line $y = 2ax - a$ in the range $0 < x < 1$.

The line $y = 2ax - a = 2a\left(x - \dfrac{1}{2}\right)$ passes through a point $\left(\dfrac{1}{2}, 0\right)$.

Since the graph of $y = x^2$ is always above the line $y = 2ax - a$, the line $y = 2ax - a$ is in-between the line ① that is passes through $\left(\dfrac{1}{2}, 0\right)$ and the line ② ; i. e., $y = 0$ (x-axis).

i) Since the line ① and the graph of $y = x^2$ intersect at one point,

the equation $x^2 = 2ax - a$ (i.e., $x^2 - 2ax + a = 0$) has the discriminant $D = 0$.

$\quad \therefore \; D = a^2 - a = a(a-1) = 0 \qquad \therefore \; a = 0$ or $a = 1$

Since ① is a line with slope, $a \neq 0$. $\qquad \therefore \; a = 1$

ii) From the line $y = 0$ (x-axis), $a = 0$

Therefore, by i) and ii), $0 \leq a < 1$

> If $a = 1$, then
> $y = x^2$ and $y = 2x - 1$ intersect at one point.
> To get $x^2 \gneqq 2x - 1$, $\; 0 \leq a \lneqq 1$

(10) There exists the value of x such that $a(x^2 + x + 1) > x$.

$a(x^2 + x + 1) > x \Rightarrow ax^2 + (a-1)x + a > 0$

Let $f(x) = ax^2 + (a-1)x + a$

Then $f(x) > 0$

To have x such that $f(x) > 0$,

$f(x)$ is one of the following Figures.

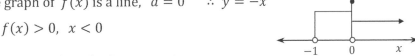

i)   If the graph of $f(x)$ is a parabola that opens upwards, $a > 0$

ii) If the graph of $f(x)$ is a parabola that opens downwards, $a < 0$

In this case, the parabola must have two different intersection points with x-axis

to satisfy $f(x) > 0$.

Thus, the discriminant D of the equation $ax^2 + (a-1)x + a = 0$ is $D > 0$.

$\therefore\ D = (a-1)^2 - 4a^2 = -3a^2 - 2a + 1 = -(3a^2 + 2a - 1) = -(3a-1)(a+1) > 0$

$\therefore\ (3a-1)(a+1) < 0 \qquad \therefore\ -1 < a < \dfrac{1}{3}$

Since $a < 0$, $-1 < a < 0$

iii) If the graph of $f(x)$ is a line, $a = 0 \quad \therefore\ y = -x$

Since $f(x) > 0$, $x < 0$

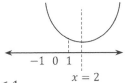

Therefore, by i), ii), and iii), $\quad a > -1$

(11) An inequality $x^2 - 4x \geq a^2 - 4a$ is always true in the range $-1 \leq x \leq 1$.

$x^2 - 4x \geq a^2 - 4a \Rightarrow x^2 - 4x - a^2 + 4a \geq 0$

Let $f(x) = x^2 - 4x - a^2 + 4a$

Then, $f(x) = (x-2)^2 - 4 - a^2 + 4a$

To have the graph of $f(x)$ such that $f(x) \geq 0$, in the range $-1 \leq x \leq 1$,

the minimum value of $y = f(x)$ must be equal to zero or positive.

Since $f(1)$ is the minimum value in the range $-1 \leq x \leq 1$,

$f(1) = 1 - 4 - a^2 + 4a = -a^2 + 4a - 3 = -(a^2 - 4a + 3) = -(a-3)(a-1) \geq 0$

$\therefore\ (a-3)(a-1) \leq 0 \quad \therefore\ 1 \leq a \leq 3$

(12) For any real number x, $ax^2 + 2\sqrt{2}x + a - 1 \geq 0$ is always true.

Let $f(x) = ax^2 + 2\sqrt{2}x + a - 1$

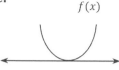

Then, i) $a > 0$

ii) The discriminant D of the equation $ax^2 + 2\sqrt{2}x + a - 1 = 0$ is $D \leq 0$.

$\therefore D = \left(2\sqrt{2}\right)^2 - 4a(a-1) = 8 - 4a^2 + 4a = -4(a^2 - a - 2)$

$\qquad = -4(a-2)(a+1) \leq 0$

$\therefore (a-2)(a+1) \geq 0 \qquad \therefore a \geq 2$ or $a \leq -1$

By i) and ii), $a \geq 2$

(13) When the two roots of the equation $x^2 + 2ax + a + 6 = 0$ are α and β,

find the range of a such that $0 < \alpha < 1 < \beta$.

Let $f(x) = x^2 + 2ax + a + 6$

i) Since the equation $x^2 + 2ax + a + 6 = 0$ has two different roots, the discriminant D of the equation is $D > 0$.

$\therefore D = a^2 - (a+6) = a^2 - a - 6 = (a-3)(a+2) > 0$

$\therefore a > 3$ or $a < -2$

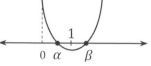

ii) Since $f(0) > 0$ and $f(1) < 0$, $a + 6 > 0$ and $1 + 2a + a + 6 < 0$

$\therefore a > -6$ and $a < -\dfrac{7}{3}$; i.e., $-6 < a < -\dfrac{7}{3}$

iii) Since $f(x) = x^2 + 2ax + a + 6 = (x+a)^2 - a^2 + a + 6$,

the axis of symmetry is $x = -a$ and $x > 0$. $\therefore a < 0$

By i), ii), and iii), $-6 < a < -\dfrac{7}{3}$

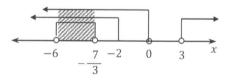

#45 Find the value.

(1) The graph of $y = x^2 - 2(a+1)x + a^2 + 4a$ and the line $y = mx + n$ intersect at one

point, not depending on the value of the real number a. Find the value of $m + n$.

For the equation $x^2 - 2(a+1)x + a^2 + 4a = mx + n$

; i.e., $x^2 - (2a + 2 + m)x + a^2 + 4a - n = 0$, the discriminant D of the equation is $D = 0$.

$\therefore D = (2a + 2 + m)^2 - 4(a^2 + 4a - n)$

$\qquad = (4a^2 + 4 + m^2 + 8a + 4am + 4m) - 4a^2 - 16a + 4n$

$\qquad = 4 + m^2 - 8a + 4am + 4m + 4n = 4 + m^2 + 4m + 4n + (4m - 8)a = 0 \cdots\cdots ①$

Since ① is always true, not depending on the value of a,

$4m - 8 = 0$ and $4 + m^2 + 4m + 4n = 0$.

$\therefore$ $m = 2$ and $4 + 4 + 8 + 4n = 0$; $n = -4$

Therefore, $m + n = 2 - 4 = -2$

(2) A parabola $y = x^2 + 2ax + a$ and the x-axis have two intersection points A and B.

Find the positive number a such that $\overline{AB} = \sqrt{3}$.

Let $A = A(\alpha, 0)$, $B = B(\beta, 0)$

Then, $\alpha + \beta = -2a$ and $\alpha\beta = a$, by the relationship between the roots and coefficients.

Since $|\alpha - \beta| = \sqrt{3}$, $(\alpha - \beta)^2 = 3$

$\therefore$ $3 = (\alpha - \beta)^2 = (\alpha + \beta)^2 - 4\alpha\beta = (-2a)^2 - 4a = 4a^2 - 4a$

$\therefore$ $4a^2 - 4a - 3 = 0$ $\qquad \therefore$ $(2a + 1)(2a - 3) = 0$ $\qquad \therefore$ $a = \dfrac{3}{2}$ or $a = -\dfrac{1}{2}$

Since $a > 0$, $a = \dfrac{3}{2}$

(3) When the leading coefficient of a quadratic function $y = f(x)$ is 1, the graph of the function and a line $y = a$ have the two intersection points $(1, a), (5, a)$. When the solution of the inequality $f(x) < f(-1) + 2$ is $m < x < n$, find the value of mn.

Since $y = f(x)$ passes through the points $(1, a)$ and $(5, a)$, $f(1) = f(5) = a$

$\therefore$ $f(x) = 1 \cdot (x - 1)(x - 5) + a = x^2 - 6x + 5 + a$

$f(x) < f(-1) + 2$ $\Rightarrow$ $x^2 - 6x + 5 + a < 1 + 6 + 5 + a + 2$

$\therefore$ $x^2 - 6x - 9 < 0$

Let α and β be the roots of the equation $x^2 - 6x - 9 = 0$.

Then $x^2 - 6x - 9 < 0$ $\Leftrightarrow$ $(x - \alpha)(x - \beta) < 0$ $\qquad \therefore$ $\alpha < x < \beta$

By the relationship between the roots and coefficients, $\alpha + \beta = 6$ and $\alpha\beta = -9$

Therefore, $mn = -9$

(4)

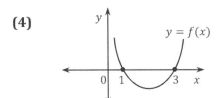

When the graph of a function $f(x) = ax^2 + bx + c$ is shown as the Figure, the inequality $bx^2 + cx + a > 0$ has solution $m < x < n$. Find the value of $m - n$.

Since the parabola opens upwards, $a > 0$

Since $x = 1$ and $x = 3$ are the two solutions of the equation $ax^2 + bx + c = 0$,

$f(x) = a(x - 1)(x - 3) = ax^2 - 4ax + 3a$

$\therefore$ $-4a = b$, $3a = c$ $\cdots\cdots$ ①

Substituting ① into the inequality $bx^2 + cx + a > 0$, $-4ax^2 + 3ax + a > 0$

Since $a > 0$, $-4x^2 + 3x + 1 > 0$ (Divide both sides by a)

$\therefore 4x^2 - 3x - 1 < 0$ $\therefore (4x + 1)(x - 1) < 0$ $\therefore -\dfrac{1}{4} < x < 1$

Therefore, $m - n = -\dfrac{1}{4} - 1 = -\dfrac{5}{4}$

(5) When an inequality $x + 2a \leq x^2 \leq 2x + b$ is always true in the range $-1 \leq x \leq 1$, find minimum value of $b - a$.

i) $x^2 \geq x + 2a$

$\Rightarrow x^2 - x - 2a \geq 0$

$\therefore$ The discriminant D of the equation $x^2 - x - 2a = 0$ is $D \leq 0$.

$\therefore D = 1 + 8a \leq 0$ $\therefore a \leq -\dfrac{1}{8}$

ii) $x^2 \leq 2x + b$

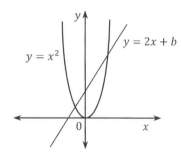

When $x = -1$, $1 \leq -2 + b$ $\therefore b \geq 3$

To have minimum value of $b - a$,

we need to have minimum value of b and

maximum value of a.

$\therefore$ Minimum value of $b - a$ is $3 - \left(-\dfrac{1}{8}\right) = \dfrac{25}{8}$

Alternative Approach:

① $x^2 \geq x + 2a$ $\Rightarrow x^2 - x - 2a \geq 0$

$\Rightarrow \left(x - \dfrac{1}{2}\right)^2 - \dfrac{1}{4} - 2a \geq 0$

Let $f(x) = \left(x - \dfrac{1}{2}\right)^2 - \dfrac{1}{4} - 2a$.

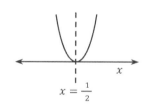

Then, the minimum value of $f(x)$ is $f\left(\dfrac{1}{2}\right) \geq 0$

$\therefore -\dfrac{1}{4} - 2a \geq 0$ $\therefore a \leq -\dfrac{1}{8}$

② $x^2 \leq 2x + b$ $\Rightarrow x^2 - 2x - b \leq 0$ $\Rightarrow (x - 1)^2 - 1 - b \leq 0$

Let $f(x) = (x - 1)^2 - 1 - b$.

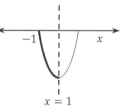

Then, maximum value of $f(x)$ should be less than or equal to zero.

Since the maximum value of $f(x)$ in the range $[-1, 1]$ such that $f(x) \leq 0$ is $f(-1)$,

$3 - b \leq 0$ $\therefore b \geq 3$

Since the minimum value of $b - a$ is the difference of minimum value of b and maximum

value of a, $3 - \left(-\dfrac{1}{8}\right) = \dfrac{25}{8}$

(6) The graph of a function $y = x^2 + 2ax + b$ intersects the lines $y = -x + 3$ and $y = 3x + 5$ at one point, respectively. Find the value of $a + b$.

i) $x^2 + 2ax + b = -x + 3$

$\Rightarrow x^2 + (2a + 1)x + b - 3 = 0$

Since the discriminant D_1 of the equation is $D_1 = 0$,

$D_1 = (2a + 1)^2 - 4(b - 3)$

$\quad = 4a^2 + 4a + 1 - 4b + 12 = 0 \cdots\cdots ①$

ii) $x^2 + 2ax + b = 3x + 5$

$\Rightarrow x^2 + (2a - 3)x + b - 5 = 0$

Since the discriminant D_2 of the equation is $D_2 = 0$,

$D_2 = (2a - 3)^2 - 4(b - 5)$

$\quad = 4a^2 - 12a + 9 - 4b + 20 = 0 \cdots\cdots ②$

$① - ②: \ 16a - 8 - 8 = 0 \qquad \therefore \ a = 1$

Substituting $a = 1$ into $①$, $\ -4b + 21 = 0$

$\therefore b = \dfrac{21}{4}$

Therefore, $a + b = 1 + \dfrac{21}{4} = \dfrac{25}{4}$

(7) When the graph of a function $y = x^2 + ax - 2$ and a line $y = -2x + 3$ intersect at two points A and B, the midpoint of the segment $\overline{AB}$ is $(4, -2)$. Find the value of the constant a.

For the equation $x^2 + ax - 2 = -2x + 3$; i.e., $x^2 + (a + 2)x - 5 = 0$,

let α and β be the two roots of the equation.

Then, by the relationship between the roots and coefficients,

$\alpha + \beta = -(a + 2), \ \alpha\beta = -5$

Since α and β are x-coordinates of A and B,

the x-coordinate of the midpoint of $\overline{AB}$ is $\dfrac{\alpha + \beta}{2} = 4$.

$\therefore \ \alpha + \beta = -(a + 2) = 8$

Therefore, $a = -10$

(8)

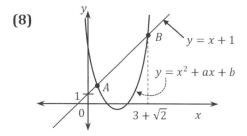

When the graph of a function $y = x^2 + ax + b$ and a line $y = x + 1$ intersect at two points A and B, the x-coordinate of the point B is $3 + \sqrt{2}$ as shown in Figure.

For rational numbers a and b, find the value of $a + b$.

$x^2 + ax + b = x + 1 \implies x^2 + (a-1)x + b - 1 = 0 \cdots\cdots ①$

Since a and b are rational numbers, and one of the roots of the equation ① is $3 + \sqrt{2}$, the other root is $3 - \sqrt{2}$.

By the relationship between the roots and coefficients,

$(3 + \sqrt{2}) + (3 - \sqrt{2}) = -(a-1)$ and $(3 + \sqrt{2})(3 - \sqrt{2}) = b - 1$

$\therefore \ a = -5$ and $b = 8$

Therefore, $a + b = 3$

(9) When the graph of a function $f(x) = x^2 - ax + 4$ and the x-axis intersect at two points A and B, length of the segment $\overline{AB}$ is $2\sqrt{2}$.

Find minimum value of $f(x)$ in the range $-\sqrt{6} \le x \le \sqrt{6}$.

The equation of the x-axis is $y = 0$.

Let $A = A(\alpha, 0), \ B = B(\beta, 0)$.

Then, α and β are roots of the equation $x^2 - ax + 4 = 0$.

$\therefore \ \alpha + \beta = a, \ \alpha\beta = 4$ by the relationship between the roots and coefficients.

Since $|\alpha - \beta| = 2\sqrt{2}$, $(\alpha - \beta)^2 = (\alpha + \beta)^2 - 4\alpha\beta = a^2 - 16 = 8$

$\therefore \ a^2 = 24 \qquad \therefore \ a = \pm 2\sqrt{6}$

$\therefore \ f(x) = x^2 \mp 2\sqrt{6}\,x + 4 = \left(x \mp \sqrt{6}\right)^2 - 6 + 4 = \left(x \mp \sqrt{6}\right)^2 - 2$

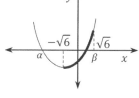

In the range $-\sqrt{6} \le x \le \sqrt{6}$,

$f(x) = \left(x + \sqrt{6}\right)^2 - 2$ has minimum value $f\left(-\sqrt{6}\right) = -2$ and

$f(x) = \left(x - \sqrt{6}\right)^2 - 2$ has minimum value $f\left(\sqrt{6}\right) = -2$

Therefore, the minimum value of $f(x)$ is -2.

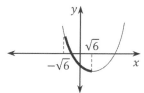

(10)

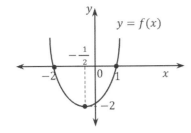

The graph of a function $y = f(x)$ is shown as the Figure. Find the sum of all three different real number solutions of the equation

$$f(f(x)) = 0.$$

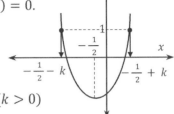

Since $f(x) = 0$ has two roots -2 and 1, $f(-2) = 0$ and $f(1) = 0$.

Note that : $f(f(x)) = 0 \Rightarrow f(x) = -2$ or $f(x) = 1$

When $f(x) = -2$, $x = -\dfrac{1}{2}$ (From the graph)

When $f(x) = 1$, $x = -\dfrac{1}{2} + k \ (k > 0)$ or $x = -\dfrac{1}{2} - k \ (k > 0)$

Therefore, the sum of three different real number roots of $f(f(x))$ is

$$-\dfrac{1}{2} + \left(-\dfrac{1}{2} + k\right) + \left(-\dfrac{1}{2} - k\right) = -\dfrac{3}{2}$$

(11)

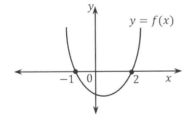

The graph of a function $y = f(x)$ is shown as the Figure. When the inequality $f\left(\dfrac{x+a}{2}\right) \leq 0$ has solution $-3 \leq x \leq 3$, find the value of a constant a.

From the graph, $x = -1$ and $x = 2$ are two roots of $f(x) = 0$.

To have $f(x) \leq 0$, $-1 \leq x \leq 2$

Let $\dfrac{x+a}{2} = k$. Then, $f(k) \leq 0$

∴ The solution of $f(k) \leq 0$ is $-1 \leq \dfrac{x+a}{2} \leq 2$.

∴ $-2 \leq x + a \leq 4$ ∴ $-2 - a \leq x \leq 4 - a$

Since the solution of $f\left(\dfrac{x+a}{2}\right) \leq 0$ is $-3 \leq x \leq 3$, $-2 - a = -3$ and $4 - a = 3$

Therefore, $a = 1$

(12) When the graph of a function $y = 2x^2$ **and a line** $y = ax + b$ **(where** a **and** b **are constants) intersect at one point, the line will be a tangent line of a circle** $x^2 + (y+1)^2 = 1$. **Find the value of** $a^2 + b$ **when** $b < 0$.

i) $2x^2 = ax + b \Rightarrow 2x^2 - ax - b = 0$

The discriminant D of the equation $2x^2 - ax - b = 0$ is $D = 0$

$\therefore D = a^2 + 8b = 0 \qquad \therefore a^2 = -8b \cdots\cdots ①$

ii) Since the line $y = ax + b$ is a tangent line of the circle $x^2 + (y+1)^2 = 1$,

the distance between the center $(0, -1)$ of the circle and the line $ax - y + b = 0$ is the length of radius of the circle.

That is, $\frac{|a \cdot 0 + (-1)\cdot(-1) + b|}{\sqrt{a^2+1}} = 1 \qquad \therefore \frac{|1+b|}{\sqrt{a^2+1}} = 1 \qquad \therefore |1+b| = \sqrt{a^2+1}$

Squaring both sides, $1 + 2b + b^2 = a^2 + 1$

By ①, $1 + 2b + b^2 = -8b + 1 \qquad \therefore b^2 + 10b = b(b+10) = 0$

$\therefore b = 0$ or $b = -10$

Since $b < 0$, $b = -10$ and $a^2 = 80$

Therefore, $a^2 + b = 80 - 10 = 70$

#46 For each given fractional function, draw the graph, and then state the range of $f(x)$ that satisfies the given domain.

(1) $f(x) = \frac{x-1}{x-2} \quad (3 \leq x \leq 4)$

$f(x) = \frac{x-1}{x-2} = \frac{x-2+1}{x-2} = 1 + \frac{1}{x-2}$

$\therefore$ The equations of the asymptotes are $x = 2$ and $y = 1$.

Since the points of intersections with the axes are $\begin{cases} (1, 0) : x - \text{axis} \\ \left(0, \frac{1}{2}\right) : y - \text{axis} \end{cases}$,

the graph is shown as Figure.

Since $f(3) = 2$ and $f(4) = \frac{3}{2}$,

the range is $\frac{3}{2} \leq f(x) \leq 2$

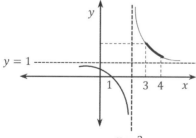

(2) $f(x) = \frac{-2x+4}{x-3}$ $\quad (1 \leq x \leq 5,\ x \neq 3)$

$$f(x) = \frac{-2x+4}{x-3} = \frac{-2(x-3)-2}{x-3} = -2 - \frac{2}{x-3}$$

∴ The equations of the asymptotes are $x = 3$ and $y = -2$.

Since the points of intersections with the axes are $\begin{cases} (2,0) : x - \text{axis} \\ \left(0, -\frac{4}{3}\right) : y - \text{axis} \end{cases}$,

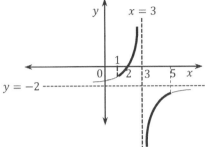

the graph is shown as Figure.

Since $f(1) = -1$ and $f(5) = -3$,

the range is $f(x) \geq -1$ or $f(x) \leq -3$

(3) $f(x) = \frac{3x}{3x-2}$ $\quad (x \geq 0,\ x \neq \frac{2}{3})$

$$f(x) = \frac{3x}{3x-2} = \frac{3x-2+2}{3x-2} = 1 + \frac{2}{3x-2}$$

∴ The equations of the asymptotes are $x = \frac{2}{3}$ and $y = 1$.

Since the point of intersection with the axes is $(0, 0)$,

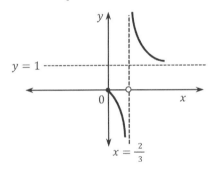

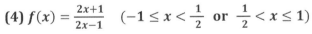

the graph is shown as Figure.

Since $f(0) = 0$,

the range is $f(x) \leq 0$ or $f(x) > 1$

(4) $f(x) = \frac{2x+1}{2x-1}$ $\quad (-1 \leq x < \frac{1}{2}$ or $\frac{1}{2} < x \leq 1)$

$$f(x) = \frac{2x+1}{2x-1} = \frac{2x-1+2}{2x-1} = 1 + \frac{2}{2x-1} = 1 + \frac{2}{2\left(x-\frac{1}{2}\right)}$$

$$= 1 + \frac{1}{x-\frac{1}{2}}$$

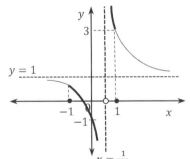

∴ The equations of the asymptotes are $x = \frac{1}{2}$ and $y = 1$.

Since the points of intersections with the axes are $\begin{cases} \left(-\frac{1}{2}, 0\right) : x - \text{axis} \\ (0, -1) : y - \text{axis} \end{cases}$,

the graph is shown as Figure.

Since $f(-1) = \frac{1}{3}$ and $f(1) = 3$, the range is $f(x) \leq \frac{1}{3}$ or $f(x) \geq 3$

#47 Maximum and minimum values

(1) Given the fractional function $f(x) = 2x + \dfrac{1}{x}$ $\left(\dfrac{1}{2} \leq x \leq 4\right)$, **draw the graph of the function and find maximum value.**

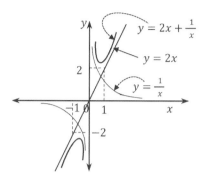

Asymptotes: $x = 0$ and $y = 2x$

$$f\left(\frac{1}{2}\right) = 2 \cdot \frac{1}{2} + \frac{1}{\frac{1}{2}} = 1 + 2 = 3$$

$$f(4) = 2 \cdot 4 + \frac{1}{4} = 8 + \frac{1}{4} = \frac{33}{4}$$

From the graph, maximum value is $f(4) = \dfrac{33}{4}$

(2) Given the fractional function $f(x) = x - \dfrac{2}{x}$ $\left(\dfrac{1}{2} \leq x \leq 4\right)$, **draw the graph of the function and find minimum value.**

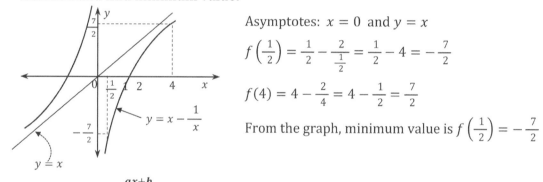

Asymptotes: $x = 0$ and $y = x$

$$f\left(\frac{1}{2}\right) = \frac{1}{2} - \frac{2}{\frac{1}{2}} = \frac{1}{2} - 4 = -\frac{7}{2}$$

$$f(4) = 4 - \frac{2}{4} = 4 - \frac{1}{2} = \frac{7}{2}$$

From the graph, minimum value is $f\left(\dfrac{1}{2}\right) = -\dfrac{7}{2}$

(3) A function $y = \dfrac{ax+b}{x+c}$ **is symmetric with respect to a point** $(2, 1)$, **and passes through a point** $(3, 3)$. **Find maximum and minimum values in the range** $-1 \leq x \leq 1$.

Since $y = \dfrac{ax+b}{x+c}$ is symmetric with respect to a point $(2, 1)$, the axes of symmetries are

$x = 2$ and $y = 1$. $\therefore y = \dfrac{k}{x-2} + 1 = \dfrac{k+(x-2)}{x-2}$

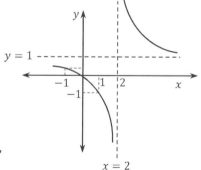

Since the function y passes through a point $(3, 3)$,

$$3 = \frac{k+(3-2)}{3-2} = k + 1$$

$\therefore k = 2$ $\therefore y = \dfrac{2}{x-2} + 1$

Since $f(-1) = \dfrac{2}{-1-2} + 1 = \dfrac{1}{3}$ and $f(1) = \dfrac{2}{1-2} + 1 = -1$,

maximum value is $f(-1) = \dfrac{1}{3}$ and minimum value is $f(1) = -1$

#48 For each of the following fractional function, state how each graph has been translated from the graph of $y = \dfrac{1}{x}$.

(1) $y = \dfrac{1}{x-2}$

The equations of the asymptotes are $x = 2$ and $y = 0$.

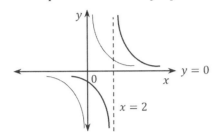   Translation: 2 units along the x-axis

(2) $y = \dfrac{1}{x} - 2$

The equations of the asymptotes are $x = 0$ and $y = -2$.

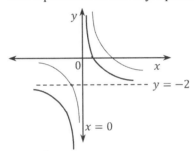   Translation: -2 units along the y-axis

(3) $y = \dfrac{1}{x-1} + 2$

The equations of the asymptotes are $x = 1$ and $y = 2$.

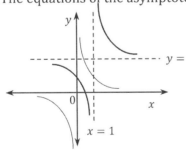   Translation: 1 unit along the x-axis and

2 units along the y-axis

(4) $y = \dfrac{1}{x+2} - 1$

The equations of the asymptotes are $x = -2$ and $y = -1$.

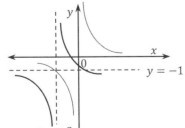   Translation: -2 units along the x-axis and

-1 unit along the y-axis

#49 **For each of the following fractional function, state how each graph has been translated from the graph of** $y = \dfrac{1}{2x}$ **.**

(1) $y = \dfrac{1}{2x-2}$

$$y = \frac{1}{2x-2} = \frac{1}{2(x-1)}$$

The equations of the asymptotes are $x = 1$ and $y = 0$

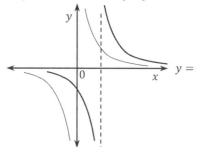

Translation: 1 unit along the x-axis

(2) $y = \dfrac{1}{2x} - 1$

The equations of the asymptotes are $x = 0$ and $y = -1$

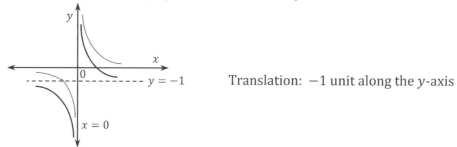

Translation: -1 unit along the y-axis

(3) $y = \dfrac{-2x-1}{2x+2}$

$$y = \frac{-2x-1}{2x+2} = \frac{-(2x+2)+1}{2x+2} = \frac{1}{2(x+1)} - 1$$

The equations of the asymptotes are $x = -1$ and $y = -1$

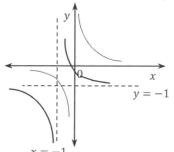

Translation: -1 unit along the x-axis and

-1 unit along the y-axis

(4) $y = \dfrac{2x-3}{2x-4}$

$y = \dfrac{2x-3}{2x-4} = \dfrac{2x-4+1}{2x-4} = \dfrac{1}{2(x-2)} + 1$

The equations of the asymptotes are $x = 2$ and $y = 1$

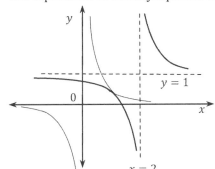

Translation: 2 units along the x-axis and

1 unit along the y-axis

#50 Find the equations of the curves obtained when the graph of $y = \sqrt{-2x}$ is translated as follows:

(1) -2 units along the y-axis

$(y + 2) = \sqrt{-2x}$ $\therefore\ y = \sqrt{-2x} - 2$

(2) 3 units along the x-axis

$y = \sqrt{-2(x-3)}$

(3) $-\dfrac{1}{2}$ units along the x-axis and 4 units along the y-axis

$(y - 4) = \sqrt{-2(x + \frac{1}{2})}$ $\therefore\ y = \sqrt{-2(x + \frac{1}{2})} + 4$

#51 Solve the following fractional equations.

(1) $\dfrac{2}{x-1} - \dfrac{1}{x+1} = \dfrac{x}{x-1}$

Multiplying both sides of the equation by the least common multiple, LCM, $(x + 1)(x - 1)$,

$2(x + 1) - (x - 1) = x(x + 1)$ where $x \neq 1,\ x \neq -1$

$\therefore\ x^2 + x - x - 3 = 0\ ;\ x^2 - 3 = 0$ $\therefore\ x^2 = 3$ $\therefore\ x = \pm\sqrt{3}$

Therefore, $x = \sqrt{3}$ or $x = -\sqrt{3}$

(2) $\dfrac{x+3}{x+1} + \dfrac{x}{x-2} = -\dfrac{1}{x-2}$

Multiplying both sides of the equation by the least common multiple, LCM, $(x + 1)(x - 2)$,

$(x + 3)(x - 2) + x(x + 1) = -(x + 1)$ where $x \neq -1,\ x \neq 2$

$$\therefore \ x^2 + x - 6 + x^2 + x + x + 1 = 0 \ ; \quad 2x^2 + 3x - 5 = 0$$

$$\therefore \ x = \frac{-3 \pm \sqrt{9+40}}{4} = \frac{-3 \pm 7}{4} \qquad \therefore \ x = 1 \ \text{ or } \ x = -\frac{10}{4} = -\frac{5}{2}$$

Therefore, $x = 1$ or $x = -\dfrac{5}{2}$

(3) $\dfrac{2}{x-2} - \dfrac{x}{x^2-3x+2} = -1$

$$\frac{2}{x-2} - \frac{x}{x^2-3x+2} = -1$$

$\Rightarrow$ Multiplying both sides of the equation by the least common multiple, LCM, $(x-2)(x-1)$,

$2(x-1) - x = -(x-2)(x-1)$ where $x \neq 1, \ x \neq 2$

$$\therefore \ x - 2 + x^2 - 3x + 2 = 0 \ ; \quad x^2 - 2x = 0 \ ; \ x(x-2) = 0 \qquad \therefore \ x = 0 \ \text{ or } \ x = 2$$

Since $x \neq 2$, $x = 0$

#52 Solve the following inequalities by using graphs.

(1) $\dfrac{2}{x-2} \leq x - 1$

Solving the original inequality as an equation,

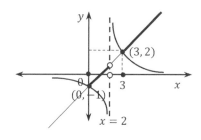

$$\frac{2}{x-2} = x - 1 \quad \Rightarrow \quad (x-1)(x-2) = 2 \, , \ x \neq 2$$

$$\Rightarrow \ x^2 - 3x = 0 \qquad \Rightarrow \ x(x-3) = 0$$

$$\therefore \ x = 0 \ \text{ or } \ x = 3$$

From the graph, $0 \leq x < 2, \ x \geq 3$

Alternative Approach:

$\dfrac{2}{x-2} \leq x - 1 \quad \Rightarrow \quad \dfrac{2}{x-2} - x + 1 \leq 0 \quad \Rightarrow \quad \dfrac{2-(x-1)(x-2)}{x-2} \leq 0$

$$\Rightarrow \ \{2 - (x-1)(x-2)\}(x-2) \leq 0 \, , \ x \neq 2$$

$$\Rightarrow \ 2(x-2) - (x-1)(x-2)^2 \leq 0 \, , \ x \neq 2$$

$$\Rightarrow \ (x-2)\{2 - (x-1)(x-2)\} \leq 0 \, , \ x \neq 2$$

$$\Rightarrow \ (x-2)(-x^2 + 3x) \leq 0 \, , \ x \neq 2$$

$$\Rightarrow \ -x(x-2)(x-3) \leq 0 \, , \ x \neq 2 \qquad \Rightarrow \ x(x-2)(x-3) \geq 0 \, , \ x \neq 2$$

$$\therefore \ 0 \leq x < 2, \ x \geq 3$$

(2) $\dfrac{x+1}{x-1} \geq x+1$

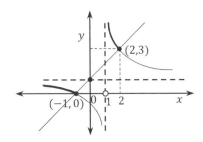

Solving the original inequality as an equation,

$$\frac{x+1}{x-1} = x+1 \;\Rightarrow\; (x+1)(x-1) = x+1 \,,\; x \neq 1$$

$$\Rightarrow\; x^2 - x - 2 = 0 \quad \Rightarrow\; (x-2)(x+1) = 0$$

$$\therefore\; x = 2 \;\text{ or }\; x = -1$$

From the graph, $x \leq -1,\; 1 < x \leq 2$

Alternative Approach:

$$\frac{x+1}{x-1} \geq x+1 \;\Rightarrow\; \{(x+1) - (x+1)(x-1)\}(x-1) \geq 0,\; x \neq 1$$

$$\frac{x+1}{x-1} = \frac{x-1+2}{x-1} = 1 + \frac{2}{x-1}$$

$$\Rightarrow\; (x+1)(x-1) - (x+1)(x-1)^2 \geq 0,\; x \neq 1$$

$$\Rightarrow\; (x+1)(x-1)\{1 - (x-1)\} \geq 0,\; x \neq 1$$

$$\Rightarrow\; (x+1)(x-1)(-x+2) \geq 0,\; x \neq 1$$

$$\Rightarrow\; -(x+1)(x-1)(x-2) \geq 0,\; x \neq 1$$

$$\Rightarrow\; (x+1)(x-1)(x-2) \leq 0,\; x \neq 1$$

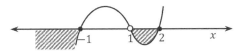

$$\therefore\; x \leq -1,\; 1 < x \leq 2$$

(3) $-1 < \dfrac{4-3x}{x-2} \leq 1$

i) $\quad -1 < \dfrac{4-3x}{x-2}$

$$\Rightarrow\; -1 < \frac{-3(x-2)-2}{x-2} = \frac{-2}{x-2} - 3$$

$$\Rightarrow\; \frac{2}{x-2} + 2 < 0$$

From the graph, $1 < x < 2$

ii) $\dfrac{4-3x}{x-2} \leq 1$

$$\Rightarrow\; \frac{-3(x-2)-2}{x-2} \leq 1$$

$$\Rightarrow\; -3 - \frac{2}{x-2} \leq 1$$

$$\Rightarrow\; \frac{2}{x-2} + 4 \geq 0$$

From the graph, $x \leq \dfrac{3}{2},\; x > 2$

By i) and ii),

 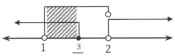

$$\therefore\; 1 < x \leq \frac{3}{2}$$

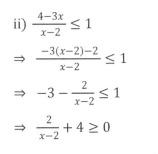

Alternative Approach:

i) $-1 < \frac{4-3x}{x-2}$ $\Rightarrow$ $\frac{4-3x+(x-2)}{x-2} > 0$

$\Rightarrow$ $\frac{-2x+2}{x-2} > 0$

$\Rightarrow$ $-2(x-1)(x-2) > 0$, $x \neq 2$

$\Rightarrow$ $2(x-1)(x-2) < 0$, $x \neq 2$

$\therefore$ $1 < x < 2$

ii) $\frac{4-3x}{x-2} \leq 1$ $\Rightarrow$ $\frac{4-3x-(x-2)}{x-2} \leq 0$

$\Rightarrow$ $\frac{-4x+6}{x-2} \leq 0$

$\Rightarrow$ $(-4x+6)(x-2) \leq 0$, $x \neq 2$ $\Rightarrow$ $-4(x-\frac{6}{4})(x-2) \leq 0$, $x \neq 2$

$\Rightarrow$ $4(x-\frac{3}{2})(x-2) \geq 0$, $x \neq 2$

$\therefore$ $x \leq \frac{3}{2}$, $x > 2$

By i) and ii),

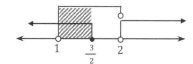

$\therefore$ $1 < x \leq \frac{3}{2}$

(4) $\frac{4}{x-2} < x+1$, $x > 2$

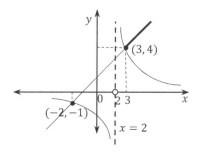

Solving the original inequality as an equation,

$\frac{4}{x-2} = x+1$ $\Rightarrow$ $(x+1)(x-2) - 4 = 0$, $x \neq 2$

$\Rightarrow$ $x^2 - x - 6 = 0$ $\Rightarrow$ $(x-3)(x+2) = 0$

$\therefore$ $x = 3$ or $x = -2$

From the graph, $x > 3$

Alternative Approach:

$\frac{4}{x-2} < x+1$ $\Rightarrow$ $\{4-(x+1)(x-2)\}(x-2) < 0$, $x \neq 2$

$\Rightarrow$ $4(x-2) - (x+1)(x-2)^2 < 0$, $x \neq 2$

$\Rightarrow$ $(x-2)\{4-(x+1)(x-2)\} < 0$, $x \neq 2$

$\Rightarrow$ $(x-2)(-x^2+x+6) < 0$, $x \neq 2$

$\Rightarrow$ $-(x-2)(x-3)(x+2) < 0$, $x \neq 2$

$\Rightarrow$ $(x-2)(x-3)(x+2) > 0$, $x \neq 2$

$\therefore$ $-2 < x < 2$, $x > 3$

Since $x > 2$, $x > 3$

#53 Draw graph of each radical function, and find maximum and minimum values within the given domain.

(1) $f(x) = \sqrt{x+1} - 2$ $(0 \leq x \leq 2)$

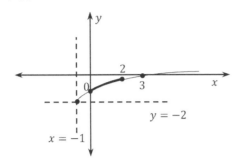

From the graph, maximum value is

$f(2) = \sqrt{2+1} - 2 = \sqrt{3} - 2$ and

minimum value is $f(0) = \sqrt{0+1} - 2 = -1$

(2) $f(x) = -\sqrt{x+2} + 1$ $(-1 \leq x \leq 2)$

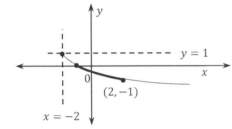

From the graph, maximum value is

$f(-1) = -\sqrt{-1+2} + 1 = 0$ and

minimum value is $f(2) = -\sqrt{2+2} + 1 = -1$

(3) $f(x) = \sqrt{3-x} - 2$ $(-1 \leq x \leq 2)$

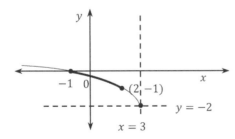

From the graph, maximum value is

$f(-1) = \sqrt{3-(-1)} - 2 = 0$ and

minimum value is $f(2) = \sqrt{3-2} - 2 = -1$

(4) $f(x) = -\sqrt{3-2x} + 1$ $(-1 \leq x \leq 1)$

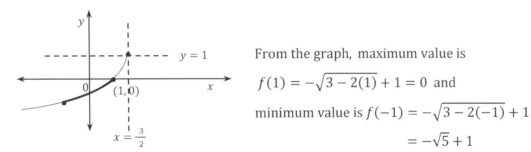

From the graph, maximum value is

$f(1) = -\sqrt{3-2(1)} + 1 = 0$ and

minimum value is $f(-1) = -\sqrt{3-2(-1)} + 1$

$$= -\sqrt{5} + 1$$

(5) $f(x) = 2 - \sqrt{x+1} \quad (0 \le x \le 3)$

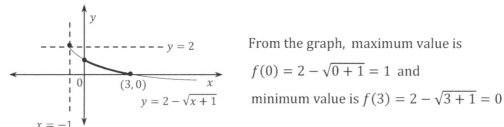

From the graph, maximum value is

$f(0) = 2 - \sqrt{0+1} = 1$ and

minimum value is $f(3) = 2 - \sqrt{3+1} = 0$

#54 Draw the graph and use it to solve the given radical equation.

(1) $\sqrt{3-x} = x - 3$

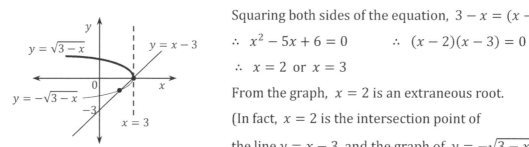

Squaring both sides of the equation, $3 - x = (x-3)^2$

$\therefore \ x^2 - 5x + 6 = 0 \qquad \therefore \ (x-2)(x-3) = 0$

$\therefore \ x = 2$ or $x = 3$

From the graph, $x = 2$ is an extraneous root.

(In fact, $x = 2$ is the intersection point of

the line $y = x - 3$ and the graph of $y = -\sqrt{3-x}$)

Therefore, $x = 3$

(2) $\sqrt{2x+1} = -2x + 1$

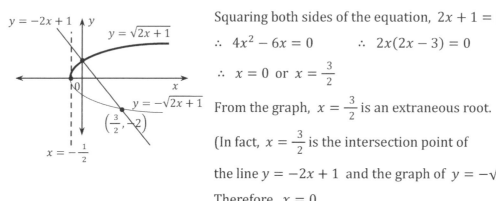

Squaring both sides of the equation, $2x + 1 = (-2x+1)^2$

$\therefore \ 4x^2 - 6x = 0 \qquad \therefore \ 2x(2x-3) = 0$

$\therefore \ x = 0$ or $x = \dfrac{3}{2}$

From the graph, $x = \dfrac{3}{2}$ is an extraneous root.

(In fact, $x = \dfrac{3}{2}$ is the intersection point of

the line $y = -2x + 1$ and the graph of $y = -\sqrt{2x+1}$)

Therefore, $x = 0$

Alternative Approach:

When we have $x = 0$ or $x = \dfrac{3}{2}$, substitute the solutions into the original equation.

If $x = 0$, then $LHS = RHS = 1$

If $x = \dfrac{3}{2}$, then $LHS = 2$, $RHS = -2 \qquad \therefore \ LHS \ne RHS$

$\therefore \ x = \dfrac{3}{2}$ is an extraneous solution.   Therefore, $x = 0$

#55 Solve the following radical inequalities by using graphs.

(1) $\sqrt{2x-3} > x-3$

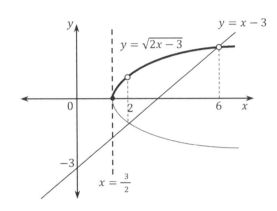

Let $\sqrt{2x-3} = x-3$

Squaring both sides, $2x-3 = (x-3)^2$

$\therefore\ x^2 - 8x + 12 = 0 \quad \therefore\ (x-6)(x-2) = 0$

$\therefore\ x = 2$ or $x = 6$

From the graph, $x = 2$ is an extraneous root.

Thus, $x = 6$

Therefore, $\dfrac{3}{2} \le x < 2,\ 2 < x < 6$

Alternative Approach:

Squaring both sides, $2x - 3 > (x-3)^2$

$\therefore\ x^2 - 8x + 12 < 0 \quad \therefore\ (x-6)(x-2) < 0 \qquad \therefore\ 2 < x < 6$

Since $2x - 3 \ge 0$, $x \ge \dfrac{3}{2}$ $\qquad$ Therefore, $\dfrac{3}{2} \le x < 2,\ 2 < x < 6$

(2) $\sqrt{2x+1} < \dfrac{x+2}{2}$

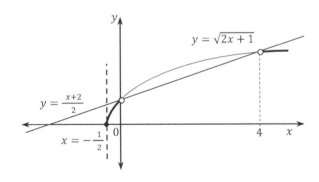

Let $\sqrt{2x+1} = \dfrac{x+2}{2}$; $2\sqrt{2x+1} = x+2$

Squaring both sides, $4(2x+1) = (x+2)^2$

$\therefore\ x^2 + 4x + 4 - 8x - 4 = x^2 - 4x$

$\qquad\qquad\qquad\qquad = x(x-4) = 0$

$\therefore\ x = 0$ or $x = 4$

From the graph, $-\dfrac{1}{2} \le x < 0$, $x > 4$

Alternative Approach:

$\sqrt{2x+1} < \dfrac{x+2}{2}$; $2\sqrt{2x+1} < x+2$

Squaring both sides, $4(2x+1) < (x+2)^2$

$\therefore\ x^2 - 4x > 0 \quad \therefore\ x(x-4) > 0 \qquad \therefore\ x > 4$ or $x < 0$

Since $2x + 1 \ge 0$, $x \ge -\dfrac{1}{2}$ $\qquad$ Therefore, $-\dfrac{1}{2} \le x < 0$, $x > 4$

(3) $\sqrt{x} - 1 < \sqrt{5 - x}$

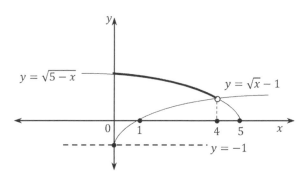

Let $\sqrt{x} - 1 = \sqrt{5 - x}$

Squaring both sides, $x - 2\sqrt{x} + 1 = 5 - x$

$\therefore 2\sqrt{x} = 2x - 4$; $\sqrt{x} = x - 2$

Squaring both sides, $x = x^2 - 4x + 4$

$\therefore x^2 - 5x + 4 = (x - 1)(x - 4) = 0$

$\therefore x = 1$ or $x = 4$

From the graph, $0 \le x < 4$

(4) $\sqrt{x} - 1 > \sqrt{5 - x}$

From the graph of (3), $x > 4$ ......①

Since $x \ge 0$ and $5 - x \ge 0$, $0 \le x \le 5$ ......②

By ① and ②, $4 < x \le 5$

(5) $\sqrt{2 - x} < x < \sqrt{x + 6}$

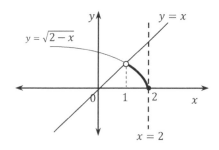

i) $\quad \sqrt{2 - x} < x$

$\Rightarrow$ Let $\sqrt{2 - x} = x$.

Squaring both sides, $2 - x = x^2$

$\therefore x^2 + x - 2 = (x + 2)(x - 1) = 0$

$\therefore x = -2$ or $x = 1$

From the graph, $1 < x \le 2$

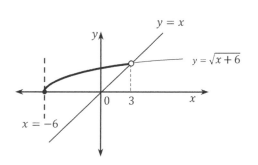

ii) $\quad x < \sqrt{x + 6}$

$\Rightarrow$ Let $x = \sqrt{x + 6}$.

Squaring both sides, $x^2 = x + 6$

$\therefore x^2 - x - 6 = (x - 3)(x + 2) = 0$

$\therefore x = 3$ or $x = -2$

From the graph, $x = -2$ is an extraneous root. $\quad \therefore -6 \le x < 3$

By i) and ii),

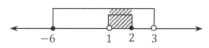

$1 < x \le 2$

Alternative Approach:

i) $\sqrt{2-x} < x \quad \Rightarrow \quad 2-x < x^2, \; x \le 2 \quad \Rightarrow x^2 + x - 2 = (x+2)(x-1) > 0$

$\therefore \; x > 1$ or $x < -2$

$\therefore \; x < -2, \; 1 < x \le 2$

ii) $x < \sqrt{x+6} \; \Rightarrow \; x^2 < x + 6 \quad, \; x \ge -6 \quad \Rightarrow \; x^2 - x - 6 = (x-3)(x+2) < 0$

$\therefore \; -2 < x < 3$

$\therefore \; -2 < x < 3$

By i) and ii),

$\therefore \; 1 < x \le 2$

(6) $\sqrt{3-2x} > -\dfrac{2x+1}{\sqrt{2}}$

$\sqrt{3-2x} > -\dfrac{2x+1}{\sqrt{2}} \quad \Rightarrow \quad \sqrt{2} \cdot \sqrt{3-2x} > -(2x+1) \quad \Rightarrow \quad \sqrt{6-4x} > -2x - 1$

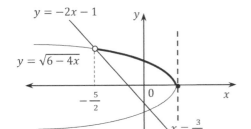

Let $\sqrt{6-4x} = -2x - 1$.

Squaring both sides, $6 - 4x = 4x^2 + 4x + 1$

$\therefore \; 4x^2 + 8x - 5 = 0$

$\therefore \; (2x+5)(2x-1) = 0$

$\therefore \; x = -\dfrac{5}{2}$ or $x = \dfrac{1}{2}$

From the graph, $-\dfrac{5}{2} < x \le \dfrac{3}{2}$

$x = \dfrac{1}{2}$ is an extraneous root.

(In fact, $x = \dfrac{1}{2}$ is the intersection point of

the line $y = -2x - 1$ and the graph of $y = -\sqrt{6-4x}$)

#56 Solve the following inequalities by using graphs.

(1) $(2x - 3)(x^2 - 4x - 5) < 0$

$\Rightarrow (2x - 3)(x - 5)(x + 1) < 0$

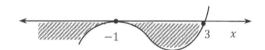

$\therefore\ x < -1,\ \ \dfrac{3}{2} < x < 5$

(2) $(3 - x)(x^2 - 2x - 15) \geq 0$

$\Rightarrow (3 - x)(x - 5)(x + 3) \geq 0$

$\Rightarrow -(x - 3)(x - 5)(x + 3) \geq 0$

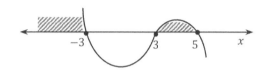

$\therefore\ x \leq -3,\ 3 \leq x \leq 5$

(3) $(x + 1)(x^2 - 2x - 3) \leq 0$

$\Rightarrow (x + 1)(x - 3)(x + 1) \leq 0 \qquad \Rightarrow (x + 1)^2(x - 3) \leq 0$

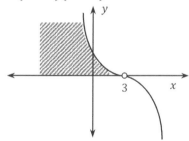

$\therefore\ x \leq -1,\ -1 \leq x \leq 3 \qquad \therefore\ x \leq 3$

(4) $(3 - x)(x^2 - 6x + 9) > 0$

$\Rightarrow (3 - x)(x - 3)^2 > 0 \qquad \Rightarrow -(x - 3)(x - 3)^2 > 0 \qquad \Rightarrow -(x - 3)^3 > 0$

$\therefore\ x < 3$

(5) $x^3 - 3x^2 + 4 \leq 0$

Let $f(x) = x^3 - 3x^2 + 4$

Then, $f(-1) = 0$

$\therefore\ f(x)$ has $(x + 1)$ as a factor.

Using long division,

$x^3 - 3x^2 + 4 = (x + 1)(x^2 - 4x + 4) = (x + 1)(x - 2)^2$

Thus, $x^3 - 3x^2 + 4 = (x + 1)(x - 2)^2 \leq 0$

Therefore, $x \leq -1$

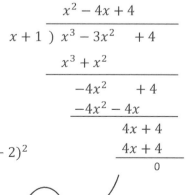

(6) $(2x - 1)(1 - x)(x^2 - 2x - 15) \leq 0$

$\Rightarrow$ $(2x - 1)(1 - x)(x - 5)(x + 3) \leq 0$

$\Rightarrow$ $-(2x - 1)(x - 1)(x - 5)(x + 3) \leq 0$

$\therefore$ $x \leq -3, \ \dfrac{1}{2} \leq x \leq 1, \ x \geq 5$

(7) $(x + 2)(1 - x)(x - 3)^2 \geq 0$

$\Rightarrow$ $-(x + 2)(x - 1)(x - 3)^2 \geq 0$

$\therefore$ $-2 \leq x \leq -1, \ x = 3$

(8) $(x^2 + 2x - 3)(x^2 + 2x + 6) \leq 0$

$x^2 + 2x + 6 = (x + 1)^2 - 1 + 6 = (x + 1)^2 + 5$

Since $(x + 1)^2 \geq 0, \ (x + 1)^2 + 5 > 0$

$\therefore$ $x^2 + 2x + 6$ is always positive.

(Or, the discriminant D of the equation $x^2 + 2x + 6 = 0$ is $D = 4 - 24 = -20 < 0$

$\therefore$ $x^2 + 2x + 6 > 0$ for all x.)

Thus, $(x^2 + 2x - 3)(x^2 + 2x + 6) \leq 0$

$\Rightarrow$ $(x^2 + 2x - 3) \leq 0$

$\Rightarrow$ $(x + 3)(x - 1) \leq 0$

$\Rightarrow$ $-3 \leq x \leq 1$

(9) $(x^2 - x + 2)(x^2 - 3x - 4)(x^3 + 4x) \leq 0$

$x^2 - x + 2 = \left(x - \dfrac{1}{2}\right)^2 - \dfrac{1}{4} + 2 = \left(x - \dfrac{1}{2}\right)^2 + \dfrac{7}{4} > 0$

(Or, the discriminant D of the equation $x^2 - x + 2 = 0$ is $D = 1 - 8 = -7 < 0$

$\therefore$ $x^2 - x + 2 > 0$ for all x.)

$x^2 - 3x - 4 = (x - 4)(x + 1)$

$x^3 + 4x = x(x^2 + 4)$ Note that $x^2 + 4 > 0$ for all x

Thus, $(x^2 - x + 2)(x^2 - 3x - 4)(x^3 + 4x) \leq 0$

$\Rightarrow$ $(x - 4)(x + 1)x \leq 0$

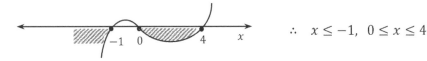

$\therefore$ $x \leq -1, \ 0 \leq x \leq 4$

(10) $(x^2 - 3x - 4)^2 > 0$

$\Rightarrow \{(x-4)(x+1)\}^2 > 0$

$\Rightarrow (x-4)^2(x+1)^2 > 0$

All numbers except $x = -1$ and $x = 4$

$\therefore \quad x < -1, \quad -1 < x < 4, \quad x > 4$

(11) $(2 - x)(x - 3)(x^2 - 6x + 9) \leq 0$

Since $x^2 - 6x + 9 = (x-3)^2 \geq 0$,

$(2-x)(x-3)(x^2-6x+9) \leq 0 \Rightarrow (2-x)(x-3) \leq 0 \Rightarrow -(x-2)(x-3) \leq 0$

$\therefore \quad x \leq 2 \quad$ or $\quad x \geq 3$

#57 For a function $y = \sqrt{1 - 2x} + 3$, find the inverse function and determine the domain and range of the inverse function.

$y = \sqrt{1 - 2x} + 3$

Since $1 - 2x \geq 0$, $x \leq \dfrac{1}{2}$ and $y \geq 3$

Since $y - 3 = \sqrt{1 - 2x}$, squaring both sides, $(y - 3)^2 = 1 - 2x$

Switching x and y, $(x - 3)^2 = 1 - 2y$

Solving for y, $y = \dfrac{1}{2} - \dfrac{(x-3)^2}{2}$

Thus, the inverse function is $y = \dfrac{1}{2} - \dfrac{(x-3)^2}{2}$.

The domain is $x \geq 3$ and the range is $y \leq \dfrac{1}{2}$.

#58 Find the value.

(1) For two functions $f(x) = \dfrac{x+4}{x-3}$ and $g(x) = \sqrt{2x - 4}$ with the same domain

$\{x \mid x > 3\}$, find the value of $(f \circ (g \circ f)^{-1} \circ f)(4)$.

$(f \circ (g \circ f)^{-1} \circ f)(4) = (f \circ f^{-1} \circ g^{-1} \circ f)(4) = (g^{-1} \circ f)(4)$

$= g^{-1}(f(4)) = g^{-1}\left(\dfrac{4+4}{4-3}\right) = g^{-1}(8)$

Let $g^{-1}(8) = k$. Then, $g(k) = 8$

$\therefore \quad \sqrt{2k - 4} = 8 \qquad \therefore \quad 2k - 4 = 64 \qquad \therefore \quad k = 34$

Therefore, $(f \circ (g \circ f)^{-1} \circ f)(4) = 34$

(2) For a function $f(x) = \sqrt{2x-1}$, there is a function $g(x)$ such that $(g \circ f)(x) = \frac{2x-1}{x+1}$. Find the value of $g(3)$.

$(g \circ f)(x) = g(f(x)) = \frac{2x-1}{x+1}$

$f(x) = 3 \Rightarrow \sqrt{2x-1} = 3 \qquad \therefore \ 2x - 1 = 9 \qquad \therefore \ x = 5$

Therefore, $g(3) = g(f(5)) = \frac{2 \cdot 5 - 1}{5 + 1} = \frac{9}{6} = \frac{3}{2}$

#59 Minimum and maximum values.

(1) When an inequality $ax \leq \frac{x+1}{x-2} \leq bx + 1$ is always true in the interval $3 \leq x \leq 4$, find the maximum value of a and minimum value of b (where a and b are real numbers).

Let $y = \frac{x+1}{x-2}$. Then, $y = \frac{x+1}{x-2} = \frac{x-2+3}{x-2} = 1 + \frac{3}{x-2}$

When $x = 3$, $y = 4$

When $x = 4$, $y = \frac{5}{2}$

To have $y = 1 + \frac{3}{x-2} \geq ax$, $3 \leq x \leq 4$,

the graph of $y = 1 + \frac{3}{x-2}$ intersects the line $y = ax$ at $(4, \frac{5}{2})$, or lies above the line.

$y = ax$ at $(4, \frac{5}{2}) \Rightarrow \frac{5}{2} = 4a \qquad \therefore \ a = \frac{5}{8}$ \qquad Therefore, maximum value of a is $\frac{5}{8}$.

To get the minimum value of b such that $\frac{x+1}{x-2} \leq bx + 1$ in $3 \leq x \leq 4$,

the graph of $y = \frac{x+1}{x-2}$ intersects the line $y = bx + 1$ at $(3, 4)$, or lies below the line.

$\therefore \ 4 = 3b + 1 \qquad \therefore \ b = 1$ \qquad Therefore, minimum value of b is 1.

(2) For a point $P(0, 2)$ in a coordinate plane, a point Q lies on a graph of $y = \frac{4}{x} + 2$ in the same plane. Find minimum value of the length of the segment $\overline{PQ}$.

Let $Q = Q(a, \frac{4}{a} + 2)$

Then, $\overline{PQ} = \sqrt{(a-0)^2 + (\frac{4}{a} + 2 - 2)^2} = \sqrt{a^2 + \frac{16}{a^2}}$

Since $a^2 > 0$ and $\frac{16}{a^2} > 0$, by the relationship between arithmetic and geometric means,

$a^2 + \frac{16}{a^2} \geq 2\sqrt{a^2 \cdot \frac{16}{a^2}} = 2\sqrt{16} = 8$ \ (when $a^2 = \frac{16}{a^2}$, $LHS = RHS$)

$\therefore \ \overline{PQ} = \sqrt{a^2 + \frac{16}{a^2}} \geq \sqrt{8} = 2\sqrt{2}$ \qquad Therefore, minimum value of $\overline{PQ}$ is $2\sqrt{2}$.

#60 When the minimum value of a function $y = \sqrt{|x| + 1}$ is a point A and the two intersection points of the graph of the function and a line $y = 2$ are B and C, find the area of the triangle $\triangle ABC$.

$x \geq 0 \;\Rightarrow\; y = \sqrt{x + 1}$

$x < 0 \;\Rightarrow\; y = \sqrt{-x + 1}$

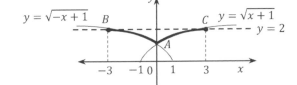

From the graph, $A = (0, 1)$

Since $2 = \sqrt{x + 1} \Rightarrow 4 = x + 1 \Rightarrow x = 3$ and

$\qquad 2 = \sqrt{-x + 1} \Rightarrow 4 = -x + 1 \Rightarrow x = -3, \quad B = (-3, 2)$ and $C = (3, 2)$

Therefore, the area of the triangle $\triangle ABC$ is $\dfrac{1}{2} \cdot 6 \cdot 1 = 3 \;(\text{unit}^2)$

#61 Find the value.

(1) When the graph of $f(x) = \dfrac{ax}{x+2}$ is symmetric with respect to the line $y = x$, find the real number a.

Since the graph of $f(x) = \dfrac{ax}{x+2}$ is symmetric with respect to the line $y = x$,

$f(x)$ and its inverse function $f^{-1}(x)$ are equivalent.

Let $y = \dfrac{ax}{x+2}$. Switching x and y gives $x = \dfrac{ay}{y+2} \qquad \therefore xy + 2x = ay$

Solving for y, $y(x - a) = -2x \qquad \therefore y = \dfrac{-2x}{x-a}$

Therefore, the inverse function is $f^{-1}(x) = \dfrac{-2x}{x-a}$

Since $\dfrac{ax}{x+2} = \dfrac{-2x}{x-a}$, $a = -2$

(2) If the inverse functions of rational functions $f(x) = \dfrac{ax+b}{cx+d} \; (d > 0)$ and $g(x) = \dfrac{x-1}{2x+3}$ exist, then $(f \circ g)(x) = x$. When the axes of symmetries for a rational function $y = f(x)$ are $x = m$ and $y = n$, find the value of $m - n$.

Since $(f \circ g)(x) = x$, g is an inverse function of f.

$g(x) = \dfrac{x-1}{2x+3} = \dfrac{x + \frac{3}{2} - \frac{5}{2}}{2\left(x + \frac{3}{2}\right)} = \dfrac{1}{2} - \dfrac{\frac{5}{2}}{2\left(x + \frac{3}{2}\right)}$

$\therefore$ The axes of symmetries for $g(x)$ are $x = -\dfrac{3}{2}$ and $y = \dfrac{1}{2}$

Thus, the axes of symmetries for $f(x) = \dfrac{ax+b}{cx+d}$ are $x = \dfrac{1}{2}, \; y = -\dfrac{3}{2}$

Therefore, $m - n = \dfrac{1}{2} - \left(-\dfrac{3}{2}\right) = 2$

(3)

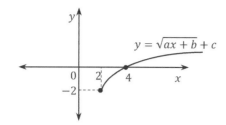

$y = \sqrt{ax + b} + c$

When the graph of a function $y = \sqrt{ax + b} + c$

is shown as the Figure,

find the value of $a + b + c$.

From the graph, $y = \sqrt{a(x - 2)} - 2$

Since the graph passes through a point $(4, 0)$, $0 = \sqrt{a(4 - 2)} - 2$

$\therefore \sqrt{2a} = 2 \qquad \therefore 2a = 4 \qquad \therefore a = 2$

Thus, $y = \sqrt{2(x - 2)} - 2 = \sqrt{2x - 4} - 2$

Therefore, $a = 2$, $b = -4$, and $c = -2 \qquad$ Hence, $a + b + c = 2 - 4 - 2 = -4$

(4) For a function $y = \frac{4x + a}{x + 2}$, the function has maximum value 3 and minimum value 2

in the interval $b \leq x \leq -3$ (where a and b are real numbers). Find the value of $a + b$.

$y = \frac{4x + a}{x + 2} = \frac{4(x + 2) - 8 + a}{x + 2} = 4 + \frac{-8 + a}{x + 2}$

$\therefore$ The axes of symmetries are $x = -2$ and $y = 4$

Since the maximum value in the interval $b \leq x \leq -3$ is 3, $a - 8 > 0$; i.e., $a > 8$

Since the function has minimum value when $x = -3$, $\frac{4(-3) + a}{(-3) + 2} = 2 \quad \therefore a = 10$

Since the maximum value is 3 when $x = b$, $\frac{4b + a}{b + 2} = 3 \quad \therefore b = 6 - a = -4$

Therefore, $a + b = 10 - 4 = 6$

(5) The graph of a fractional function $y = \frac{-4x + 3}{2x - 1}$ is symmetric with respect to the line

$y = ax + b$, $a \neq 0$. Find the real number $a^2 - b^2$.

$y = \frac{-4x + 3}{2x - 1} = \frac{-2(2x - 1) + 1}{2x - 1} = -2 + \frac{1}{2x - 1} \quad \cdots\cdots \text{①}$

$\therefore$ The axes of symmetries are $x = \frac{1}{2}$, $y = -2$

The graph of ① is symmetric with respect to the line

that passes through a point $(\frac{1}{2}, -2)$ and has slope ± 1.

Thus, the line is $y = \pm\left(x - \frac{1}{2}\right) - 2$

$\therefore y = x - \frac{5}{2}$ or $y = -x - \frac{3}{2}$

By the graph, $y = -x - \frac{3}{2} \qquad \therefore a = -1$, $b = -\frac{3}{2}$

Therefore, $a^2 - b^2 = 1 - \frac{9}{4} = -\frac{5}{4}$

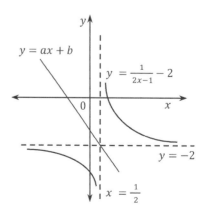

(6) When the graph of a function $y = \sqrt{x+1}$ is translated with 3 units along the x-axis, -1unit along the y-axis, we reflect the translated graph through the line $x = 1$. Then, it will be the same as the graph of $y = \sqrt{ax+b} + c$. Find the value of $a + b + c$.

Translation: $y + 1 = \sqrt{(x-3)+1}$ $\quad \therefore y = \sqrt{x-2} - 1$

Reflection: $y = \sqrt{-x} - 1$ (From the graph)

$\therefore \sqrt{-x} - 1 = \sqrt{ax+b} + c$

Thus, $a = -1$, $b = 0$, and $c = -1$

Therefore, $a + b + c = -2$

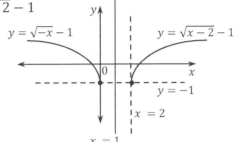

#62 Find the range.

(1) When an equation $\sqrt{x+2} = x + a$ has two different real number solutions, find the range of a.

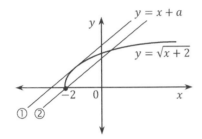

To get the line ①,

the line $y = x + a$ and the graph of $y = \sqrt{x+2}$

intersect at one point.

$\sqrt{x+2} = x + a \implies x + 2 = (x+a)^2$

$\implies x + 2 = x^2 + 2ax + a^2$

$\implies x^2 + (2a-1)x + a^2 - 2 = 0 \cdots\cdots ❶$

Let D be the discriminant of the equation ❶.

Then, $D = (2a-1)^2 - 4(a^2 - 2) = 0$ $\quad \therefore -4a + 9 = 0$ $\quad \therefore a = \dfrac{9}{4}$

To get the line ②, the line $y = x + a$ passes through a point $(-2, 0)$.

$\therefore 0 = -2 + a$ $\quad \therefore a = 2$

Therefore, $2 \le a < \dfrac{9}{4}$

(2) When the graphs of two functions $y = \dfrac{x+2}{x-2}$ and $y = \sqrt{x+a}$ intersect at two different points, find the range of a.

$y = \dfrac{x+2}{x-2} = \dfrac{x-2+4}{x-2} = 1 + \dfrac{4}{x-2}$

The axes of symmetries are $x = 2$, $y = 1$

Note that the graph of $y = \sqrt{x+a}$ is translated

from the graph of $y = \sqrt{x}$, $-a$ units along the x-axis.

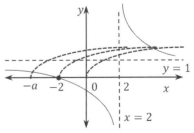

From the graph, if $-a \le -2$, then the graphs of two functions intersect at two different points. Therefore, $a \ge 2$

(3) When the graph of $y = \frac{a}{x-2} + 1$ passes all 4 quadrants in a coordinate plane,

find the range of the real number a.

If $a > 0$, then the graph of y is shown as the Figure 1.

If $a < 0$, then the graph of y is shown as the Figure 2.

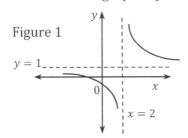 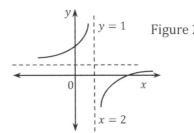

Figure 1

$y = 1$

$x = 2$

$y = 1$ Figure 2

$x = 2$

From the graph, when $a < 0$, the graph cannot pass the third quadrant.

Let $f(x) = \frac{a}{x-2} + 1$.

When $a > 0$, $f(0) < 0$ to satisfy the given condition.

$\therefore \ \frac{a}{-2} + 1 < 0 \ ; \ \frac{a}{-2} < -1 \ ; \ \frac{a}{2} > 1$

Therefore, $a > 2$

(4) Let the axes of symmetries of a function $y = \frac{x+1}{2x-1}$ be $x = m, \ y = n$. When the graph

of y and a line $y = ax + \dfrac{m}{n}$ intersect at the range: $a \leq \alpha$ or $a \geq \beta$,

find the value of $(\alpha - \beta)^2$ (where α and β are constants).

$y = \dfrac{x+1}{2x-1} = \dfrac{\frac{1}{2}(2x-1) + \frac{3}{2}}{2x-1} = \dfrac{\frac{3}{2}}{2x-1} + \dfrac{1}{2}$

$\therefore$ The axes of symmetries are $x = \dfrac{1}{2}, \ y = \dfrac{1}{2}$

$\therefore \ m = \dfrac{1}{2}, \ n = \dfrac{1}{2}$

$\therefore \ y = ax + \dfrac{m}{n} = ax + 1$

Since the graphs intersect, the equation $\dfrac{x+1}{2x-1} = ax + 1$ has real number solutions.

; i.e., $(2x - 1)(ax + 1) = x + 1 \ ; \ 2ax^2 - ax + 2x - 1 = x + 1 \ ; \ 2ax^2 + (1 - a)x - 2 = 0$,

Let D be the discriminant of the equation. Then, $D = (1 - a)^2 + 16a \geq 0$

$\therefore \ a^2 + 14a + 1 \geq 0$

Let α and β $(\alpha < \beta)$ be the roots of the equation $a^2 + 14a + 1 = 0$

Then, $a \leq \alpha$ or $a \geq \beta$

By the relationship between roots and coefficients, $\alpha + \beta = -14, \ \alpha\beta = 1$

Therefore, $(\alpha - \beta)^2 = (\alpha + \beta)^2 - 4\alpha\beta = 196 - 4 = 192$

(5) When the graphs of two functions $f(x) = \frac{1}{3}x^2 + a$ and $g(x) = \sqrt{3x - 3a}$ $(x \geq 0,$

$a \geq 0)$ intersect at two different points, find the range of real number a.

Let $y = \frac{1}{3}x^2 + a$.

When $x \geq 0$, $x^2 = 3y - 3a$ $\qquad \therefore x = \sqrt{3y - 3a}$, $3y - 3a \geq 0$

Switching x and y, $y = \sqrt{3x - 3a}$

Thus, $y = \sqrt{3x - 3a}$ is an inverse of $y = \frac{1}{3}x^2 + a$

$\therefore$ The intersection points of the graphs of $y = f(x)$ and $y = g(x)$ are equal to the

intersection points of the graph of $y = f(x)$ and a line $y = x$.

$\therefore \frac{1}{3}x^2 + a = x$; $x^2 - 3x + 3a = 0$ $\cdots\cdots$ ①

Since $3x - 3a \geq 0$, $x \geq a$

Since $a \geq 0$, $x \geq 0$

Thus, the equation ① has two different real number solutions which are greater than or

equal to zero.

i) Let D be the discriminant of ①. Then, $D > 0$

$\qquad \therefore D = (-3)^2 - 12a > 0$ $\qquad \therefore a < \frac{3}{4}$

ii) The product of the two solutions is greater than or equal to zero ; i.e., $3a \geq 0$.

$\qquad \therefore a \geq 0$

$\qquad$ By i) and ii), $0 \leq a < \frac{3}{4}$

(6) For two points $A = A(2, 3)$ and $B = B(3, 2)$, the segment $\overline{AB}$ intersects the graph of

$y = \sqrt{ax + 2}$, $a > 0$. Find the range of a.

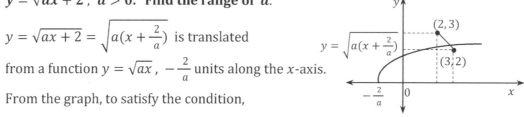

$y = \sqrt{ax + 2} = \sqrt{a(x + \frac{2}{a})}$ is translated

from a function $y = \sqrt{ax}$, $-\frac{2}{a}$ units along the x-axis.

From the graph, to satisfy the condition,

a point $A(2, 3)$ must lie on the graph of $y = \sqrt{ax + 2}$ or above the graph.

$\therefore 3 \geq \sqrt{2a + 2}$; $9 \geq 2a + 2$ $\qquad \therefore a \leq \frac{7}{2}$

Since $a > 0$, $0 < a \leq \frac{7}{2}$ $\cdots\cdots$ ①

Also, the other point $B = B(3, 2)$ must lie on the graph of y or below the graph.

$\therefore 2 \leq \sqrt{3a + 2}$; $4 \leq 3a + 2$ $\qquad \therefore a \geq \frac{2}{3}$ $\cdots\cdots$ ②

By ① and ②, $\frac{2}{3} \leq a \leq \frac{7}{2}$

#63 Find the value.

(1) For a fractional equation $\dfrac{x^2+2x}{x-1} - \dfrac{16(x-1)}{x(x+2)} - 6 = 0$, find the sum of all real number

solutions.

$\dfrac{x^2+2x}{x-1} - \dfrac{16(x-1)}{x(x+2)} - 6 = 0 \Rightarrow \dfrac{x(x+2)}{x-1} - \dfrac{16(x-1)}{x(x+2)} - 6 = 0$

Let $\dfrac{x(x+2)}{x-1} = X$.

Then, $X - \dfrac{16}{X} - 6 = 0$

$\therefore\ X^2 - 6X - 16 = 0 \quad \therefore\ (X-8)(X+2) = 0 \quad \therefore\ X = 8$ or $X = -2$

i)   When $\dfrac{x(x+2)}{x-1} = 8$, $\ x^2 + 2x = 8x - 8 \quad \therefore\ x^2 - 6x + 8 = (x-2)(x-4) = 0$

   $\therefore\ x = 2$ or $x = 4$

ii) When $\dfrac{x(x+2)}{x-1} = -2$, $\ x^2 + 2x = -2x + 2 \quad \therefore\ x^2 + 4x - 2 = 0$

   $\therefore\ x = \dfrac{-2\pm\sqrt{4+2}}{1} = -2 \pm \sqrt{6}$

   Therefore, the sum of all real number solutions is

   $2 + 4 + \left(-2 + \sqrt{6}\right) + \left(-2 - \sqrt{6}\right) = 2$

(2) For a fractional equation $\dfrac{x^2+x+1}{x-2} - \dfrac{x+2}{x-1} = \dfrac{3}{(x-1)(x-2)} - 2$, find the sum of all real

number solutions.

Multiplying both sides by the LCM, $(x-1)(x-2)$,

$(x^2 + x + 1)(x-1) - (x+2)(x-2) = 3 - 2(x-1)(x-2)$

$\therefore\ x^3 - x^2 + x^2 - x + x - 1 - x^2 + 4 = 3 - 2x^2 + 6x - 4$

$\therefore\ x^3 + x^2 - 6x + 4 = 0$

Let $f(x) = x^3 + x^2 - 6x + 4$

Then, $f(1) = 0$

$$
\begin{array}{r|rrrr}
1 & 1 & 1 & -6 & 4 \\
  &   & 1 & 2 & -4 \\
\hline
  & 1 & 2 & -4 & 0
\end{array}
$$
$\qquad \therefore\ f(x) = (x-1)(x^2 + 2x - 4)$

$\therefore\ x = 1$ or $x^2 + 2x - 4 = 0$

Since $x = 1$ makes the denominators of the given fractional equation zero,

$x = 1$ is an extraneous root.   Thus, $x = 1$ is rejected. $\therefore\ x^2 + 2x - 4 = 0 \cdots\cdots ①$

Let α and β be roots of ①.   Then, $\alpha + \beta = -2$, $\alpha\beta = -4$

Therefore, the sum of all roots is -2.

(3) When the fractional equation $\dfrac{2x}{x^2-4} - \dfrac{1}{x+2} = a$ has no solution, find the value of a.

Multiplying both sides by the LCM, $(x+2)(x-2)$, $2x - (x-2) = a(x+2)(x-2)$

$\therefore\ x + 2 = ax^2 - 4a \qquad \therefore\ x + 2 = a(x+2)(x-2) \qquad \therefore\ (x+2)(a(x-2)-1) = 0$

$\therefore\ x = -2\ $ or $\ ax - 2a - 1 = 0$

Since $x = -2$ is an extraneous root, it is rejected. Thus, $ax - 2a - 1 = 0$

To satisfy the condition, $ax - 2a - 1 = 0$ has no solution or makes the denominator zero.

i) $ax - 2a - 1 = 0$

   If $a = 0$, then there is no solution.

ii) $ax - 2a - 1 = 0$

   $\Rightarrow\ x = \dfrac{2a+1}{a},\ a \neq 0$

   If $x = -2$ or $x = 2$, then the denominators will be zero.

   $\therefore\ \dfrac{2a+1}{a} = -2\ $ or $\ \dfrac{2a+1}{a} = 2$

   $\therefore\ 4a = -1 \qquad \therefore\ a = -\dfrac{1}{4}$

Therefore, by i) and ii), $a = 0$ or $a = -\dfrac{1}{4}$

(4) For a fractional equation $x^2 + \dfrac{1}{x^2} = \dfrac{x^2+1}{x}$, find the value of the real number solution.

$x^2 + \dfrac{1}{x^2} = \dfrac{x^2+1}{x} \quad \Rightarrow \quad x^2 + \dfrac{1}{x^2} = x + \dfrac{1}{x}$

$\Rightarrow\ \left(x + \dfrac{1}{x}\right)^2 - 2 = x + \dfrac{1}{x}$

Let $x + \dfrac{1}{x} = X$

Then, $X^2 - 2 = X \qquad \therefore\ X^2 - X - 2 = (X-2)(X+1) = 0 \qquad \therefore\ X = 2$ or $X = -1$

i) $X = 2$

   $\Rightarrow\ x + \dfrac{1}{x} = 2 \qquad \therefore\ x^2 - 2x + 1 = 0 \qquad \therefore\ (x-1)^2 = 0 \qquad \therefore\ x = 1$

ii) $X = -1$

   $\Rightarrow\ x + \dfrac{1}{x} = -1 \qquad \therefore\ x^2 + x + 1 = 0 \qquad \therefore\ x = \dfrac{-1 \pm \sqrt{1-4}}{2} = \dfrac{-1 \pm \sqrt{3}\,i}{2}$

Therefore, the real number solution is $x = 1$.

(5) For a fractional equation $\dfrac{1}{(x-1)x} + \dfrac{1}{x(x+1)} + \dfrac{1}{(x+1)(x+2)} = \dfrac{3}{4}$, let α and β be the two

roots. Find the value of $\dfrac{\beta}{\alpha} + \dfrac{\alpha}{\beta}$.

$\dfrac{1}{(x-1)x} + \dfrac{1}{x(x+1)} + \dfrac{1}{(x+1)(x+2)} = \dfrac{3}{4}$

$\Rightarrow\ \left(\dfrac{1}{x-1} - \dfrac{1}{x}\right) + \left(\dfrac{1}{x} - \dfrac{1}{x+1}\right) + \left(\dfrac{1}{x+1} - \dfrac{1}{x+2}\right) = \dfrac{3}{4}$

$$\Rightarrow \quad \frac{1}{x-1} - \frac{1}{x+2} = \frac{3}{4}$$

Multiplying both sides by LCM, $(x-1)(x+2)$, $(x+2) - (x-1) = \frac{3}{4}(x-1)(x+2)$

$\therefore \quad 3(x-1)(x+2) = 12$

$\therefore \quad x^2 + x - 2 = 4 \quad \therefore \quad x^2 + x - 6 = 0$

$\therefore \quad \alpha + \beta = -1$ and $\alpha\beta = -6$ by the relationship between roots and coefficients

$\therefore \quad \frac{\beta}{\alpha} + \frac{\alpha}{\beta} = \frac{\alpha^2 + \beta^2}{\alpha\beta} = \frac{(\alpha+\beta)^2 - 2\alpha\beta}{\alpha\beta} = \frac{1+12}{-6} = -\frac{13}{6}$

(6) **There is a $a\%$ salt solution containing 30 ounces of salt. If 50 ounces of water is evaporated, then it will be a $(a+2)\%$ salt solution. Find the value of a.**

Let the beginning salt water be x ounces.

Then, $\frac{30}{x} = a\%$; i.e., $\frac{30}{x} = \frac{a}{100} \quad \therefore \quad x = \frac{3000}{a} \quad \cdots\cdots$ ①

If 50 ounces of water is evaporated, $\frac{30}{x-50} = (a+2)\%$; i.e., $\frac{30}{x-50} = \frac{a+2}{100}$

$\therefore \quad (x-50)(a+2) = 3000 \quad \therefore \quad x = \frac{3000}{a+2} + 50 \quad \cdots\cdots$ ②

By ① and ②, $\frac{3000}{a} = \frac{3000}{a+2} + 50$

$\therefore \quad 3000(a+2) = 3000a + 50a(a+2)$

$\therefore \quad 60(a+2) = 60a + a(a+2)$

$\therefore \quad a^2 + 2a - 120 = (a+12)(a-10) = 0 \quad \therefore \quad a = -12 \text{ or } a = 10$

Since $a > 0$, $a = 10$

(7) **For a fractional equation $f(x) = \frac{-x+3}{x-2}$ and its inverse function $g(x)$, find the two different real number roots such that $f(x) + g(x) = 0$.**

Let $y = \frac{-x+3}{x-2}$

Switching x and y, $x = \frac{-y+3}{y-2}$; $xy - 2x = -y + 3$

Solving for y, $y = \frac{2x+3}{x+1}$

$\therefore \quad g(x) = \frac{2x+3}{x+1}$

Since $f(x) + g(x) = 0$, $\frac{-x+3}{x-2} + \frac{2x+3}{x+1} = 0$

$\therefore \quad (-x+3)(x+1) + (2x+3)(x-2) = 0$

$\therefore \quad -x^2 + 2x + 3 + 2x^2 - x - 6 = x^2 + x - 3 = 0$

$\therefore \quad x = \frac{-1 \pm \sqrt{1+12}}{2} = \frac{-1 \pm \sqrt{13}}{2}$

Therefore, $x = \frac{-1+\sqrt{13}}{2}$ or $x = \frac{-1-\sqrt{13}}{2}$

(8) When a fractional equation $\dfrac{1}{x-1} - \dfrac{1}{x} - \dfrac{1}{x+3} + \dfrac{1}{x+4} = 0$ **has a root** α,

find the value of $2\alpha + 3$.

$$\frac{1}{x-1} - \frac{1}{x} - \frac{1}{x+3} + \frac{1}{x+4} = 0$$

$$\Rightarrow \frac{1}{x-1} - \frac{1}{x} = \frac{1}{x+3} - \frac{1}{x+4}$$

$$\therefore \frac{1}{x(x-1)} = \frac{1}{(x+3)(x+4)}$$

Multiplying both sides by LCM, $x(x-1)(x+3)(x+4)$, $(x+3)(x+4) = x(x-1)$

$$\therefore \ 7x + 12 = -x \qquad \therefore \ 8x = -12 \qquad \therefore \ x = -\frac{3}{2}$$

Since $x = -\dfrac{3}{2}$ does not make the denominators zero, $x = -\dfrac{3}{2}$ is a root.

$$\therefore \ \alpha = -\frac{3}{2}$$

Therefore, $2\alpha + 3 = 2\left(-\dfrac{3}{2}\right) + 3 = 0$

(9) When a fractional equation $\dfrac{x^2 + 2x + a - 11}{x-2} = a$ **has no real number solution,**

find the sum of all possible integers for a.

Multiplying both sides by LCM, $x - 2$, $x^2 + 2x + a - 11 = a(x - 2)$

$$\therefore \ x^2 + (2 - a)x + 3a - 11 = 0 \ \cdots\cdots \text{①}$$

For the equation ① to have no real number solution, the equation ① has imaginary

solutions or $x = 2$ (extraneous root) as a double root.

i) Let D be the discriminant of the equation ①.

If $D < 0$, then ① has imaginary solutions.

$$\therefore \ D = (2 - a)^2 - 4(3a - 11) = a^2 - 4a + 4 - 12a + 44$$

$$= a^2 - 16a + 48 = (a - 12)(a - 4) < 0$$

$$\therefore \ 4 < a < 12$$

ii) If $x = 2$ (extraneous root) as a double root, $4 + (2 - a)2 + 3a - 11 = 0$

$$\therefore \ a = 3$$

By i) and ii), $a = 3, 5, 6, 7, 8, 9, 10, 11$

Therefore, the sum is 59.

(10)

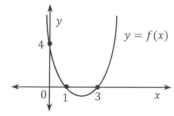

When the graph of a function $y = f(x)$ **is**

shown as the Figure, $g(x) = \dfrac{1}{x} - \dfrac{2}{x+4}$.

Find the number of real roots

such that $g(f(x)) = 0$.

Let $f(x) = t$. Then, $g(f(x)) = g(t) = 0$

$\therefore \dfrac{1}{t} - \dfrac{2}{t+4} = 0$ $\qquad \therefore \dfrac{-t+4}{t(t+4)} = 0$ $\qquad \therefore t = 4$ $\qquad \therefore f(x) = 4$

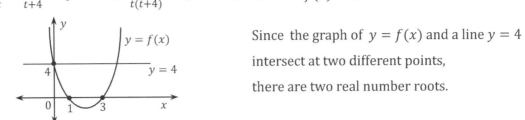

Since the graph of $y = f(x)$ and a line $y = 4$

intersect at two different points,

there are two real number roots.

(11) For a function $f(x) = x^2 - 5x + 4$, the fractional equation $\dfrac{f(x)}{x+a} - \dfrac{f(x)}{x+b} = 0$ has no

solution. Find the value of $(a - b)^2$ (where a and b are real numbers).

$\dfrac{f(x)}{x+a} - \dfrac{f(x)}{x+b} = 0 \ \Rightarrow \ \dfrac{f(x)(x+b) - f(x)(x+a)}{(x+a)(x+b)} = 0 \ \Rightarrow \ \dfrac{f(x)(b-a)}{(x+a)(x+b)} = 0 \ \cdots\cdots ①$

If $a = b$, $f(x) = 0$ from ①.

For ① to have no solution, $x = -a$ and $x = -b$ are the solutions of $f(x) = 0$.

By the relationship between roots and coefficients, $-a - b = 5$, $(-a)(-b) = 4$

That is, $a + b = -5$ and $ab = 4$

Therefore, $(a - b)^2 = (a + b)^2 - 4ab = 25 - 16 = 9$

#64 Find the value.

(1) Find the value of real number a so that the radical equation $\sqrt{1 - x^2} = x + a$ has a

real number solution.

Let $y = \sqrt{1 - x^2}$, $y = x + a$

From the graph,

i) When the line $y = x + a$ and the graph of $y = \sqrt{1 - x^2}$ intersect at one point,

$\sqrt{1 - x^2} = x + a \Rightarrow 1 - x^2 = (x + a)^2$ (Square both sides)

$\Rightarrow 2x^2 + 2ax + a^2 - 1 = 0 \cdots\cdots ①$

Since the discriminant D of the equation ① is $D = 0$,

$D = a^2 - 2(a^2 - 1) = -a^2 + 2 = 0$

$\therefore a = \pm\sqrt{2}$

Since $a > 0$, $a = \sqrt{2}$

ii) When the line $y = x + a$ passes through a point $(1, 0)$,

$0 = 1 + a \qquad \therefore a = -1$

Therefore, by i) and ii), $-1 \leq a \leq \sqrt{2}$

(2) When a radical equation $x^2 - 2x + \sqrt{x^2 - 2x + a} = 4$ has a root 3, find the value of the other root.

Since $x = 3$ is a root, $9 - 6 + \sqrt{9 - 6 + a} = 4$

$\therefore \sqrt{3 + a} = 1 \qquad \therefore 3 + a = 1 \qquad \therefore a = -2$

Thus, the given function is $x^2 - 2x + \sqrt{x^2 - 2x - 2} = 4 \ \cdots\cdots ①$

Let $\sqrt{x^2 - 2x - 2} = t$. Then, $t \geq 0$ and $x^2 - 2x - 2 = t^2$; $x^2 - 2x = t^2 + 2$

From ①, $t^2 + 2 + t = 4 \qquad \therefore t^2 + t - 2 = (t + 2)(t - 1) = 0 \qquad \therefore t = -2$ or $t = 1$

Since $t \geq 0$, $t = 1$

Thus, $\sqrt{x^2 - 2x - 2} = 1$; $x^2 - 2x - 2 = 1$; $x^2 - 2x - 3 = 0$; $(x - 3)(x + 1) = 0$

$\therefore x = -1$ or $x = 3$ (Note that these roots are not extraneous roots.)

Therefore, the other root is $x = -1$.

(3) For a radical equation $\sqrt{(a + 2)x + a + 3} = x$, find minimum value of real number a so that the equation has a real number root.

$$\sqrt{(a + 2)x + a + 3} = x \quad \Rightarrow \quad (a + 2)x + a + 3 = x^2 \quad \Rightarrow \quad x^2 - (a + 2)x - a - 3 = 0$$

$$\Rightarrow \quad (x + 1)(x - a - 3) = 0 \quad \Rightarrow \quad x = -1 \text{ or } x = a + 3$$

Since $\sqrt{(a + 2)x + a + 3} \geq 0$, $x \geq 0$

$\therefore x = -1$ is an extraneous root, and is rejected. $\qquad \therefore x = a + 3$

$\therefore \sqrt{(a + 2)(a + 3) + a + 3} = a + 3$; $\sqrt{(a + 3)\big((a + 2) + 1\big)} = a + 3$

$\therefore \sqrt{(a + 3)^2} = a + 3 \qquad \therefore |a + 3| = a + 3$

Since $a + 3 \geq 0$, $a \geq -3$ $\qquad$ Therefore, minimum value of a is -3.

(4) When a radical equation $\sqrt{x + 5} = x + a$ has an extraneous root – 1, find the real number root of the equation (where a is a constant).

Since $x = -1$ is an extraneous root,

the equation $-\sqrt{x + 5} = x + a$ has a solution $x = -1$.

$\therefore -\sqrt{(-1) + 5} = (-1) + a \quad \therefore -2 = a - 1 \quad \therefore a = -1$

Thus, $\sqrt{x + 5} = x - 1$; $x + 5 = (x - 1)^2 = x^2 - 2x + 1$

$\therefore x^2 - 3x - 4 = (x - 4)(x + 1) = 0 \quad \therefore x = 4$ or $x = -1$

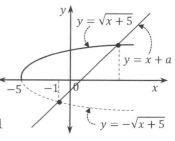

Since $x = -1$ is an extraneous root, the real number root is $x = 4$.

(5) For a radical equation $\sqrt{x^2 - 3x + 1} - \sqrt{x^2 - 3x} = \frac{1}{2}$, find the sum of all roots.

Let $\sqrt{x^2 - 3x} = t$. Then, $t \geq 0$ and $x^2 - 3x = t^2$

$\therefore \sqrt{t^2 + 1} - t = \frac{1}{2}$ $\therefore \sqrt{t^2 + 1} = \frac{1}{2} + t$ $\therefore t^2 + 1 = (\frac{1}{2} + t)^2 = t^2 + t + \frac{1}{4}$

$\therefore t = \frac{3}{4}$ Thus, $\sqrt{x^2 - 3x} = \frac{3}{4}$

Squaring both sides, $x^2 - 3x = \frac{9}{16}$ $\therefore 16x^2 - 48x - 9 = 0$

$\therefore x = \frac{24 \pm \sqrt{(24)^2 + 16 \cdot 9}}{16} = \frac{24 \pm \sqrt{720}}{16} = \frac{24 \pm 12\sqrt{5}}{16} = \frac{6 \pm 3\sqrt{5}}{4}$

Since $x^2 - 3x \geq 0$, $x(x - 3) \geq 0$ $\therefore x \geq 3$ or $x \leq 0$

Since $x = \frac{6 + 3\sqrt{5}}{4}$ and $x = \frac{6 - 3\sqrt{5}}{4}$ are not extraneous roots, these are roots.

Therefore, the sum is $\frac{6 + 3\sqrt{5}}{4} + \frac{6 - 3\sqrt{5}}{4} = \frac{12}{4} = 3$

(6) For a function $f(x) = \sqrt{2x + 1} + x$,

find all real number solutions such that $f(f(x)) = 1$.

Let $f(x) = t$.

Then, $f(f(x)) = f(t) = 1$

From the given function $f(x) = \sqrt{2x + 1} + x$, $f(t) = \sqrt{2t + 1} + t = 1$ $\therefore \sqrt{2t + 1} = 1 - t$

Squaring both sides, $2t + 1 = (1 - t)^2$ $\therefore t^2 - 4t = t(t - 4) = 0$ $\therefore t = 0$ or $t = 4$

i)   When $t = 0$; i.e., $f(x) = \sqrt{2x + 1} + x = 0$,

$\sqrt{2x + 1} = -x$ $\therefore 2x + 1 = x^2$ $\therefore x^2 - 2x - 1 = 0$

$\therefore x = \frac{1 \pm \sqrt{1 + 1}}{1} = 1 \pm \sqrt{2}$

Since $\sqrt{2x + 1} = -x \geq 0$, $x \leq 0$

$\therefore x = 1 + \sqrt{2}$ is an extraneous root. $\therefore x = 1 - \sqrt{2}$

ii)   When $t = 4$; i.e., $f(x) = \sqrt{2x + 1} + x = 4$,

$\sqrt{2x + 1} = 4 - x$ $\therefore 2x + 1 = x^2 - 8x + 16$ $\therefore x^2 - 10x + 15 = 0$

$\therefore x = \frac{5 \pm \sqrt{25 - 15}}{1} = 5 \pm \sqrt{10}$

Since $\sqrt{2x + 1} = 4 - x \geq 0$, $x \leq 4$

$\therefore x = 5 + \sqrt{10}$ is an extraneous root. $\therefore x = 5 - \sqrt{10}$

Therefore, by i) and ii), $x = 1 - \sqrt{2}$, $x = 5 - \sqrt{10}$

#65 Find the range.

(1) For a radical equation $\sqrt{x+1} = |x+a|$, find the range of a so that the product of the two real number roots of the equation is negative.

$\sqrt{x+1} = |x+a| \Rightarrow x+1 = (x+a)^2 \quad \therefore x^2 + (2a-1)x + a^2 - 1 = 0$

Since the eqaution has two different real number solutions, the discriminant D of the equation is $D > 0$.

$\therefore (2a-1)^2 - 4(a^2-1) = -4a + 5 > 0 \quad \therefore a < \dfrac{5}{4} \cdots\cdots ①$

Since the product of the two roots is negative,

$a^2 - 1 < 0$ (by the relationship between roots and coefficient)

$\therefore a^2 < 1 \quad \therefore -1 < a < 1 \cdots\cdots ②$

Therefore, by ① and ②, $-1 < a < 1$

(2) For a positive real number a, find the range of a so that the radical equation $\sqrt{2a+x} + \sqrt{2a-x} = 4a$ has a real number solution.

From the given equation, $2a+x \geq 0$, $2a-x \geq 0$. $\quad \therefore -2a \leq x \leq 2a$

$\sqrt{2a+x} + \sqrt{2a-x} = 4a \Rightarrow (2a+x) + 2 \cdot \sqrt{2a+x} \cdot \sqrt{2a-x} + (2a-x) = 16a^2$

$\therefore \sqrt{2a+x} \cdot \sqrt{2a-x} = 8a^2 - 2a \cdots\cdots ①$

Squaring both sides, $(2a+x)(2a-x) = 64a^4 - 32a^3 + 4a^2$

$\therefore 4a^2 - x^2 = 64a^4 - 32a^3 + 4a^2 \; ; \; x^2 = -64a^4 + 32a^3 = 32a^3(1-2a) \cdots\cdots ②$

From ①, $8a^2 - 2a \geq 0 \quad \therefore 2a(4a-1) \geq 0 \quad \therefore a \geq \dfrac{1}{4}$ or $a \leq 0$

Since $a > 0$, $a \geq \dfrac{1}{4}$

From ②, $1 - 2a \geq 0 \quad (\because a^3 > 0) \quad \therefore a \leq \dfrac{1}{2}$

Therefore, $\dfrac{1}{4} \leq a \leq \dfrac{1}{2}$

#66 A container is filled with 100 ounces of 8% salt water solution. Suppose 6 ounces of water is evaporated per day. How many days do we need to have more than 10% salt water solution?

Let x be the amount of salt. Then, $\dfrac{x}{100} = 8\% \quad \therefore x = 8$

Thus, 100 ounces of 8% salt water solution contains 8 ounces of salt.

Since 6 ounces of water is evaporated per day, $6y$ ounces of water will be evaporated after y days.

$\therefore \dfrac{8}{100-6y} \geq 10\%$; $\dfrac{8}{100-6y} \geq \dfrac{10}{100}$ $\quad \therefore \dfrac{4}{50-3y} \geq \dfrac{1}{10}$; $\dfrac{40-(50-3y)}{10(50-3y)} \geq 0$; $\dfrac{-10+3y}{10(50-3y)} \geq 0$

$\therefore \dfrac{10-3y}{10(50-3y)} \leq 0$; $\dfrac{10-3y}{50-3y} \leq 0$ $\quad \therefore (10-3y)(50-3y) \leq 0$, $y \neq \dfrac{50}{3}$

$\therefore \dfrac{10}{3} \leq y < \dfrac{50}{3}$

Therefore, at least 4 days are needed.

#67 Solve each inequality.

(1) $x^4 + x^3 \leq x^2 + x$

$x^4 + x^3 \leq x^2 + x \quad \Rightarrow \quad x^4 + x^3 - x^2 - x \leq 0 \quad \Rightarrow \quad x(x^3 + x^2 - x - 1) \leq 0$

Let $f(x) = x^3 + x^2 - x - 1$

$\qquad = x^2(x+1) - (x+1) = (x+1)(x^2-1) = (x+1)^2(x-1)$

Then, $x(x+1)^2(x-1) \leq 0$

Since $(x+1)^2 \geq 0$, $x(x-1) \leq 0$ or $x = -1$

$\therefore 0 \leq x \leq 1$ or $x = -1$

(2) $\dfrac{2}{x-1} + \dfrac{3}{x-2} + \dfrac{2}{(x-1)(x-2)} + 3 \leq 0$

$\dfrac{2}{x-1} + \dfrac{3}{x-2} + \dfrac{2}{(x-1)(x-2)} + 3 \leq 0 \quad \Rightarrow \quad \dfrac{2(x-2)+3(x-1)+2+3(x-1)(x-2)}{(x-1)(x-2)} \leq 0$

$\Rightarrow \dfrac{3x^2-4x+1}{(x-1)(x-2)} \leq 0 \quad \Rightarrow \quad \dfrac{(3x-1)(x-1)}{(x-1)(x-2)} \leq 0$

$\therefore (3x-1)(x-2) \leq 0$, $x \neq 1$, $x \neq 2$

$\therefore \dfrac{1}{3} \leq x \leq 2$, $x \neq 1$, $x \neq 2$

Therefore, $\dfrac{1}{3} \leq x < 1$ or $1 < x < 2$

(3) $\dfrac{x^2-x-30}{|x(x-1)|} \leq 0$

Since $|x(x-1)| \geq 0$, the given inequality is $x^2 - x - 30 \leq 0$, $x \neq 0$, $x \neq 1$

$\therefore (x-6)(x+5) \leq 0$, $x \neq 0$, $x \neq 1$

$\therefore -5 \leq x \leq 6$, $x \neq 0$, $x \neq 1$

$\therefore -5 \leq x < 0$, $0 < x < 1$, $1 < x \leq 6$

(4) $(|x| - 1)(x^2 - 2|x| - 15) \leq 0$

$(|x| - 1)(x^2 - 2|x| - 15) \leq 0 \Rightarrow (|x| - 1)(|x|^2 - 2|x| - 15) \leq 0$

$\Rightarrow (|x| - 1)(|x| - 5)(|x| + 3) \leq 0$

Since $|x| \geq 0$, $|x| + 3 \geq 0$

$\therefore (|x| - 1)(|x| - 5) \leq 0 \qquad \therefore 1 \leq |x| \leq 5$

$\therefore 1 \leq x \leq 5$ or $1 \leq -x \leq 5$

$\therefore 1 \leq x \leq 5$ or $-5 \leq x \leq -1$

(5) $\begin{cases} (x-1)(x-2)^2(x-4) \geq 0 \\ \dfrac{2x-3}{x+2} \leq 1 \end{cases}$

Since $(x-2)^2 \geq 0$, $(x-1)(x-2)^2(x-4) \geq 0 \Leftrightarrow (x-1)(x-4) \geq 0$ or $x = 2$

$\therefore x \geq 4$ or $x \leq 1$ or $x = 2 \cdots\cdots$①

$\dfrac{2x-3}{x+2} \leq 1 \Rightarrow \dfrac{2x-3-(x+2)}{x+2} \leq 0 \Rightarrow \dfrac{x-5}{x+2} \leq 0$

$\therefore (x-5)(x+2) \leq 0$, $x \neq -2$

$\therefore -2 < x \leq 5 \cdots\cdots$②

By ① and ②,

Therefore, $-2 < x \leq 1$ or $4 \leq x \leq 5$ or $x = 2$

#68 Find the range.

(1) For any real number x, an inequality $x^4 - x^3 - x^2 - 2x < 0$ is always true. Find the range of the real number a so that a function $f(x) = -x^2 + x + a$ has always positive values.

$x^4 - x^3 - x^2 - 2x < 0 \Rightarrow x(x^3 - x^2 - x - 2) < 0$

Let $g(x) = x^3 - x^2 - x - 2$

Then, $g(2) = 8 - 4 - 2 - 2 = 0$

$$\begin{array}{c|cccc} 2 & 1 & -1 & -1 & -2 \\ & & 2 & 2 & 2 \\ \hline & 1 & 1 & 1 & 0 \end{array} \qquad g(x) = (x-2)(x^2 + x + 1)$$

$\therefore x(x-2)(x^2 + x + 1) < 0$

Since $x^2 + x + 1 = \left(x + \dfrac{1}{2}\right)^2 - \dfrac{1}{4} + 1 = \left(x + \dfrac{1}{2}\right)^2 + \dfrac{3}{4} > 0$, $x(x-2) < 0$

$\therefore 0 < x < 2$

Thus, $f(x) = -x^2 + x + a > 0$ for all x in the interval $0 < x < 2$.

$\therefore\ f(x) = -x^2 + x + a = -\left(x - \dfrac{1}{2}\right)^2 + \dfrac{1}{4} + a > 0$

Since $f(0) = a \geq 0$ and $f(2) = -2 + a \geq 0$, $a \geq 2$

Therefore, $a \geq 2$

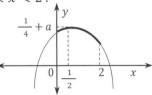

(2) When an inequality $(x + 1)(x^2 + ax + a) < 0$ has solution $x < -1$,

find the range of the real number a.

When $x^2 + ax + a \geq 0$ for any real number x,

$(x + 1)(x^2 + ax + a) < 0 \Rightarrow (x + 1) < 0 \qquad \therefore\ x < -1$

Thus, $x^2 + ax + a \geq 0$ is true.

Let D be the discriminant of the equation $x^2 + ax + a = 0$. Then, $D \leq 0$

$\therefore\ a^2 - 4a \leq 0$; $a(a - 4) \leq 0 \qquad \therefore\ 0 \leq a \leq 4$

(3) For any x, $(x - 1)(x - 2)(x - 4)^2 \leq 0$. Find the range of a such that $x^2 - 4x + a \leq 0$

Since $(x - 4)^2 \geq 0$,

$(x - 1)(x - 2)(x - 4)^2 \leq 0 \iff (x - 1)(x - 2) \leq 0$ or $x = 4$

$\therefore\ 1 \leq x \leq 2$ or $x = 4$

Let $f(x) = x^2 - 4x + a$. Then, $f(x) = (x - 2)^2 - 4 + a$

The axis of symmetry is $x = 2$

To have the inequality $x^2 - 4x + a \leq 0$ in the range: $(1 \leq x \leq 2$ or $x = 4)$, $f(4) \leq 0$

Since $f(4) = a$, $a \leq 0$

(4) For a fractional function $f(x) = \dfrac{x}{x+2}$, $(f \circ f)(x) \geq \dfrac{1}{2}$. Find the range of x.

$(f \circ f)(x) = f\big(f(x)\big) = \dfrac{\frac{x}{x+2}}{\frac{x}{x+2}+2} = \dfrac{\frac{x}{x+2}}{\frac{x+2x+4}{x+2}} = \dfrac{x}{3x+4} \geq \dfrac{1}{2}$

$\therefore\ \dfrac{x}{3x+4} - \dfrac{1}{2} \geq 0$; $\dfrac{2x-(3x+4)}{2(3x+4)} \geq 0 \qquad \therefore\ (-x - 4)(3x + 4) \geq 0,\ x \neq -\dfrac{4}{3}$

$\therefore\ (x + 4)(3x + 4) \leq 0,\ x \neq -\dfrac{4}{3} \qquad \therefore\ -4 \leq x < -\dfrac{4}{3}$

Since $x \neq -2$, $-4 \leq x < -2$ or $-2 < x < -\dfrac{4}{3}$

(5) When there are three integers for x so that a system of inequalities

$\begin{cases} x^3 + x^2 - 2x - 2 > 0 \\ x^2 - (a + 1)x + a \leq 0 \end{cases}$ is always true, find the range of real number a.

$x^3 + x^2 - 2x - 2 > 0 \iff x^2(x + 1) - 2(x + 1) > 0$

$\iff (x + 1)(x^2 - 2) > 0 \iff (x + 1)(x + \sqrt{2})(x - \sqrt{2}) > 0$

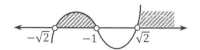 $\therefore \ -\sqrt{2} < x < -1 \ $ or $\ x > \sqrt{2}$

$x^2 - (a+1)x + a \le 0 \iff (x-1)(x-a) \le 0$

i) If $a > 1$, then $1 \le x \le a$

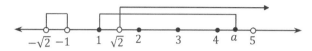

To have three integers, $4 \le a < 5$

ii) If $a \le 1$, then $a \le x \le 1$

In this case, there is no integer x such that the system.

Therefore, $4 \le a < 5$

(6) When there are two integers for x such that $x(x-a)(x-3)^2 \le 0$, find the range of real number a.

Consider following cases:

$a < 0, \quad a = 0, \quad 0 < a < 3, \quad a = 3, \quad a > 3$

i) When $a < 0$,

 $\therefore \ a \le x \le 0 \ $ or $\ x = 3$

To have two integers, $\boxed{-1 < a < 0}$

ii) When $a = 0$,

$x(x-a)(x-3)^2 \le 0 \iff x^2(x-3)^2 \le 0 \qquad \therefore \ x = 0 \text{ or } x = 3$

Since $x = 0$ and $x = 3$ are two integers, $\boxed{a = 0}$

iii) When $0 < a < 3$,

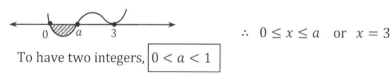 $\therefore \ 0 \le x \le a \ $ or $\ x = 3$

To have two integers, $\boxed{0 < a < 1}$

iv) When $a = 3$,

$x(x-a)(x-3)^2 \le 0 \iff x(x-3)^3 \le 0 \iff x(x-3) \le 0$

$\therefore \ 0 \le x \le 3$

Since $x = 0, 1, 2, 3$ (4 integers), $\boxed{a \ne 3}$

v)   When $a > 3$,

$\therefore$ $0 \leq x \leq a$

Since there are more than two integers, $\boxed{a \not> 3}$

Therefore, by i), ii), iii), iv), and v),

$\therefore$ $-1 < a < 1$

(7) If an inequality $(|x| + 1)(x + a) < 0$ is true for all x, then $x^3 - x^2 - 4x + 4 < 0$ is also true. Find the range of real number a.

Since $|x| + 1 > 0$,

$(|x| + 1)(x + a) < 0 \Rightarrow x + a < 0$ $\therefore$ $x < -a$

$x^3 - x^2 - 4x + 4 < 0 \Leftrightarrow x^2(x - 1) - 4(x - 1) < 0 \Leftrightarrow (x - 1)(x^2 - 4) < 0$

$\Leftrightarrow (x - 1)(x + 2)(x - 2) < 0$

$\therefore$ $x < -2$ or $1 < x < 2$

Since ($x < -a \Rightarrow x < -2$ or $1 < x < 2$),

the range $(-\infty, -a)$ is included in the range $(-\infty, -2) \cup (1, 2)$ $\therefore$ $-a \leq -2$

Therefore, $a \geq 2$

(8) For two quadratic expressions $f(x)$ and $g(x)$ with leading coefficients 1, their GCF and LCM are $x + 2$ and $x(x + 2)(x - 4)$, respectively.

Find the range of x such that $\dfrac{1}{f(x)} + \dfrac{1}{g(x)} \leq 0$

Since GCF is $x + 2$,

$f(x) = (x + 2)A$ and $g(x) = (x + 2)B$ where A and B are disjoint linear expressions.

Since LCM is $x(x + 2)(x - 4)$, $AB(x + 2) = x(x + 2)(x - 4)$

$\therefore$ $f(x) = x(x + 2), \; g(x) = (x + 2)(x - 4)$ or $f(x) = (x + 2)(x - 4), \; g(x) = x(x + 2)$

$\dfrac{1}{f(x)} + \dfrac{1}{g(x)} \leq 0 \Leftrightarrow \dfrac{1}{x(x+2)} + \dfrac{1}{(x+2)(x-4)} \leq 0 \Leftrightarrow \dfrac{x-4+x}{x(x+2)(x-4)} \leq 0$

$\Leftrightarrow 2(x - 2)x(x + 2)(x - 4) \leq 0, \; x \neq 0, \; x \neq -2, \; x \neq 4$

$\therefore$ $-2 < x < 0$ or $2 \leq x < 4$

(9)

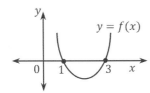

The graph of a function $y = f(x)$ is shown as the Figure.

Find the range of x such that $\dfrac{f(x-2)}{f(x)} \le 0$

From the graph, $f(x) = a(x-1)(x-3)$, $a > 0$

$\therefore$ $f(x-2) = a(x-3)(x-5)$, $a > 0$

$\therefore$ $\dfrac{f(x-2)}{f(x)} \le 0 \iff \dfrac{a(x-3)(x-5)}{a(x-1)(x-3)} \le 0 \iff \dfrac{(x-3)(x-5)}{(x-1)(x-3)} \le 0$

$\qquad\qquad \iff (x-1)(x-3)^2(x-5) \le 0$, $x \ne 1$, $x \ne 3$

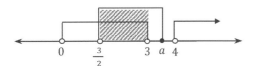

$\therefore$ $1 < x < 3$ or $3 < x \le 5$

(10) For two inequalities $x^3 - 7x^2 + 12x > 0$ and $\dfrac{x-a}{2x-3} \le 0$, the common(intersection) range of their solutions is $\dfrac{3}{2} < x < 3$. Find the range of the constant a.

$x^3 - 7x^2 + 12x > 0 \iff x(x^2 - 7x + 12) > 0 \iff x(x-3)(x-4) > 0$

$\therefore$ $0 < x < 3$ or $x > 4$ ······ ①

$\dfrac{x-a}{2x-3} \le 0 \iff (x-a)(2x-3) \le 0$, $x \ne \dfrac{3}{2}$

i) If $a < \dfrac{3}{2}$, then $a \le x < \dfrac{3}{2}$ ······ ②

$\therefore$ ① ∩ ② is $0 < x < \dfrac{3}{2}$ or $a \le x < \dfrac{3}{2}$

Since ① ∩ ② $\iff \dfrac{3}{2} < x < 3$, $a \not< \dfrac{3}{2}$

ii) If $a \ge \dfrac{3}{2}$, then $\dfrac{3}{2} < x \le a$ ······ ③

Since ① ∩ ③ $\iff \dfrac{3}{2} < x < 3$, $3 \le a < 4$

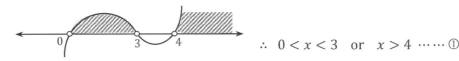

Therefore, $3 \le a < 4$

(11) For a positive number a, there are 6 integers so that the system of inequalities
$$\begin{cases} x(x+a)(x-2a) < 0 \\ x^2 + ax - 2a^2 \le 0 \end{cases}$$ **is always true. Solve the system and state the integers.**

$x(x+a)(x-2a) < 0$

$\Rightarrow$ Since $a > 0$,

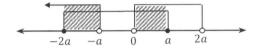

$\therefore \quad x < -a \quad$ or $\quad 0 < x < 2a \quad \cdots\cdots$ ①

$x^2 + ax - 2a^2 \le 0 \Leftrightarrow (x+2a)(x-a) \le 0$

Since $a > 0$, $\quad -2a \le x \le a \quad \cdots\cdots$ ②

By ① and ②,

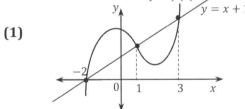

$\therefore -2a \le x < -a \quad$ or $\quad 0 < x \le a$

Since $|a - 0| = |-a - (-2a)| = a$, each interval contains 3 integers.

Therefore, $3 \le a < 4$ and the integers are $1, 2, 3, -6, -5, -4$.

#69 Find the value.

(1)

$y = f(x)$

$y = x + 1$

The graph of a function $y = f(x)$ and a line $y = x + 1$ intersect at $x = -2$, $x = 1$, and $x = 3$. Find maximum and minimum values of x such that $\dfrac{x}{f(2x)-1} \ge \dfrac{1}{2}$.

$\dfrac{x}{f(2x)-1} \ge \dfrac{1}{2} \Leftrightarrow \dfrac{2x}{f(2x)-1} \ge 1$

Let $2x = t$ Then, $\dfrac{2x}{f(2x)-1} \ge 1 \Leftrightarrow \dfrac{t}{f(t)-1} \ge 1$

i) When $f(t) > 1$; i.e., $f(t) - 1 > 0$,

Multiplying both sides by $f(t) - 1$, $\dfrac{t}{f(t)-1}(f(t) - 1) \ge f(t) - 1$

$\therefore \quad t \ge f(t) - 1 \qquad \therefore \quad f(t) \le t + 1$

From the graph, $1 \le t \le 3 \quad \cdots\cdots$ ①

ii) When $f(t) < 1$; i.e., $f(t) - 1 < 0$,

Multiplying both sides by $f(t) - 1$, $\dfrac{t}{f(t)-1}(f(t) - 1) \le f(t) - 1$

$\therefore\ t \le f(t) - 1 \qquad \therefore\ f(t) \ge t + 1$

Let $f(a) = 1$. Then, from the graph, $\quad -2 \le t < a$ ······ ②

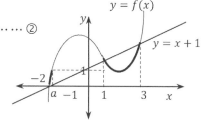

By ① and ②, $1 \le t \le 3$ or $-2 \le t < a$

$\therefore\ 1 \le 2x \le 3$ or $-2 \le 2x < a$

$\therefore\ \dfrac{1}{2} \le x \le \dfrac{3}{2}$ or $-1 \le x < \dfrac{a}{2}$

Therefore, maximum value of x is $\dfrac{3}{2}$ and minimum value of x is -1.

(2) **For constants a and b, the two inequalities $\dfrac{x^2-a^2}{x^2+x+1} < 0$ and $\dfrac{1}{x+2b} < \dfrac{1}{x+1}$ have the same solution. Find the value of $a + b$ $(a > 0)$.**

For any real number x, $x^2 + x + 1 = \left(x + \dfrac{1}{2}\right)^2 - \dfrac{1}{4} + 1 = \left(x + \dfrac{1}{2}\right)^2 + \dfrac{3}{4} > 0$

$\therefore\ \dfrac{x^2-a^2}{x^2+x+1} < 0 \iff x^2 - a^2 < 0 \iff (x+a)(x-a) < 0$

$\therefore\ -a < x < a\ (\because a > 0)$ ······ ①

$\dfrac{1}{x+2b} < \dfrac{1}{x+1} \iff \dfrac{1}{x+2b} - \dfrac{1}{x+1} < 0 \iff \dfrac{x+1-x-2b}{(x+2b)(x+1)} < 0 \iff \dfrac{1-2b}{(x+2b)(x+1)} < 0$

$\qquad \iff (1 - 2b)(x + 2b)(x + 1) < 0,\ x \ne -2b,\ x \ne -1$ ······ ②

Since ① and ② have the same solution, $1 - 2b > 0$ and $(x + 2b)(x + 1) < 0$

Since $2b < 1$, $-1 < x < -2b$ ······ ③

Since ③ $\iff$ ①, $-a = -1$ and $a = -2b \qquad \therefore\ a = 1,\ b = -\dfrac{1}{2}$

Therefore, $a + b = 1 - \dfrac{1}{2} = \dfrac{1}{2}$

(3) **When the two inequalities $\sqrt{x - 2} < 4 - x$ and $\dfrac{x+a}{x+b} \le 0$ have the same solution, find the value of ab (where a and b are constants).**

Let $\sqrt{x - 2} = 4 - x$

Squaring both sides, $x - 2 = (4 - x)^2 \qquad \therefore\ x^2 - 9x + 18 = 0 \qquad \therefore\ (x - 3)(x - 6) = 0$

$\therefore\ x = 3$ or $x = 6$

From the graph, $x = 6$ is an extraneous root, and is rejected.

To have $\sqrt{x - 2} < 4 - x$, $2 \le x < 3$

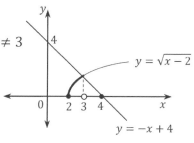

$\iff (x - 2)(x - 3) \le 0,\ x \ne 3$

$\iff \dfrac{x-2}{x-3} \le 0$

$\iff \dfrac{x+a}{x+b} \le 0$

$\therefore\ a = -2,\ b = -3 \qquad$ Therefore, $ab = 6$

#70 For two inequalities $\frac{(x+1)(x-4)}{(x-1)^2} \le 0$ and $\frac{x^2+3x-4}{x^3+1} \ge 0$, the common range of their

solutions is the solution of an inequality $f(x) \le 0$. Find the expression for $f(x)$.

$\frac{(x+1)(x-4)}{(x-1)^2} \le 0 \Leftrightarrow (x+1)(x-4)(x-1)^2 \le 0, \ x \ne 1$

$\therefore -1 \le x < 1 \ \text{ or } \ 1 < x \le 4 \ \cdots\cdots \ ①$

$\frac{x^2+3x-4}{x^3+1} \ge 0 \Leftrightarrow \frac{(x+4)(x-1)}{(x+1)(x^2+x+1)} \ge 0$

Since $x^2 + x + 1 = (x+\frac{1}{2})^2 - \frac{1}{4} + 1 = \left(x+\frac{1}{2}\right)^2 + \frac{3}{4} > 0$,

$\frac{(x+4)(x-1)}{(x+1)(x^2+x+1)} \ge 0 \Leftrightarrow (x+4)(x-1)(x+1) \ge 0, \ x \ne -1$

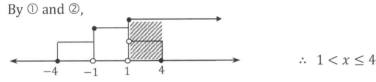

$\therefore -4 \le x < -1 \ \text{ or } \ x \ge 1 \ \cdots\cdots \ ②$

By ① and ②,

$\therefore 1 < x \le 4$

$1 < x \le 4 \ \Leftrightarrow a(x-1)(x-4) \le 0, \ x \ne 1, \ a > 0 \Leftrightarrow \frac{a(x-4)}{x-1} \le 0, \ a > 0$

$\Leftrightarrow f(x) \le 0, \ a > 0$

Therefore, $f(x) = \frac{a(x-4)}{x-1}, \ a > 0$

#71 For any two real numbers a and b, two inequalities $x(x+1)(x+2) > 0$ and $x^2 + ax + b \le 0$ have the union (sum) range of their solutions, $x > -2$, and the intersection (common) range of their solutions, $0 < x \le 2$. Find the value of $a + b$.

$x(x+1)(x+2) > 0 \Rightarrow$

$\therefore -2 < x < -1 \ \text{ or } \ x > 0 \ \cdots\cdots ①$

Let $x^2 + ax + b = 0$.

If α and β $(\alpha \le \beta)$ are two roots of the equation, then the solution of the inequality $x^2 + ax + b \le 0$ is $\alpha \le x \le \beta$ $\cdots\cdots ②$

Since ① ∪ ② is $x > -2$ and ① ∩ ② is $0 < x \le 2$, ② is $-1 \le x \le 2$

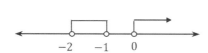 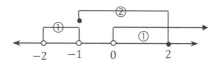

∴ $\alpha = -1$ and $\beta = 2$

That is, $(x + 1)(x - 2) \le 0$

∴ $x^2 - x - 2 \le 0 \iff x^2 + ax + b \le 0$

∴ $a = -1$, $b = -2$

Therefore, $a + b = -3$

#72 For any real number x ($x \ne 0$), the inequality $x^4 - 2x^3 + ax^2 - 2x + 1 \ge 0$ is always true. Find minimum value of a.

Dividing both sides by x^2, $x^2 - 2x + a - \frac{2}{x} + \frac{1}{x^2} \ge 0$

∴ $\left(x^2 + \frac{1}{x^2}\right) - 2\left(x + \frac{1}{x}\right) + a \ge 0$

∴ $\left\{\left(x + \frac{1}{x}\right)^2 - 2\right\} - 2\left(x + \frac{1}{x}\right) + a \ge 0$

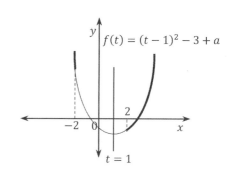

Let $x + \frac{1}{x} = t$ Then, $t^2 - 2 - 2t + a \ge 0$

If $x > 0$, then $x + \frac{1}{x} \ge 2\sqrt{x \cdot \frac{1}{x}} = 2$ ∴ $t \ge 2$

(by the relationship between arithmetic and geometric means)

If $x < 0$, then $(-x) + (-\frac{1}{x}) \ge 2\sqrt{(-x) \cdot (-\frac{1}{x})} = 2$ ∴ $-t \ge 2$; i.e., $t \le -2$

(by the relationship between arithmetic and geometric means)

Thus, the inequality $t^2 - 2 - 2t + a \ge 0$ is always true in the ranges $t \ge 2$ or $t \le -2$.

Since the axis of symmetry is $t = 1$,

$f(t) = t^2 - 2 - 2t + a = (t - 1)^2 - 3 + a$ has minimum value when $t = 2$.

Since $f(2) = -2 + a \ge 0$, $a \ge 2$

Therefore, minimum value of a is 2.

Chapter 6. Exponential and Logarithmic Functions

#1 Simplify the following given expressions.

(1) $a^5 \times a^3 = a^{5+3} = a^8$

(2) $a^2 \div a^6 = a^{2-6} = a^{-4} = \frac{1}{a^4}$

(3) $a^3 \div a^3 = a^{3-3} = a^0 = 1$

(4) $(a^2)^3 = a^{2\times3} = a^6$

(5) $(ab)^2 = a^2 b^2$

(6) $(a^2 b^5)^3 = a^{2\times3} b^{5\times3} = a^6 b^{15}$

(7) $(-3a^2 b^3)^3 = (-3)^3 (a^2)^3 (b^3)^3 = -27 a^6 b^9$

(8) $(a^3)^2 a^5 = a^6 a^5 = a^{6+5} = a^{11}$

(9) $\left(\frac{ab^2}{c^3}\right)^4 = \frac{a^4 (b^2)^4}{(c^3)^4} = \frac{a^4 b^8}{c^{12}}$

(10) $\left(\frac{-3ab^2}{2c^3}\right)^5 = \frac{(-3)^5 a^5 (b^2)^5}{2^5 (c^3)^5} = \frac{-243 a^5 b^{10}}{32 c^{15}}$

#2 Evaluate each expression.

(1) $3^{-2} = \frac{1}{3^2} = \frac{1}{9}$

(2) $2^{-4} = \frac{1}{2^4} = \frac{1}{16}$

(3) $4^0 = 1$

(4) $5^{-3} = \frac{1}{5^3} = \frac{1}{125}$

(5) $(2^3)^0 = 2^{3\times0} = 2^0 = 1$

(6) $2^{-1} \div 2^{-3} = 2^{(-1)-(-3)} = 2^2 = 4$

(7) $2^2 \div 2^{-3} = 2^{2-(-3)} = 2^5 = 32$

(8) $3^0 \div 3^4 = 3^{0-4} = 3^{-4} = \frac{1}{3^4} = \frac{1}{81}$

(9) $2^{-10} \div 2^{-2} \times 2^5 = 2^{-10-(-2)+5} = 2^{-3} = \frac{1}{2^3} = \frac{1}{8}$

(10) $8^3 \times 16^{-2} \div 32 = (2^3)^3 \times (2^4)^{-2} \div 2^5 = 2^9 \times 2^{-8} \div 2^5 = 2^{9+(-8)-5} = 2^{-4} = \frac{1}{2^4} = \frac{1}{16}$

(11) $(3 \times 2^{-3})^{-2} = 3^{-2} \times (2^{-3})^{-2} = \frac{1}{3^2} \times 2^6 = \frac{64}{9}$

(12) $(-2)^2 (2^{-3} \times 3^2)^{-1} = 2^2 (2^{-3})^{-1} (3^2)^{-1} = 2^2 2^3 3^{-2} = \frac{2^{2+3}}{3^2} = \frac{2^5}{3^2} = \frac{32}{9}$

(13) $(-10)^0 = 1$

#3 Evaluate each expression.

(1) $\sqrt[3]{64} = \sqrt[3]{4^3} = 4$

(2) $\sqrt[3]{-64} = \sqrt[3]{-(4^3)} = -\sqrt[3]{4^3} = -4$

(3) $\sqrt[4]{81} = \sqrt[4]{3^4} = 3$

(4) $\sqrt[5]{-32} = \sqrt[5]{-(2^5)} = -\sqrt[5]{2^5} = -2$

(5) $\sqrt[4]{2} \cdot \sqrt[4]{8} = \sqrt[4]{2 \cdot 8} = \sqrt[4]{16} = \sqrt[4]{2^4} = 2$

(6) $\frac{\sqrt[3]{2}}{\sqrt[3]{16}} = \sqrt[3]{\frac{2}{16}} = \sqrt[3]{\frac{1}{8}} = \sqrt[3]{\left(\frac{1}{2}\right)^3} = \frac{1}{2}$

(7) $\frac{\sqrt[4]{64}}{\sqrt[4]{4}} = \sqrt[4]{\frac{64}{4}} = \sqrt[4]{16} = \sqrt[4]{2^4} = 2$

(8) $\sqrt[3]{8^2} = (\sqrt[3]{8})^2 = (\sqrt[3]{2^3})^2 = 2^2 = 4$

(9) $\sqrt[3]{\left(\frac{8}{27}\right)^2} = \left(\sqrt[3]{\frac{8}{27}}\right)^2 = \left(\sqrt[3]{\left(\frac{2}{3}\right)^3}\right)^2 = \left(\frac{2}{3}\right)^2 = \frac{4}{9}$

(10) $(-\sqrt[3]{3})^9 = -\left(\sqrt[3]{3}\right)^9 = -\sqrt[3]{3^9} = -\sqrt[3]{(3^3)^3} = -3^3 = -27$

(11) $\sqrt[4]{(-2)^4} = \sqrt[4]{2^4} = 2$

(12) $\sqrt[3]{0.0001} \cdot \sqrt[3]{10} = \sqrt[3]{0.0001 \cdot 10} = \sqrt[3]{0.001} = \sqrt[3]{(0.1)^3} = 0.1$

(13) $\sqrt[3]{\sqrt[4]{8}} \cdot \sqrt{\sqrt[4]{64}} = \sqrt[4]{\sqrt[3]{8}} \cdot \sqrt[4]{\sqrt{64}} = \sqrt[4]{\sqrt[3]{2^3}} \cdot \sqrt[4]{\sqrt{8^2}} = \sqrt[4]{2} \cdot \sqrt[4]{8} = \sqrt[4]{2 \cdot 8} = \sqrt[4]{16} = \sqrt[4]{2^4} = 2$

(14) $\sqrt[9]{3^3} \cdot \sqrt[12]{3^8} = \sqrt[3 \times 3]{3^{1 \times 3}} \cdot \sqrt[3 \times 4]{3^{2 \times 4}} = \sqrt[3]{3} \cdot \sqrt[3]{3^2} = \sqrt[3]{3 \cdot 3^2} = \sqrt[3]{3^3} = 3$

(15) $\dfrac{\sqrt[4]{400}}{\sqrt{10}} = \dfrac{400^{\frac{1}{4}}}{10^{\frac{1}{2}}} = \dfrac{(2^2 \cdot 10^2)^{\frac{1}{4}}}{10^{\frac{1}{2}}} = \dfrac{2^{\frac{1}{2}} \cdot 10^{\frac{1}{2}}}{10^{\frac{1}{2}}} = 2^{\frac{1}{2}} = \sqrt{2}$

#4 Evaluate following expressions.

(1) $\sqrt{16} = \sqrt{4^2} = 4$

(2) The square roots of 16 $= \pm 4$

(3) $\sqrt[4]{16} = \sqrt[4]{2^4} = 2$

(4) $\sqrt[4]{-16}$: Undefined ($\because \sqrt[n]{\text{(negative number)}}$ is not defined. $n = 4, 6, 8, \cdots \cdots$)

(5) The 4th root of 16 $= \pm 2$

(6) Cube root of 8 $= 2$

(7) Cube root of $-8 = -2$

(8) $\sqrt[3]{-8} = -\sqrt[3]{8} = -\sqrt[3]{2^3} = -2$

(9) The 5th root of 32 $= 2$

(10) The 5th root of $-32 = -2$

#5 Evaluate each expression.

(1) $16^{\frac{1}{2}} = \sqrt[2]{16} = \sqrt[2]{4^2} = 4$

(2) $4^{\frac{3}{2}} = \sqrt[2]{4^3} = (\sqrt[2]{4})^3 = (\sqrt[2]{2^2})^3 = 2^3 = 8$

(3) $32^{\frac{4}{5}} = \sqrt[5]{32^4} = (\sqrt[5]{32})^4 = (\sqrt[5]{2^5})^4 = 2^4 = 16$

(4) $8^{-\frac{2}{3}} = \dfrac{1}{8^{\frac{2}{3}}} = \dfrac{1}{\sqrt[3]{8^2}} = \dfrac{1}{(\sqrt[3]{8})^2} = \dfrac{1}{(\sqrt[3]{2^3})^2} = \dfrac{1}{2^2} = \dfrac{1}{4}$

(5) $64^{-\frac{3}{2}} = \dfrac{1}{64^{\frac{3}{2}}} = \dfrac{1}{\sqrt[2]{64^3}} = \dfrac{1}{(\sqrt[2]{64})^3} = \dfrac{1}{(\sqrt[2]{8^2})^3} = \dfrac{1}{8^3} = \dfrac{1}{512}$

(6) $(-64)^{\frac{2}{3}} = \sqrt[3]{(-64)^2} = \sqrt[3]{64^2} = \sqrt[3]{(4^3)^2} = \sqrt[3]{(4^2)^3} = 4^2 = 16$

(7) $\left\{\left(\frac{9}{16}\right)^{-\frac{3}{4}}\right\}^{\frac{2}{3}} = \left(\frac{9}{16}\right)^{-\frac{3}{4} \cdot \frac{2}{3}} = \left(\frac{9}{16}\right)^{-\frac{1}{2}} = \left\{\left(\frac{3}{4}\right)^2\right\}^{-\frac{1}{2}} = \left(\frac{3}{4}\right)^{-1} = \frac{4}{3}$

(8) $\sqrt[3]{16} \div \sqrt[3]{2^{10}} = \dfrac{\sqrt[3]{16}}{\sqrt[3]{2^{10}}} = \sqrt[3]{\dfrac{16}{2^{10}}} = \sqrt[3]{\dfrac{2^4}{2^{10}}} = \sqrt[3]{\dfrac{1}{2^6}} = \sqrt[3]{\left(\dfrac{1}{2^2}\right)^3} = \dfrac{1}{2^2} = \dfrac{1}{4}$

(9) $16^{\frac{1}{3}} \div 24^{\frac{2}{3}} \div 18^{\frac{1}{3}} = (2^4)^{\frac{1}{3}} \div (3 \cdot 2^3)^{\frac{2}{3}} \div (2 \cdot 3^2)^{\frac{1}{3}}$

$= 2^{\frac{4}{3}} \div \left(3^{\frac{2}{3}} \cdot 2^2\right) \div \left(2^{\frac{1}{3}} \cdot 3^{\frac{2}{3}}\right) = 2^{\frac{4}{3}} \times 3^{-\frac{2}{3}} \times 2^{-2} \times 2^{-\frac{1}{3}} \times 3^{-\frac{2}{3}}$

$= 2^{\frac{4}{3}-2-\frac{1}{3}} \times 3^{-\frac{2}{3}-\frac{2}{3}} = 2^{-1} \times 3^{-\frac{4}{3}} = \dfrac{1}{2 \cdot 3^{\frac{4}{3}}} = \dfrac{1}{2 \cdot \sqrt[3]{3^4}} = \dfrac{1}{2 \cdot 3 \sqrt[3]{3}} = \dfrac{1}{6 \sqrt[3]{3}}$

(10) $2^{\frac{1}{3}} \times 3^{\frac{1}{2}} \div 6^{\frac{1}{6}} \div \left(\dfrac{3}{2}\right)^{\frac{1}{3}} = 2^{\frac{1}{3}} \times 3^{\frac{1}{2}} \times 6^{-\frac{1}{6}} \times \left(\dfrac{3}{2}\right)^{-\frac{1}{3}}$

$= 2^{\frac{1}{3}} \times 3^{\frac{1}{2}} \times (2 \times 3)^{-\frac{1}{6}} \times \left(\dfrac{3}{2}\right)^{-\frac{1}{3}} = 2^{\frac{1}{3}} \times 3^{\frac{1}{2}} \times 2^{-\frac{1}{6}} \times 3^{-\frac{1}{6}} \times \left(\dfrac{2}{3}\right)^{\frac{1}{3}}$

$= 2^{\frac{1}{3}} \times 3^{\frac{1}{2}} \times 2^{-\frac{1}{6}} \times 3^{-\frac{1}{6}} \times 2^{\frac{1}{3}} \times 3^{-\frac{1}{3}} = 2^{\frac{1}{3}-\frac{1}{6}+\frac{1}{3}} \times 3^{\frac{1}{2}-\frac{1}{6}-\frac{1}{3}} = 2^{\frac{1}{2}} \times 3^0 = 2^{\frac{1}{2}} = \sqrt{2}$

(11) $\sqrt{2^3 \cdot 3} \div \sqrt[4]{2^2 \cdot 3^5} \times \sqrt[6]{2 \cdot 3^{10}} = \left(2^{\frac{3}{2}} \times 3^{\frac{1}{2}}\right) \div \left(2^{\frac{2}{4}} \times 3^{\frac{5}{4}}\right) \times \left(2^{\frac{1}{6}} \times 3^{\frac{10}{6}}\right)$

$= \left(2^{\frac{3}{2}} \times 3^{\frac{1}{2}}\right) \times \left(2^{-\frac{2}{4}} \times 3^{-\frac{5}{4}}\right) \times \left(2^{\frac{1}{6}} \times 3^{\frac{10}{6}}\right) = 2^{\frac{3}{2}-\frac{2}{4}+\frac{1}{6}} \times 3^{\frac{1}{2}-\frac{5}{4}+\frac{10}{6}}$

$= 2^{\frac{18-6+2}{12}} \times 3^{\frac{6-15+20}{12}} = 2^{\frac{14}{12}} \times 3^{\frac{11}{12}} = 2^{\frac{7}{6}} \times 3^{\frac{11}{12}}$

#6 (Magnitude) Compare the following numbers.

(1) $\sqrt[3]{3}$, $\sqrt[4]{9}$, $\sqrt[5]{27}$

Convert the given numbers so that all numbers are written with the same base, 3 .

$\sqrt[3]{3} = 3^{\frac{1}{3}}$, $\sqrt[4]{9} = 3^{\frac{2}{4}}$, $\sqrt[5]{27} = 3^{\frac{3}{5}}$

Since the base is greater than 1, the numbers are in the order of the powers.

Comparing the powers, $\dfrac{1}{3} < \dfrac{2}{4} < \dfrac{3}{5}$

Therefore, $\sqrt[3]{3} < \sqrt[4]{9} < \sqrt[5]{27}$

(2) $\left(\dfrac{1}{2}\right)^{\frac{1}{3}}$, $\left(\dfrac{1}{2}\right)^{-2}$, $\sqrt{\dfrac{1}{2}}$, $\dfrac{1}{4}$

Convert the given numbers so that all numbers are written with the same base, $\dfrac{1}{2}$.

Since the base is less than 1, the numbers are in the reverse order of the powers.

Comparing the powers, $-2 < \dfrac{1}{3} < \dfrac{1}{2} < 2 \qquad \therefore \left(\dfrac{1}{2}\right)^{-2} > \left(\dfrac{1}{2}\right)^{\frac{1}{3}} > \left(\dfrac{1}{2}\right)^{\frac{1}{2}} > \left(\dfrac{1}{2}\right)^{2}$

Therefore, $\dfrac{1}{4} < \sqrt{\dfrac{1}{2}} < \left(\dfrac{1}{2}\right)^{\frac{1}{3}} < \left(\dfrac{1}{2}\right)^{-2}$

(3) $\sqrt[6]{16}$, $\sqrt[12]{81}$, $\sqrt[9]{64}$

$\sqrt[6]{16} = \sqrt[6]{4^2} = 4^{\frac{2}{6}} = 4^{\frac{1}{3}}$, $\quad \sqrt[12]{81} = \sqrt[12]{3^4} = 3^{\frac{4}{12}} = 3^{\frac{1}{3}}$, $\quad \sqrt[9]{64} = \sqrt[9]{4^3} = 4^{\frac{3}{9}} = 4^{\frac{1}{3}}$

Note that : when a and b are positive, $a < b \iff a^n < b^n$ for positive n.

Therefore, $\sqrt[12]{81} < \sqrt[6]{16} = \sqrt[9]{64}$

(4) $\sqrt{2}$, $\sqrt[3]{3}$, $\sqrt[4]{5}$, $\sqrt[6]{6}$

Note that : when a and b are positive, $a < b \iff \sqrt[n]{a} < \sqrt[n]{b}$.

$\sqrt{2} = \sqrt[2\cdot6]{2^{1\cdot6}} = \sqrt[12]{2^6} = \sqrt[12]{64}$, $\quad \sqrt[3]{3} = \sqrt[3\cdot4]{3^{1\cdot4}} = \sqrt[12]{3^4} = \sqrt[12]{81}$,

$\sqrt[4]{5} = \sqrt[4\cdot3]{5^{1\cdot3}} = \sqrt[12]{5^3} = \sqrt[12]{125}$, $\quad \sqrt[6]{6} = \sqrt[6\cdot2]{6^{1\cdot2}} = \sqrt[12]{6^2} = \sqrt[12]{36}$

Therefore, $\sqrt[6]{6} < \sqrt{2} < \sqrt[3]{3} < \sqrt[4]{5}$

(5) $\sqrt[3]{3}$, $\sqrt[4]{5}$, $\sqrt{\sqrt[3]{9}}$

$\sqrt[3]{3} = \sqrt[3\cdot4]{3^{1\cdot4}} = \sqrt[12]{3^4} = \sqrt[12]{81}$, $\quad \sqrt[4]{5} = \sqrt[4\cdot3]{5^{1\cdot3}} = \sqrt[12]{5^3} = \sqrt[12]{125}$,

$\sqrt{\sqrt[3]{9}} = \sqrt[2\cdot3]{9} = \sqrt[2\cdot3\cdot2]{9^{1\cdot2}} = \sqrt[12]{81}$

Therefore, $\sqrt[3]{3} = \sqrt{\sqrt[3]{9}} < \sqrt[4]{5}$

(6) $\left(\sqrt{2\sqrt[3]{3}}\right)^2$, $\left(\sqrt[3]{3\sqrt{2}}\right)^2$, $\left(\sqrt[3]{2\sqrt{3}}\right)^2$

$\left(\sqrt{2\sqrt[3]{3}}\right)^2 = \left(\sqrt{\sqrt[3]{2^3\cdot3}}\right)^2 = \left(\sqrt{\sqrt[3]{24}}\right)^2 = \left(\sqrt[6]{24}\right)^2 = 24^{\frac{2}{6}} = 24^{\frac{1}{3}} = \sqrt[3]{24}$

$\left(\sqrt[3]{3\sqrt{2}}\right)^2 = \left(\sqrt[3]{\sqrt{18}}\right)^2 = \left(\sqrt{\sqrt[3]{18}}\right)^2 = \left(\sqrt[6]{18}\right)^2 = 18^{\frac{2}{6}} = 18^{\frac{1}{3}} = \sqrt[3]{18}$

$\left(\sqrt[3]{2\sqrt{3}}\right)^2 = \left(\sqrt[3]{\sqrt{12}}\right)^2 = \left(\sqrt{\sqrt[3]{12}}\right)^2 = \left(\sqrt[6]{12}\right)^2 = 12^{\frac{2}{6}} = 12^{\frac{1}{3}} = \sqrt[3]{12}$

Since $\sqrt[3]{12} < \sqrt[3]{18} < \sqrt[3]{24}$, $\quad \left(\sqrt[3]{2\sqrt{3}}\right)^2 < \left(\sqrt[3]{3\sqrt{2}}\right)^2 < \left(\sqrt{2\sqrt[3]{3}}\right)^2$

(7) $\sqrt{3\sqrt{3}}$, $\sqrt{3^{\sqrt{3}}}$, $(\sqrt{3})^{\sqrt{3}}$

$\sqrt{3\sqrt{3}} = 3^{\frac{1}{4}} \cdot 3^{\frac{1}{2}} = 3^{\frac{1}{4}+\frac{1}{2}} = 3^{\frac{3}{4}}$, $\quad \sqrt{3^{\sqrt{3}}} = (3^{\sqrt{3}})^{\frac{1}{2}} = 3^{\frac{\sqrt{3}}{2}}$, $\quad (\sqrt{3})^{\sqrt{3}} = (3^{\frac{1}{2}})^{\sqrt{3}} = 3^{\frac{\sqrt{3}}{2}}$

Since $\dfrac{3}{4} = \dfrac{\sqrt{9}}{4}$ and $\dfrac{\sqrt{3}}{2} = \dfrac{2\sqrt{3}}{4} = \dfrac{\sqrt{12}}{4}$, $\quad \sqrt{3\sqrt{3}} < \sqrt{3^{\sqrt{3}}} = (\sqrt{3})^{\sqrt{3}}$

(8) **For positive integers a and b such that $1 < a^{b-3} < b^{a-5}$,**

compare the numbers A, B, C: $\quad A = a^{\frac{1}{a-5}} \cdot b^{\frac{1}{b-3}}$, $\quad B = a^{-\frac{1}{a-5}} \cdot b^{\frac{1}{b-3}}$, $\quad C = a^{\frac{1}{a-5}} \cdot b^{-\frac{1}{b-3}}$

Since $a^{b-3} > 1$ and $1 = a^0$, $b - 3 > 0$

Since $b^{a-5} > 1$, $a - 5 > 0$

Since $b^{a-5} > a^{b-3}$, $\dfrac{b^{a-5}}{a^{b-3}} > 1$

$$\frac{A}{B} = \frac{a^{\frac{1}{a-5}} \cdot b^{\frac{1}{b-3}}}{a^{-\frac{1}{a-5}} \cdot b^{\frac{1}{b-3}}} = \frac{a^{\frac{1}{a-5}}}{a^{-\frac{1}{a-5}}} = a^{\frac{2}{a-5}}$$

Since $a - 5 > 0$, $\dfrac{2}{a-5} > 0$ $\quad \therefore a^{\frac{2}{a-5}} > a^0$ $\quad \therefore a^{\frac{2}{a-5}} > 1$ $\quad \therefore \dfrac{A}{B} > 1$ $\quad \therefore A > B$

$$\frac{A}{C} = \frac{a^{\frac{1}{a-5}} \cdot b^{\frac{1}{b-3}}}{a^{\frac{1}{a-5}} \cdot b^{-\frac{1}{b-3}}} = \frac{b^{\frac{1}{b-3}}}{b^{-\frac{1}{b-3}}} = b^{\frac{2}{b-3}}$$

Since $b - 3 > 0$, $\dfrac{2}{b-3} > 0$ $\quad \therefore b^{\frac{2}{b-3}} > b^0$ $\quad \therefore b^{\frac{2}{b-3}} > 1$ $\quad \therefore \dfrac{A}{C} > 1$ $\quad \therefore A > C$

$$\frac{B}{C} = \frac{a^{-\frac{1}{a-5}} \cdot b^{\frac{1}{b-3}}}{a^{\frac{1}{a-5}} \cdot b^{-\frac{1}{b-3}}} = \frac{b^{\frac{2}{b-3}}}{a^{\frac{2}{a-5}}} = \left(\frac{b^{a-5}}{a^{b-3}}\right)^{\frac{2}{(a-5)(b-3)}}$$

Since $\dfrac{b^{a-5}}{a^{b-3}} > 1$ and $\dfrac{2}{(a-5)(b-3)} > 0$, $\quad \therefore \dfrac{B}{C} > 1$ $\quad \therefore B > C$

Therefore, $A > B > C$

#7 Simplify each expression.

(1) $\sqrt[3]{\dfrac{\sqrt[4]{a}}{\sqrt{a}}} \div \sqrt{\dfrac{\sqrt[6]{a}}{\sqrt[3]{a}}} = \dfrac{\sqrt[3]{\sqrt[4]{a}}}{\sqrt[3]{\sqrt{a}}} \div \dfrac{\sqrt{\sqrt[6]{a}}}{\sqrt{\sqrt[3]{a}}} = \dfrac{\sqrt[12]{a}}{\sqrt[6]{a}} \div \dfrac{\sqrt[12]{a}}{\sqrt[6]{a}} = 1$

(2) $\sqrt{\dfrac{\sqrt[3]{a}}{\sqrt[4]{a}}} \times \sqrt[4]{\dfrac{\sqrt{a}}{\sqrt[3]{a}}} \times \sqrt[3]{\dfrac{a}{\sqrt[4]{a}}} = \dfrac{\sqrt{\sqrt[3]{a}}}{\sqrt{\sqrt[4]{a}}} \times \dfrac{\sqrt[4]{\sqrt{a}}}{\sqrt[4]{\sqrt[3]{a}}} \times \dfrac{\sqrt[3]{a}}{\sqrt[3]{\sqrt[4]{a}}} = \dfrac{\sqrt[6]{a}}{\sqrt[8]{a}} \times \dfrac{\sqrt[8]{a}}{\sqrt[12]{a}} \times \dfrac{\sqrt[3]{a}}{\sqrt[12]{a}} = \dfrac{\sqrt[6]{a}}{\sqrt[12]{a}} \times \dfrac{\sqrt[3]{a}}{\sqrt[12]{a}}$

$= \dfrac{\sqrt[12]{a^2}}{\sqrt[12]{a}} \times \dfrac{\sqrt[12]{a^4}}{\sqrt[12]{a}} = \sqrt[12]{\dfrac{a^2}{a}} \times \sqrt[12]{\dfrac{a^4}{a}} = \sqrt[12]{a} \times \sqrt[12]{a^3} = \sqrt[12]{a \cdot a^3} = \sqrt[12]{a^4} = a^{\frac{4}{12}} = a^{\frac{1}{3}} = \sqrt[3]{a}$

(3) $\sqrt{\dfrac{a}{\sqrt{a}} \cdot \sqrt[3]{a^2}} = \left(a \div a^{\frac{1}{2}} \times a^{\frac{2}{3}}\right)^{\frac{1}{2}} = \left(a^{1-\frac{1}{2}+\frac{2}{3}}\right)^{\frac{1}{2}} = \left(a^{\frac{7}{6}}\right)^{\frac{1}{2}} = a^{\frac{7}{12}}$

(4) $\sqrt{a\sqrt{a\sqrt{a}}} = \left\{a\left(a \cdot a^{\frac{1}{2}}\right)^{\frac{1}{2}}\right\}^{\frac{1}{2}} = \left\{a\left(a^{\frac{3}{2}}\right)^{\frac{1}{2}}\right\}^{\frac{1}{2}} = \left(a \cdot a^{\frac{3}{4}}\right)^{\frac{1}{2}} = \left(a^{\frac{7}{4}}\right)^{\frac{1}{2}} = a^{\frac{7}{8}}$

Or, $\sqrt{a\sqrt{a\sqrt{a}}} = a^{\frac{1}{2}} \cdot a^{\frac{1}{4}} \cdot a^{\frac{1}{8}} = a^{\frac{1}{2}+\frac{1}{4}+\frac{1}{8}} = a^{\frac{4+2+1}{8}} = a^{\frac{7}{8}}$

(5) $\sqrt{a\sqrt{a\sqrt{a\sqrt{a}}}} = a^{\frac{1}{2}} \cdot a^{\frac{1}{4}} \cdot a^{\frac{1}{8}} \cdot a^{\frac{1}{16}} = a^{\frac{1}{2}+\frac{1}{4}+\frac{1}{8}+\frac{1}{16}} = a^{\frac{8+4+2+1}{16}} = a^{\frac{15}{16}}$

(6) $\left(a^{\frac{1}{4}} + b^{\frac{1}{4}}\right)\left(a^{\frac{1}{4}} - b^{\frac{1}{4}}\right)\left(a^{\frac{1}{2}} + b^{\frac{1}{2}}\right) = \left\{\left(a^{\frac{1}{4}}\right)^2 - \left(b^{\frac{1}{4}}\right)^2\right\}\left(a^{\frac{1}{2}} + b^{\frac{1}{2}}\right)$

$= \left(a^{\frac{1}{2}} - b^{\frac{1}{2}}\right)\left(a^{\frac{1}{2}} + b^{\frac{1}{2}}\right) = \left(a^{\frac{1}{2}}\right)^2 - \left(b^{\frac{1}{2}}\right)^2 = a - b$

Or, Letting $a^{\frac{1}{4}} = x$ and $b^{\frac{1}{4}} = y$, we have $a^{\frac{1}{2}} = x^2$ and $b^{\frac{1}{2}} = y^2$

Therefore, $\left(a^{\frac{1}{4}} + b^{\frac{1}{4}}\right)\left(a^{\frac{1}{4}} - b^{\frac{1}{4}}\right)\left(a^{\frac{1}{2}} + b^{\frac{1}{2}}\right) = (x + y)(x - y)(x^2 + y^2)$

$$= (x^2 - y^2)(x^2 + y^2) = x^4 - y^4 = \left(a^{\frac{1}{4}}\right)^4 - \left(b^{\frac{1}{4}}\right)^4 = a - b$$

(7) $(a - b) \div \left(a^{\frac{1}{3}} - b^{\frac{1}{3}}\right)$

Note that $a^3 - b^3 = (a - b)(a^2 + ab + b^2)$

Since $a - b = \left(a^{\frac{1}{3}}\right)^3 - \left(b^{\frac{1}{3}}\right)^3$, let $a^{\frac{1}{3}} = x$ and $b^{\frac{1}{3}} = y$.

Then, $(a - b) \div \left(a^{\frac{1}{3}} - b^{\frac{1}{3}}\right) = (x^3 - y^3) \div (x - y) = (x - y)(x^2 + xy + y^2) \div (x - y)$

$$= x^2 + xy + y^2 = \left(a^{\frac{1}{3}}\right)^2 + a^{\frac{1}{3}}b^{\frac{1}{3}} + \left(b^{\frac{1}{3}}\right)^2 = a^{\frac{2}{3}} + a^{\frac{1}{3}}b^{\frac{1}{3}} + b^{\frac{2}{3}}$$

(8) $(a - a^{-1}) \div \left(a^{\frac{1}{2}} - a^{-\frac{1}{2}}\right)$

Let $a^{\frac{1}{2}} = x$ and $a^{-\frac{1}{2}} = y$. Then, $a = x^2$ and $a^{-1} = y^2$

Thus, $(a - a^{-1}) \div \left(a^{\frac{1}{2}} - a^{-\frac{1}{2}}\right) = (x^2 - y^2) \div (x - y) = (x + y)(x - y) \div (x - y)$

$$= x + y = a^{\frac{1}{2}} + a^{-\frac{1}{2}}$$

#8 Find the value.

(1) When $x^{\frac{1}{2}} + x^{-\frac{1}{2}} = 3$, ① $x + x^{-1}$ ② $x^2 + x^{-2}$ ③ $x^{\frac{3}{2}} + x^{-\frac{3}{2}}$

① $x + x^{-1}$

Squaring both sides of the given equation, we obtain $\left(x^{\frac{1}{2}} + x^{-\frac{1}{2}}\right)^2 = 3^2$

$\therefore x + 2x^{\frac{1}{2}} \cdot x^{-\frac{1}{2}} + x^{-1} = 9$ $\qquad \therefore x + 2 + x^{-1} = 9$ $\qquad \therefore x + x^{-1} = 7$

② $x^2 + x^{-2}$

Since $x + x^{-1} = 7$, $(x + x^{-1})^2 = 7^2$

$\therefore x^2 + 2x \cdot x^{-1} + x^{-2} = 49$ $\qquad \therefore x^2 + 2 + x^{-2} = 49$ $\qquad \therefore x^2 + x^{-2} = 47$

③ $x^{\frac{3}{2}} + x^{-\frac{3}{2}}$

Since $x^{\frac{1}{2}} + x^{-\frac{1}{2}} = 3$, $\left(x^{\frac{1}{2}} + x^{-\frac{1}{2}}\right)^3 = 3^3$

$\therefore x^{\frac{3}{2}} + 3x \cdot x^{-\frac{1}{2}} + 3x^{\frac{1}{2}} \cdot x^{-1} + x^{-\frac{3}{2}} = 27$

$\therefore x^{\frac{3}{2}} + 3x^{\frac{1}{2}} + 3x^{-\frac{1}{2}} + x^{-\frac{3}{2}} = 27$

$\therefore x^{\frac{3}{2}} + 3\left(x^{\frac{1}{2}} + x^{-\frac{1}{2}}\right) + x^{-\frac{3}{2}} = 27$ $\qquad \therefore x^{\frac{3}{2}} + 9 + x^{-\frac{3}{2}} = 27$ $\qquad \therefore x^{\frac{3}{2}} + x^{-\frac{3}{2}} = 18$

When $x + \dfrac{1}{x} = 7$, $\sqrt{x} + \dfrac{1}{\sqrt{x}}$

Note that $\sqrt{x} + \dfrac{1}{\sqrt{x}} = x^{\frac{1}{2}} + x^{-\frac{1}{2}}$

$\therefore \left(x^{\frac{1}{2}} + x^{-\frac{1}{2}}\right)^2 = x + 2x^{\frac{1}{2}} \cdot x^{-\frac{1}{2}} + x^{-1} = x + 2 + x^{-1} = x + \dfrac{1}{x} + 2 = 7 + 2 = 9$

$\therefore x^{\frac{1}{2}} + x^{-\frac{1}{2}} = \pm 3$

Since $x > 0$, $x^{\frac{1}{2}} + x^{-\frac{1}{2}} > 0$

$\therefore x^{\frac{1}{2}} + x^{-\frac{1}{2}} = 3$ 　　　Therefore, $\sqrt{x} + \dfrac{1}{\sqrt{x}} = 3$

(2) When $e^{2x} = 5$,

① $\left(\dfrac{1}{e^2}\right)^{-3x} = (e^{-2})^{-3x} = e^{6x} = (e^{2x})^3 = 5^3 = 125$

② $\dfrac{e^x + e^{-x}}{e^x - e^{-x}}$

Multiplying both nominator and denominator by e^x, we obtain

$\dfrac{e^x + e^{-x}}{e^x - e^{-x}} = \dfrac{e^{2x} + e^0}{e^{2x} - e^0} = \dfrac{e^{2x} + 1}{e^{2x} - 1} = \dfrac{5+1}{5-1} = \dfrac{6}{4} = \dfrac{3}{2}$

③ $\dfrac{e^{3x} + e^{-3x}}{e^x + e^{-x}}$

Multiplying both nominator and denominator by e^x, we obtain

$\dfrac{e^{3x} + e^{-3x}}{e^x + e^{-x}} = \dfrac{e^{4x} + e^{-2x}}{e^{2x} + e^0} = \dfrac{(e^{2x})^2 + \frac{1}{e^{2x}}}{e^{2x} + 1} = \dfrac{5^2 + \frac{1}{5}}{5 + 1} = \dfrac{25 + \frac{1}{5}}{6} = \dfrac{\frac{126}{5}}{6} = \dfrac{21}{5}$

(3) When α and β are two roots of an equation $x^2 - 4x + 1 = 0$, $\dfrac{4^\alpha \cdot 8^\beta}{(2 \cdot 4^\alpha)^\beta}$

By the relationship between the roots and coefficients, $\alpha + \beta = 4$, $\alpha\beta = 1$

$\dfrac{4^\alpha \cdot 8^\beta}{(2 \cdot 4^\alpha)^\beta} = \dfrac{(2^2)^\alpha \cdot (2^3)^\beta}{2^\beta (2^2)^{\alpha\beta}} = \dfrac{2^{2\alpha + 3\beta}}{2^{\beta + 2\alpha\beta}} = 2^{(2\alpha + 3\beta) - (\beta + 2\alpha\beta)} = 2^{2(\alpha + \beta) - 2\alpha\beta} = 2^{2 \cdot 4 - 2 \cdot 1} = 2^6$

(4) When $\sqrt[3]{x} + \dfrac{1}{\sqrt[3]{x}} = 2$ for positive x, $\sqrt[3]{x^4} + \dfrac{1}{\sqrt[3]{x^4}}$

$\sqrt[3]{x} + \dfrac{1}{\sqrt[3]{x}} = 2 \;\Rightarrow\; x^{\frac{1}{3}} + x^{-\frac{1}{3}} = 2 \;\cdots\cdots\; ①$

Squaring both sides of ①, $x^{\frac{2}{3}} + 2 + x^{-\frac{2}{3}} = 4$ 　　$\therefore x^{\frac{2}{3}} + x^{-\frac{2}{3}} = 2 \;\cdots\cdots\; ②$

Squaring both sides of ②, $x^{\frac{4}{3}} + 2 + x^{-\frac{4}{3}} = 4$ 　　$\therefore x^{\frac{4}{3}} + x^{-\frac{4}{3}} = 2$

Therefore, $\sqrt[3]{x^4} + \dfrac{1}{\sqrt[3]{x^4}} = x^{\frac{4}{3}} + x^{-\frac{4}{3}} = 2$

(5) When $a^{\frac{1}{2}} - a^{-\frac{1}{2}} = 3 \ (a > 0)$, $\dfrac{a^{\frac{3}{2}} - a^{-\frac{3}{2}} + 5}{a + a^{-1} + 3}$

$$\left(a^{\frac{1}{2}} - a^{-\frac{1}{2}}\right)^3 = a^{\frac{3}{2}} - 3aa^{-\frac{1}{2}} + 3a^{\frac{1}{2}}a^{-1} - a^{-\frac{3}{2}} = 3^3$$

$$\therefore \ a^{\frac{3}{2}} - 3a^{\frac{1}{2}} + 3a^{-\frac{1}{2}} - a^{-\frac{3}{2}} = 27$$

$$\therefore \ a^{\frac{3}{2}} - 3\left(a^{\frac{1}{2}} - a^{-\frac{1}{2}}\right) - a^{-\frac{3}{2}} = 27$$

$$\therefore \ a^{\frac{3}{2}} - 3 \cdot 3 - a^{-\frac{3}{2}} = 27 \qquad \therefore \ a^{\frac{3}{2}} - a^{-\frac{3}{2}} = 36$$

Since $\left(a^{\frac{1}{2}} - a^{-\frac{1}{2}}\right)^2 = a - 2a^{\frac{1}{2}}a^{-\frac{1}{2}} + a^{-1} = a - 2 + a^{-1} = 3^2$, $\ a + a^{-1} = 11$

Therefore, $\dfrac{a^{\frac{3}{2}} - a^{-\frac{3}{2}} + 5}{a + a^{-1} + 3} = \dfrac{36 + 5}{11 + 3} = \dfrac{41}{14}$

(6) When $a^{2x} = 7 + 4\sqrt{3} \ (a > 0, \ a \neq 1)$, $\dfrac{a^x - a^{-x}}{a^{2x} - a^{-2x}}$

$$a^{2x} = 7 + 4\sqrt{3} \ \Rightarrow \ a^x = \sqrt{7 + 4\sqrt{3}} = \sqrt{7 + 2\sqrt{12}} = \sqrt{(4 + 3) + 2\sqrt{(4 \cdot 3)}}$$

$$= \sqrt{4} + \sqrt{3} = 2 + \sqrt{3}$$

$$a^{-x} = \frac{1}{a^x} = \frac{1}{2 + \sqrt{3}} = \frac{2 - \sqrt{3}}{(2 + \sqrt{3})(2 - \sqrt{3})} = \frac{2 - \sqrt{3}}{4 - 3} = 2 - \sqrt{3}$$

Therefore, $\dfrac{a^x - a^{-x}}{a^{2x} - a^{-2x}} = \dfrac{a^x - a^{-x}}{(a^x)^2 - (a^{-x})^2} = \dfrac{a^x - a^{-x}}{(a^x + a^{-x})(a^x - a^{-x})} = \dfrac{1}{a^x + a^{-x}} = \dfrac{1}{(2 + \sqrt{3}) + (2 - \sqrt{3})} = \dfrac{1}{4}$

(7) When $a^x = b^y = 2^z$ **and** $\dfrac{1}{x} + \dfrac{1}{y} - \dfrac{4}{z} = 0 \ (xyz \neq 0, \ a > 0, \ b > 0)$, $\dfrac{1}{ab}$

Let $a^x = b^y = 2^z = k \ (k > 0, \ k \neq 1)$. Then, $a = k^{\frac{1}{x}}$, $b = k^{\frac{1}{y}}$, $2 = k^{\frac{1}{z}}$

Thus, $ab = k^{\frac{1}{x}}k^{\frac{1}{y}} = k^{\frac{1}{x} + \frac{1}{y}}$

Since $\dfrac{1}{x} + \dfrac{1}{y} - \dfrac{4}{z} = 0$, $\dfrac{1}{x} + \dfrac{1}{y} = \dfrac{4}{z}$ $\qquad \therefore \ ab = k^{\frac{4}{z}} = \left(k^{\frac{1}{z}}\right)^4 = 2^4 = 16$

Therefore, $\dfrac{1}{ab} = \dfrac{1}{16}$

(8) When $a^x = 3$, $(ab)^y = 3^4$, $(abc)^z = 3^5$ **(**$a, b,$ **and** c **are positive numbers),** $3^{\frac{1}{x} + \frac{4}{y} - \frac{5}{z}}$

$$a^x = 3 \ \Rightarrow \ a = 3^{\frac{1}{x}} \qquad (ab)^y = 3^4 \ \Rightarrow \ ab = 3^{\frac{4}{y}} \qquad (abc)^z = 3^5 \ \Rightarrow \ abc = 3^{\frac{5}{z}}$$

Therefore, $3^{\frac{1}{x} + \frac{4}{y} - \frac{5}{z}} = 3^{\frac{1}{x}} \times 3^{\frac{4}{y}} \div 3^{\frac{5}{z}} = a \times ab \div abc = \dfrac{a^2 b}{abc} = \dfrac{a}{c}$

(9) When $a = 2^m 3^n$, $\sqrt{\dfrac{a}{2}} \times \sqrt[3]{\dfrac{a}{4}}$

$$\sqrt{\frac{a}{2}} \times \sqrt[3]{\frac{a}{4}} = \left(\frac{2^m 3^n}{2}\right)^{\frac{1}{2}} \cdot \left(\frac{2^m 3^n}{4}\right)^{\frac{1}{3}} = (2^{m-1}3^n)^{\frac{1}{2}} \cdot (2^{m-2}3^n)^{\frac{1}{3}} = 2^{\frac{m-1}{2}} \cdot 3^{\frac{n}{2}} \cdot 2^{\frac{m-2}{3}} \cdot 3^{\frac{n}{3}}$$

$$= 2^{\frac{m-1}{2}+\frac{m-2}{3}} \cdot 3^{\frac{n}{2}+\frac{n}{3}} = 2^{\frac{3m-3+2m-4}{6}} \cdot 3^{\frac{3n+2n}{6}} = 2^{\frac{5m-7}{6}} \cdot 3^{\frac{5n}{6}}$$

(10) When $f(x) = \dfrac{1+x+x^2+\cdots\cdots+x^{10}}{x^{-2}+x^{-3}+\cdots\cdots+x^{-12}}$, $f(\sqrt[24]{2})$

$$f(x) = \frac{1+x+x^2+\cdots\cdots+x^{10}}{x^{-2}+x^{-3}+\cdots\cdots+x^{-12}} = \frac{1+x+x^2+\cdots\cdots+x^{10}}{x^{-12}(1+x+x^2+\cdots\cdots+x^{10})} = \frac{1}{x^{-12}} = x^{12}$$

$$\therefore \ f(\sqrt[24]{2}) = (\sqrt[24]{2})^{12} = 2^{\frac{12}{24}} = 2^{\frac{1}{2}} = \sqrt{2}$$

(11) When $2^{3x} = 5$ **(x is a real number),** $\dfrac{2^{5x}+2^{-4x}}{2^{2x}+2^{-x}} + \dfrac{2^{2x}-2^{-4x}}{2^{-4x}+2^{-x}}$

Multiplying the numerator and denominator by 2^x,

$$\frac{2^{5x}+2^{-4x}}{2^{2x}+2^{-x}} + \frac{2^{2x}-2^{-4x}}{2^{-4x}+2^{-x}} = \frac{2^{6x}+2^{-3x}}{2^{3x}+2^0} + \frac{2^{3x}-2^{-3x}}{2^{-3x}+2^0} = \frac{\left(2^{3x}\right)^2+\left(2^{3x}\right)^{-1}}{2^{3x}+2^0} + \frac{2^{3x}-\left(2^{3x}\right)^{-1}}{\left(2^{3x}\right)^{-1}+2^0}$$

$$= \frac{5^2+5^{-1}}{5+1} + \frac{5-5^{-1}}{5^{-1}+1} = \frac{25+\frac{1}{5}}{6} + \frac{5-\frac{1}{5}}{\frac{1}{5}+1} = \frac{\frac{126}{5}}{6} + \frac{\frac{24}{5}}{\frac{6}{5}} = \frac{21}{5} + 4 = 8\frac{1}{5}$$

(12) When $x^{\frac{1}{2}} = e^{\frac{1}{2}} + e^{-\frac{1}{2}}$ **($e > 1$),** $\dfrac{x-2-\sqrt{x^2-4x}}{x-2+\sqrt{x^2-4x}}$

$$x = \left(x^{\frac{1}{2}}\right)^2 = \left(e^{\frac{1}{2}}+e^{-\frac{1}{2}}\right)^2 = e+2+e^{-1} \quad \therefore \ x-2 = e+e^{-1} \ \cdots\cdots ①$$

Squaring both sides of ①, $x^2 - 4x + 4 = e^2 + 2 + e^{-2}$ $\quad \therefore \ x^2 - 4x = e^2 + e^{-2} - 2$

Thus, $\sqrt{x^2-4x} = \sqrt{e^2+e^{-2}-2} = \sqrt{(e-e^{-1})^2} = e - e^{-1} \ \ (\because \ e > e^{-1})$

Therefore, $\dfrac{x-2-\sqrt{x^2-4x}}{x-2+\sqrt{x^2-4x}} = \dfrac{e+e^{-1}-(e-e^{-1})}{e+e^{-1}+(e-e^{-1})} = \dfrac{2e^{-1}}{2e} = \dfrac{e^{-1}}{e} = \dfrac{1}{e^2}$

#9 Find the value.

(1) When a point $P(a, b)$ lies on the line that has x-intercept 2 and y-intercept 1, find the minimum value of $\sqrt{2^a} + 2^b$.

The equation of the line is $\dfrac{x}{2} + \dfrac{y}{1} = 1 \ \cdots\cdots ①$

Since $P(a, b)$ is on the line ①, $\dfrac{a}{2} + b = 1$ $\qquad$ Since $2^a > 0$ and $2^b > 0$,

$$\sqrt{2^a} + 2^b = 2^{\frac{a}{2}} + 2^b \geq 2\sqrt{2^{\frac{a}{2}} \cdot 2^b} \quad \text{by the arithmetic and geometric means.}$$

$$= 2\sqrt{2^{\frac{a}{2}+b}} = 2\sqrt{2^1} = 2\sqrt{2} \ \left(\text{when } \frac{a}{2} = b, \text{ LHS} = \text{RHS}\right)$$

Therefore, the minimum value of $\sqrt{2^a} + 2^b$ is $2\sqrt{2}$.

(2) When α and β are two roots of an equation $x^2 + 2ax + 4 = 0$, $\dfrac{\alpha^{-1}-\beta^{-1}}{\alpha^{-2}-\beta^{-2}} = \dfrac{3}{2}$ ($\alpha > 0$, $\beta > 0$). Find the value of the constant a.

By the relationship between the roots and coefficients, $\alpha + \beta = -2a$, $\alpha\beta = 4$

$$\frac{\alpha^{-1}-\beta^{-1}}{\alpha^{-2}-\beta^{-2}} = \frac{\alpha^{-1}-\beta^{-1}}{(\alpha^{-1}+\beta^{-1})(\alpha^{-1}-\beta^{-1})} = \frac{1}{\alpha^{-1}+\beta^{-1}} = \frac{1}{\frac{1}{\alpha}+\frac{1}{\beta}} = \frac{1}{\frac{\alpha+\beta}{\alpha\beta}} = \frac{\alpha\beta}{\alpha+\beta} = \frac{4}{-2a} = -\frac{2}{a}$$

$\therefore \ -\dfrac{2}{a} = \dfrac{3}{2}$ Therefore, $a = -\dfrac{4}{3}$

(3) For real numbers x, y, and z, $2^x = 10$, $20^y = 10$, and $a^z = 10$ ($a > 0$).

When $\dfrac{1}{x} - \dfrac{1}{y} + \dfrac{1}{z} = 3$, find the value of a.

$2^x = 10 \ \Rightarrow \ 2 = 10^{\frac{1}{x}}$ $\qquad 20^y = 10 \ \Rightarrow \ 20 = 10^{\frac{1}{y}}$ $\qquad a^z = 10 \ \Rightarrow \ a = 10^{\frac{1}{z}}$

$10^{\frac{1}{x}} \div 10^{\frac{1}{y}} \times 10^{\frac{1}{z}} = 10^{\frac{1}{x}-\frac{1}{y}+\frac{1}{z}} = 10^3$

Since $2 \div 20 \times a = 10^3$, $a = 10^4$

(4) For real numbers x and y, $3^x = 5$ and $5^{\frac{y}{2}} = 27$. Find the value of xy.

$3^x = 5 \ \Rightarrow \ 3 = 5^{\frac{1}{x}} \cdots\cdots ①$ $\qquad 5^{\frac{y}{2}} = 27 \ \Rightarrow \ 5^{\frac{y}{2}} = 3^3 \cdots\cdots ②$

Sbbstituting ① into ②, $5^{\frac{y}{2}} = \left(5^{\frac{1}{x}}\right)^3 = 5^{\frac{3}{x}}$ $\quad \therefore \ \dfrac{y}{2} = \dfrac{3}{x}$ $\qquad \therefore \ xy = 6$

Alternative approach:

$3^x = 5 \ \Rightarrow \ 3 = 5^{\frac{1}{x}}$

$5^{\frac{y}{2}} = 27 \ \Rightarrow \ 5^{\frac{y}{2}} = 3^3$; $\ 5^{\frac{y}{6}} = 3$ $\quad \therefore \ 5^{\frac{y}{6}} = 5^{\frac{1}{x}}$ $\quad \therefore \ \dfrac{1}{x} = \dfrac{y}{6}$ $\quad \therefore \ xy = 6$

(5) For real numbers a, b, and c such that $abc = 12$, $2^a = 3^2$ and $3^b = 5^3$.

Find the value of 5^c.

$2^a = 3^2 \ \Rightarrow \ 2^{\frac{a}{2}} = 3$ $\qquad 3^b = 5^3 \ \Rightarrow \ 3^{\frac{b}{3}} = 5$

$\therefore \ 5^c = \left(3^{\frac{b}{3}}\right)^c = 3^{\frac{bc}{3}} = \left(2^{\frac{a}{2}}\right)^{\frac{bc}{3}} = 2^{\frac{abc}{6}}$

Since $abc = 12$, $\dfrac{abc}{6} = 2$ $\quad$ Therefore, $5^c = 2^2 = 4$

(6) When $a = 2^{\frac{2}{3}}$ and $b = 3^{\frac{1}{4}}$, find the value of mn such that $a^m b^n = 36$.

$a^m b^n = \left(2^{\frac{2}{3}}\right)^m \left(3^{\frac{1}{4}}\right)^n = 2^{\frac{2m}{3}} 3^{\frac{n}{4}} = 36$

Since $36 = 2^2 \cdot 3^2$, $\dfrac{2m}{3} = 2$ and $\dfrac{n}{4} = 2$

$\therefore\ m = 3,\ n = 8$ Therefore, $mn = 24$

(7) Find the value of the constant a such that $4^a = \left(\sqrt{2+\sqrt{3}} - \sqrt{2-\sqrt{3}}\right)^{\frac{1}{3}}$.

$\sqrt{2+\sqrt{3}} - \sqrt{2-\sqrt{3}} = \sqrt{\dfrac{4+2\sqrt{3}}{2}} - \sqrt{\dfrac{4-2\sqrt{3}}{2}}$

$= \dfrac{\sqrt{(3+1)+2\sqrt{3\cdot1}}}{\sqrt{2}} - \dfrac{\sqrt{(3+1)-2\sqrt{3\cdot1}}}{\sqrt{2}} = \dfrac{\sqrt{3}+1}{\sqrt{2}} - \dfrac{\sqrt{3}-1}{\sqrt{2}} = \dfrac{2}{\sqrt{2}} = \dfrac{2\sqrt{2}}{2} = \sqrt{2}$

$\therefore\ \left(\sqrt{2+\sqrt{3}} - \sqrt{2-\sqrt{3}}\right)^{\frac{1}{3}} = \left(\sqrt{2}\right)^{\frac{1}{3}} = \left(2^{\frac{1}{2}}\right)^{\frac{1}{3}} = 2^{\frac{1}{6}}$

Since $4^a = (2^2)^a = 2^{2a}$, $2a = \dfrac{1}{6}$ Therefore, $a = \dfrac{1}{12}$

(8) A function $y = 3^{-x^2+4x-3}$ has the maximum value when $x = a$. Find the value of a.

Let $f(x) = -x^2 + 4x - 3$

Then, $f(x) = -(x^2 - 4x) - 3 = -(x-2)^2 + 4 - 3 = -(x-2)^2 + 1$

Since $y = 3^{f(x)}$ and $3 > 1$,

y has maximum value when $f(x)$ has maximum value.

Since $f(x)$ has the maximum value 1 when $x = 2$,

y has maximum value $3^1 = 3$ when $x = 2$.   Therefore, $a = 2$.

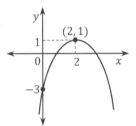

(9) When a function $y = a^{x^2-4x+6}$ has the maximum value $\dfrac{1}{100}$,

find the value of the constant a $(a > 0)$.

i) If $a > 1$, then the exponential function cannot have the maximum value $\dfrac{1}{100}$.

ii) If $a = 1$, then $y = 1$ $\therefore y$ cannot have the maximum value $\dfrac{1}{100}$.

iii) If $0 < a < 1$, then let $f(x) = x^2 - 4x + 6$

When $f(x)$ has minimum value, $y = a^{f(x)}$ has maximum value. $(\because 0 < a < 1)$

$f(x) = x^2 - 4x + 6 = (x-2)^2 - 4 + 6 = (x-2)^2 + 2 \geq 2$

$\therefore$ The minimum value of $f(x)$ is 2.

$\therefore$ The maximum value of $y = a^{f(x)}$ is a^2.

Since $a^2 = \dfrac{1}{100} = \left(\dfrac{1}{10}\right)^2$ and $a > 0$, $a = \dfrac{1}{10}$

(10) A function $y = (2^x - 1)^2 + (2^{-x} - 1)^2$ has the minimum value when $x = a$.

Find the value of a.

$y = (2^x)^2 - 2 \cdot 2^x + 1 + (2^{-x})^2 - 2 \cdot 2^{-x} + 1 = \{(2^x)^2 + 2 + (2^{-x})^2\} - 2(2^x + 2^{-x})$

$\quad = (2^x + 2^{-x})^2 - 2(2^x + 2^{-x})$

Let $X = 2^x + 2^{-x}$ $\qquad$ Since $2^x > 0$ and $2^{-x} > 0$,

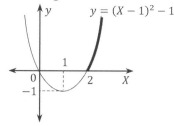

$2^x + 2^{-x} \geq 2\sqrt{2^x \cdot 2^{-x}} = 2$ by the relationship between arithmetic and geometric means

(when $2^x = 2^{-x}$; i.e., $x = 0$, LHS=RHS) $\quad \therefore X \geq 2$

$\therefore y = X^2 - 2X = (X-1)^2 - 1$, $X \geq 2$

From the graph, y has minimum value 0 when $X = 2$.

$\therefore 2^x + 2^{-x} = 2$ $\quad \therefore x = 0$ $\qquad$ Therefore, $a = 0$

#10 Compare the graph of each of the following with the graph of $f(x) = 2^x$.

$\quad$ **Identify the domain and range of each function.**

$\quad$ **(1) $g(x) = 2^{x+1}$**

$\quad$ **(2) $h(x) = 2^x - 3$**

$\quad$ **(3) $k(x) = -2^x$**

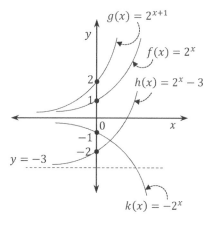

(1) Since $g(x) = 2^{x+1} = f(x+1)$, the graph of g is obtained by shifting the graph of f 1 unit to the left. The domain is $(-\infty, \infty)$ and the range is $(0, \infty)$.

(2) Since $h(x) = 2^x - 3 = f(x) - 3$, the graph of h is obtained by shifting the graph of f down 3 units. The domain is $(-\infty, \infty)$ and the range is $(-3, \infty)$.

(3) Since $k(x) = -2^x = -f(x)$, the graph of k is obtained by reflecting the graph of f in the x-axis. The domain is $(-\infty, \infty)$ and the range is $(-\infty, 0)$.

#11 For each function, find the maximum and minimum values within the given domain.

$\quad$ **(1) $y = 3 \cdot 2^x$ $(-1 \leq x \leq 1)$**

Let $2^x = X$. Then, $y = 3X$

Since $-1 \leq x \leq 1$, $2^{-1} \leq 2^x \leq 2^1$; i.e., $\dfrac{1}{2} \leq X \leq 2$

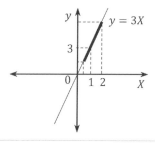

From the graph of $y = 3X$, $\dfrac{1}{2} \leq X \leq 2$,

the maximum value is 6 when $X = 2$; i.e., $x = 1$

and the minimum value is $\dfrac{3}{2}$ when $X = \dfrac{1}{2}$; i.e., $x = -1$.

(2) $y = 2^{1+2x}$ $(-1 \le x \le 1)$

$y = 2^{1+2x} = 2 \cdot 2^{2x}$

Let $2^x = X$. Then, $y = 2X^2$

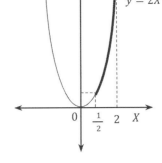
$y = 2X^2$

Since $-1 \le x \le 1$, $2^{-1} \le 2^x \le 2^1$; i.e., $\dfrac{1}{2} \le X \le 2$

From the graph of $y = 2X^2$,

the maximum value is 8 when $X = 2$; i.e., $x = 1$

and the minimum value is $\dfrac{1}{2}$ when $X = \dfrac{1}{2}$; i.e., $x = -1$.

(3) $y = 3^x \cdot 2^{-x}$ $(0 \le x \le 2)$

$y = 3^x \cdot 2^{-x} = 3^x \cdot \dfrac{1}{2^x} = \left(\dfrac{3}{2}\right)^x$, $0 \le x \le 2$

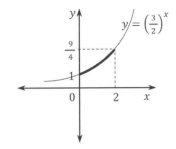
$y = \left(\dfrac{3}{2}\right)^x$

From the graph of $y = \left(\dfrac{3}{2}\right)^x$; Increasing function,

the maximum value is $\left(\dfrac{3}{2}\right)^2 = \dfrac{9}{4}$ when $x = 2$

and the minimum value is $\left(\dfrac{3}{2}\right)^0 = 1$ when $x = 0$.

(4) $y = 4^x + 2^{x+1} - 2$ $(-1 \le x \le 1)$

Let $2^x = X$. Then,

$y = (2^2)^x + 2 \cdot 2^x - 2 = X^2 + 2X - 2 = (X+1)^2 - 1 - 2 = (X+1)^2 - 3$

Since $-1 \le x \le 1$, $2^{-1} \le 2^x \le 2^1$; i.e., $\dfrac{1}{2} \le X \le 2$

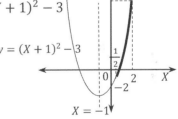
$y = (X+1)^2 - 3$

From the graph of $y = (X+1)^2 - 3$,

the maximum value is 6 when $X = 2$; i.e., $x = 1$

and the minimum value is $-\dfrac{3}{4}$ when $X = \dfrac{1}{2}$; i.e., $x = -1$.

(5) $y = -4^x + 2^{x+3}$ $(2 \le x \le 3)$

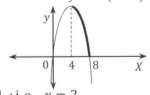
$y = -X(X-8)$

Let $2^x = X$. Then, $y = -X^2 + 8X = -(X^2 - 8X) = -X(X-8)$

Since $2 \le x \le 3$, $2^2 \le 2^x \le 2^3$; i.e., $4 \le X \le 8$

Form the graph, the maximum value is $-4(4-8) = 16$ when $X = 4$; i.e., $x = 2$

and the minimum value is 0 when $X = 8$; i.e., $x = 3$.

(6) $y = \left(\dfrac{1}{4}\right)^x - 2\left(\dfrac{1}{2}\right)^x - 1$ $(-2 \le x \le 3)$

$y = \left(\dfrac{1}{4}\right)^x - 2\left(\dfrac{1}{2}\right)^x - 1$ $\Rightarrow$ $y = \left\{\left(\dfrac{1}{2}\right)^x\right\}^2 - 2\left(\dfrac{1}{2}\right)^x - 1$

Let $\left(\dfrac{1}{2}\right)^x = X$.

Then, $X > 0$ and $y = X^2 - 2X - 1 = (X-1)^2 - 2$

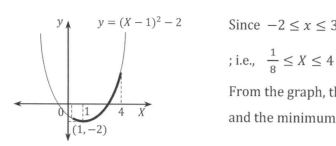

Since $-2 \leq x \leq 3$, $\left(\frac{1}{2}\right)^3 \leq \left(\frac{1}{2}\right)^x \leq \left(\frac{1}{2}\right)^{-2}$

; i.e., $\frac{1}{8} \leq X \leq 4$

From the graph, the maximum value is 7 when $X = 4$

and the minimum value is -2 when $X = 1$.

#12 Find the minimum value for each function.

(1) $y = 2^{2+x} + 2^{2-x}$

Since $2^{2+x} > 0$ and $2^{2-x} > 0$ for any real number x,

$2^{2+x} + 2^{2-x} \geq 2\sqrt{2^{2+x} \cdot 2^{2-x}}$ by the relationship between arithmetic and geometric means

$\qquad = 2\sqrt{2^{2+x+2-x}} = 2\sqrt{2^4} = 8$

Therefore, the minimum value is 8.

(2) $y = 4^x + 4^{-x} - 2(2^x + 2^{-x}) + 7$

Let $2^x + 2^{-x} = X$

Since $2^x > 0$ and $2^{-x} > 0$ for any real number x,

$2^x + 2^{-x} \geq 2\sqrt{2^x \cdot 2^{-x}}$ by the relationship between arithmetic and geometric means

$\qquad = 2 \qquad$ (when $2^x = 2^{-x}$; i.e., $x = 0$, LHS = RHS)

Since $(2^x + 2^{-x})^2 = 4^x + 2 + 4^{-x}$,

$4^x + 4^{-x} = X^2 - 2$

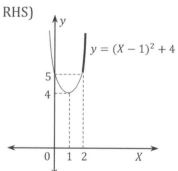

$\therefore \; y = 4^x + 4^{-x} - 2(2^x + 2^{-x}) + 7$

$\qquad = X^2 - 2 - 2X + 7 = X^2 - 2X + 5$

$\qquad = (X - 1)^2 + 4 \, , \; X \geq 2$

From the graph, the minimum value is 5 when $X = 2$.

#13 Solve for x.

(1) $\left(\frac{2}{3}\right)^{x^2-2} = \left(\frac{3}{2}\right)^{-2x-1}$

$\left(\frac{2}{3}\right)^{x^2-2} = \left(\frac{3}{2}\right)^{-2x-1} = \left\{\left(\frac{2}{3}\right)^{-1}\right\}^{-2x-1} = \left(\frac{2}{3}\right)^{2x+1}$

$\therefore \; x^2 - 2 = 2x + 1 \qquad \therefore \; x^2 - 2x - 3 = 0 \qquad \therefore \; (x - 3)(x + 1) = 0$

$\therefore \; x = 3$ or $x = -1$

(2) $x^{x^2} > x^{2x+15} \; (x > 0)$

i) When $x > 1$

$x^2 > 2x + 15 \qquad \therefore \; x^2 - 2x - 15 > 0 \qquad \therefore \; (x - 5)(x + 3) > 0 \qquad \therefore \; x > 5$ or $x < -3$

Since $x > 1,\ x > 5$ $\cdots\cdots$ ①

ii) When $0 < x < 1$

$x^2 < 2x + 15$ $\qquad \therefore\ x^2 - 2x - 15 < 0$ $\qquad\qquad \therefore\ (x-5)(x+3) < 0$ $\qquad \therefore\ -3 < x < 5$

Since $0 < x < 1,\ 0 < x < 1$ $\cdots\cdots$ ②

By ① and ②, $x > 5$ or $0 < x < 1$

(3) $\left(\frac{1}{4}\right)^x + \frac{1}{4}\left(\frac{1}{2}\right)^x > 4\left(\frac{1}{2}\right)^x + 1$

$\left(\frac{1}{4}\right)^x + \frac{1}{4}\left(\frac{1}{2}\right)^x > 4\left(\frac{1}{2}\right)^x + 1 \Rightarrow \left\{\left(\frac{1}{2}\right)^x\right\}^2 + \frac{1}{4}\left(\frac{1}{2}\right)^x > 4\left(\frac{1}{2}\right)^x + 1$

Let $\left(\frac{1}{2}\right)^x = X.$ Then $X > 0$

$\therefore\ X^2 + \frac{1}{4}X > 4X + 1$ $\qquad \therefore\ 4X^2 + X > 16X + 4$ $\qquad \therefore\ 4X^2 - 15X - 4 > 0$

$\therefore\ (4X + 1)(X - 4) > 0$ $\qquad \therefore\ X > 4$ or $X < -\frac{1}{4}$

Since $X > 0,\ X > 4$

$\therefore\ \left(\frac{1}{2}\right)^x > 4$ $\qquad \therefore\ \left(\frac{1}{2}\right)^x > \left(\frac{1}{2}\right)^{-2}$

Since $\frac{1}{2} < 1,\ x < -2$

#14 (Transformations)

(1) If the graph of a function $y = 2^{2x}$ is translated with m units along the x-axis and n units along the y-axis, then the graph will be the same as the graph of a function $y = \frac{2^{2x}-3}{16}$. Find the value of $m + n$.

$y = \frac{2^{2x}-3}{16} = \frac{2^{2x}}{2^4} - \frac{3}{16} = 2^{2(x-2)} - \frac{3}{16}$

$\therefore$ The graph of $y = \frac{2^{2x}-3}{16}$ is translated from the graph of $y = 2^{2x}$ with 2 units along the

x-axis and $-\frac{3}{16}$ units along the y-axis. Therefore, $m + n = 2 - \frac{3}{16} = \frac{29}{16}$.

(2) If the graph of a function $f(x) = 2^x$ is translated with m units along the x-axis and n units along the y-axis, then the graph will be the same as the graph of a function $y = g(x)$. By the translation, a point $P\big(1, f(1)\big)$ will be moved to a point $P'\big(4, g(4)\big)$. When the graph of $y = g(x)$ passes through a point $(0, 1)$, find the value of mn.

Since $f(x) = 2^x,\ g(x) = 2^{(x-m)} + n$

Since the point $(0, 1)$ lies on the graph of $g(x),\ 1 = 2^{-m} + n$ $\cdots\cdots$ ①

Since $P\big(1, f(1)\big) \longrightarrow P'\big(4, g(4)\big)$ and $f(1) = 2^1 = 2,\ (1 + m, 2 + n) = (4, 2^{(4-m)} + n)$

$\therefore\ m = 3$

Substituting $m = 3$ into ①, $1 = 2^{-3} + n$ $\quad \therefore \; n = 1 - \dfrac{1}{8} = \dfrac{7}{8}$

Therefore, $mn = 3 \cdot \dfrac{7}{8} = \dfrac{21}{8}$

#15 Solve exponential equation.

(1) When an equation $4^x = 2^{x+2} + a$ of x has two different real number solutions, find the range of the constant a.

$4^x = 2^{x+2} + a \;\Rightarrow\; \left(2^x\right)^2 = 4 \cdot 2^x + a$

Let $2^x = X$. Then $X > 0$

$X^2 = 4X + a \qquad \therefore \; X^2 - 4X - a = 0 \;\; \cdots\cdots ①$

Let D be the discriminant of the equation ①. Then $D > 0$

$\therefore \; 4 + a > 0 \quad \therefore \; a > -4$

Since $X > 0$, the two solutions are positive.

Thus, the product of the two roots is positive. ; i.e., $-a > 0 \quad \therefore \; a < 0$

Therefore, $-4 < a < 0$

(2) When an equation $8^x - 4^{x+1} - 2^{x+3} + 8 = 0$ has two different real roots, find the sum of the roots.

$8^x - 4^{x+1} - 2^{x+3} + 8 = 0 \;\Rightarrow\; \left(2^x\right)^3 - 4 \cdot \left(2^x\right)^2 - 8 \cdot 2^x + 8 = 0$

Let $2^x = X$. Then $X > 0$

$X^3 - 4X^2 - 8X + 8 = 0$

Let $P(X) = X^3 - 4X^2 - 8X + 8$. Then $P(-2) = -8 - 16 + 16 + 8 = 0$

$\therefore \; P(X)$ has $X + 2$ as a factor.

$$
\begin{array}{r|rrrr}
-2 & 1 & -4 & -8 & 8 \\
   &   & -2 & 12 & -8 \\
\hline
   & 1 & -6 & 4 & \boxed{0}
\end{array}
\qquad P(X) = (X+2)(X^2 - 6X + 4)
$$

$\therefore \; (X + 2)\left(X^2 - 6X + 4\right) = 0 \quad \therefore \; X = -2$ or $X = 3 \pm \sqrt{5}$

Since $X > 0$, $X = 3 \pm \sqrt{5}$

Let the two real roots of the given equation be α, β.

Then, 2^α and 2^β are the two real roots of the equation $X^2 - 6X + 4 = 0$.

By the relationship between roots and coefficients, $2^\alpha \cdot 2^\beta = 4 \quad \therefore \; 2^{\alpha+\beta} = 2^2$

$\therefore \; \alpha + \beta = 2 \qquad$ Therefore, the sum of the roots of the given equation is 2.

(3) Find the sum of all roots of an equation $2^{2x} - 2^{x+2} + 8 = 0$.

Let $2^x = X$. Then $X > 0$

$X^2 - 4X + 8 = 0$

Let the two real roots of the given equation be α, β.

Then, 2^α and 2^β are the two real roots of the equation $X^2 - 4X + 8 = 0$.

By the relationship between roots and coefficients, $2^\alpha \cdot 2^\beta = 8$ $\therefore$ $2^{\alpha+\beta} = 2^3$

$\therefore$ $\alpha + \beta = 3$ Therefore, the sum of the roots of the given equation is 3.

(4) Find the product of all roots of an equation $(x-2)^{x^2+2x-5} = (x-2)^{4x+3}$, $x > 2$

i)   When $x - 2 = 1$; i.e., $x = 3$,

   Since the base is 1, both sides of the equation are equal to 1.

   Thus, the equation is always true when $x = 3$.

ii) When $x - 2 \neq 1$; i.e., $x \neq 3$,

   $x^2 + 2x - 5 = 4x + 3$ $\Rightarrow$ $x^2 - 2x - 8 = 0$ $\therefore (x-4)(x+2) = 0$

   $\therefore$ $x = 4$ or $x = -2$

   Since $x > 2$, $x = 4$

   Therefore, the product of roots is $3 \times 4 = 12$.

#16 Solve each inequality.

(1) $\dfrac{1}{2^x-1} - \dfrac{2}{4^x+2^x+1} \leq \dfrac{5}{8^x-1}$

Let $2^x = X$. Then $X > 0$

Since $4^x = \left(2^2\right)^x = \left(2^x\right)^2 = X^2$ and $8^x = \left(2^3\right)^x = \left(2^x\right)^3 = X^3$,

$\dfrac{1}{2^x-1} - \dfrac{2}{4^x+2^x+1} \leq \dfrac{5}{8^x-1}$ $\Rightarrow$ $\dfrac{1}{X-1} - \dfrac{2}{X^2+X+1} \leq \dfrac{5}{X^3-1}$

$\Rightarrow$ $\dfrac{X^2+X+1}{(X-1)(X^2+X+1)} - \dfrac{2(X-1)}{(X-1)(X^2+X+1)} \leq \dfrac{5}{(X-1)(X^2+X+1)}$

$\Rightarrow$ $\dfrac{X^2-X-2}{(X-1)(X^2+X+1)} \leq 0$

$\Rightarrow$ $\dfrac{(X-2)(X+1)}{(X-1)(X^2+X+1)} \leq 0$

Since $X^2 + X + 1 = \left(X + \dfrac{1}{2}\right)^2 - \dfrac{1}{4} + 1 = \left(X + \dfrac{1}{2}\right)^2 + \dfrac{3}{4} > 0$ and $X + 1 > 0$,

$\dfrac{(X-2)(X+1)}{(X-1)(X^2+X+1)} \leq 0$ $\Rightarrow$ $\dfrac{X-2}{X-1} \leq 0$ $\Rightarrow$ $(X-2)(X-1) \leq 0$, $X \neq 1$

$\therefore$ $1 < X \leq 2$ $\therefore$ $1 < 2^x \leq 2$ $\therefore$ $2^0 < 2^x \leq 2^1$

Therefore, $0 < x \leq 1$

(2) $\left(\frac{1}{2}x - 1\right)^{x^2-8x} < \left(\frac{1}{2}x - 1\right)^{4x-27}$, $x > 2$

i) When $0 < \frac{1}{2}x - 1 < 1$; i.e., $1 < \frac{1}{2}x < 2$; $2 < x < 4$ ……①

$\left(\frac{1}{2}x - 1\right)^{x^2-8x} < \left(\frac{1}{2}x - 1\right)^{4x-27}$ $\Rightarrow$ $x^2 - 8x > 4x - 27$

∴ $x^2 - 12x + 27 > 0$ ∴ $(x-3)(x-9) > 0$ ∴ $x > 9$ or $x < 3$ ……②

By ① and ②, $2 < x < 3$

ii) When $\frac{1}{2}x - 1 = 1$; i.e., $\frac{1}{2}x = 2$; $x = 4$

$\left(\frac{1}{2}x - 1\right)^{x^2-8x} < \left(\frac{1}{2}x - 1\right)^{4x-27}$ $\Rightarrow$ $1 < 1$ This is not true.   ∴ $x \neq 4$

iii) When $\frac{1}{2}x - 1 > 1$; i.e., $\frac{1}{2}x > 2$; $x > 4$ ……③

$\left(\frac{1}{2}x - 1\right)^{x^2-8x} < \left(\frac{1}{2}x - 1\right)^{4x-27}$ $\Rightarrow$ $x^2 - 8x < 4x - 27$

∴ $x^2 - 12x + 27 < 0$ ∴ $(x-3)(x-9) < 0$ ∴ $3 < x < 9$ ……④

By ③ and ④, $4 < x < 9$

From i) and iii), $2 < x < 3$ or $4 < x < 9$

#17 Sketch the graph of a function.

(1)

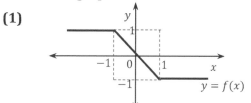

When the graph of a function $y = f(x)$ is shown as the Figure, sketch the graph of a function $y = 2^{1-f(x)}$.

From the graph, $f(x) = \begin{cases} 1 & , \ x < -1 \\ -x & , \ -1 \leq x < 1 \\ -1 & , \ x \geq 1 \end{cases}$

∴ $-f(x) = \begin{cases} -1 & , \ x < -1 \\ x & , \ -1 \leq x < 1 \\ 1 & , \ x \geq 1 \end{cases}$

∴ $1 - f(x) = \begin{cases} 0 & , \ x < -1 \\ x+1 & , \ -1 \leq x < 1 \\ 2 & , \ x \geq 1 \end{cases}$

Therefore, $2^{1-f(x)} = \begin{cases} 1 & , \ x < -1 \\ 2^{x+1} & , \ -1 \leq x < 1 \\ 4 & , \ x \geq 1 \end{cases}$

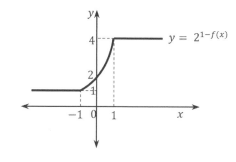

(2) The three functions $f(x)$, $g(x)$, and $h(x)$ satisfy the following conditions:

① $f(x) = f(-x)$,

② $g(-x) = -g(x)$, **and**

③ $h(x) = f(x) + g(x)$.

When $h(x) = 3^x$, sketch the graph of a function $y = f(x)$.

Since $h(x) = f(x) + g(x) = 3^x$, ⋯⋯❶

$h(-x) = f(-x) + g(-x) = 3^{-x}$

Since $f(x) = f(-x)$ and $g(-x) = -g(x)$,

$f(x) - g(x) = f(-x) + g(-x) = 3^{-x}$ ⋯⋯❷

❶+❷ : $2f(x) = 3^x + 3^{-x}$ ∴ $f(x) = \dfrac{3^x + 3^{-x}}{2}$

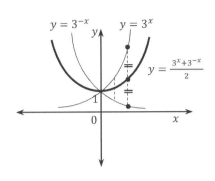

#18 Find the range.

(1) Find the range of x so that $\log_{x-1}(-3x^2 + 22x - 24)$ is defined for all real number x.

i) $x - 1 > 0$, $x - 1 \neq 1$

Since $x > 1$ and $x \neq 2$, $1 < x < 2$ or $x > 2$ ⋯⋯①

ii) $-3x^2 + 22x - 24 > 0$

Since $3x^2 - 22x + 24 < 0$, $(3x - 4)(x - 6) < 0$ ∴ $\dfrac{4}{3} < x < 6$ ⋯⋯②

By ① and ②, $\dfrac{4}{3} < x < 2$ or $2 < x < 6$

(2) Find the range of a so that $\log_{2a-1}(ax^2 + 3ax + 1 + a)$ is defined for all real number x.

i) $2a - 1 > 0$, $2a - 1 \neq 1$

Since $a > \dfrac{1}{2}$ and $a \neq 1$, $\dfrac{1}{2} < a < 1$ or $a > 1$ ⋯⋯①

ii) $ax^2 + 3ax + 1 + a > 0$

Let D be the discriminant of an equation $ax^2 + 3ax + 1 + a = 0$. Then, $D < 0$

∴ $9a^2 - 4a(1 + a) < 0$ ∴ $5a^2 - 4a = a(5a - 4) < 0$

∴ $0 < x < \dfrac{4}{5}$ ⋯⋯②

By ① and ②, $\dfrac{1}{2} < x < \dfrac{4}{5}$

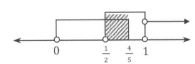

#19 Find the value of x for each expression.

(1) $\log_3 81 = x$

$\log_3 81 = x \Rightarrow 3^x = 81$

Since $81 = 3^4$, $x = 4$ ∴ $\log_3 81 = 4$

(2) $\log_8 0.25 = x$

$\log_8 0.25 = x \implies 8^x = 0.25 \qquad \therefore (2^3)^x = \dfrac{1}{4} \qquad \therefore 2^{3x} = 2^{-2} \qquad \therefore 3x = -2$

$\therefore x = -\dfrac{2}{3} \qquad \therefore \log_8 0.25 = -\dfrac{2}{3}$

(3) $\log_x 16 = 2$

$\log_x 16 = 2 \implies x^2 = 16$

Since $x > 0$ and $x \neq 1$, $x = 4$

(4) $\log_{2\sqrt{2}} 64 = x$

$\log_{2\sqrt{2}} 64 = x \implies (2\sqrt{2})^x = 64 \qquad \therefore (\sqrt{8})^x = 64 \qquad \therefore 8^{\frac{x}{2}} = 8^2 \qquad \therefore \dfrac{x}{2} = 2 \qquad \therefore x = 4$

(5) $\log_x 16 = \dfrac{2}{3}$

$\log_x 16 = \dfrac{2}{3} \implies x^{\frac{2}{3}} = 16 \qquad \therefore \left(x^{\frac{2}{3}}\right)^{\frac{3}{2}} = 16^{\frac{3}{2}} = (4^2)^{\frac{3}{2}} = 4^3 \qquad \therefore x = 4^3 = 64$

(6) $\log_9 x = 1.5$

$\log_9 x = 1.5 \implies 9^{1.5} = x$

Since $9^{1.5} = (3^2)^{1.5} = (3^2)^{\frac{3}{2}} = 3^3$, $x = 3^3 = 27$

(7) $\log_4(\log_{81} x) = -1$

$\log_4(\log_{81} x) = -1 \implies 4^{-1} = \log_{81} x \qquad \therefore \log_{81} x = \dfrac{1}{4} \qquad \therefore 81^{\frac{1}{4}} = x$

$\therefore (3^4)^{\frac{1}{4}} = x \qquad \therefore x = 3$

#20 Express each expression.

(1) When $\log_{10} 2 = a$ and $\log_{10} 3 = b$, express the following expressions with a and b.

① $\log_{10} 6$ ② $\log_{10} 5$ ③ $\log_{10} 800$ ④ $\log_{10}\left(\dfrac{3}{5}\right)^{-10}$ ⑤ $\log_{10} \sqrt[4]{500}$

① $\log_{10} 6 = \log_{10} 2 \cdot 3 = \log_{10} 2 + \log_{10} 3 = a + b$

② $\log_{10} 5 = \log_{10} \dfrac{10}{2} = \log_{10} 10 - \log_{10} 2 = 1 - \log_{10} 2 = 1 - a$

③ $\log_{10} 800 = \log_{10}(2 \cdot 4 \cdot 100) = \log_{10} 2 + \log_{10} 4 + \log_{10} 100$

$\qquad\qquad = \log_{10} 2 + \log_{10} 2^2 + \log_{10} 10^2 = \log_{10} 2 + 2\log_{10} 2 + 2\log_{10} 10$

$\qquad\qquad = a + 2a + 2 \cdot 1 = 3a + 2$

④ $\log_{10}\left(\dfrac{3}{5}\right)^{-10} = -10\log_{10}\left(\dfrac{3}{5}\right) = -10(\log_{10} 3 - \log_{10} 5) = -10\left(\log_{10} 3 - \log_{10} \dfrac{10}{2}\right)$

$\qquad\qquad = -10\{\log_{10} 3 - (\log_{10} 10 - \log_{10} 2)\}$

$\qquad\qquad = -10(\log_{10} 3 - 1 + \log_{10} 2) = -10(b - 1 + a)$

⑤ $\log_{10} \sqrt[4]{500} = \log_{10}(500)^{\frac{1}{4}} = \frac{1}{4}\log_{10} 500 = \frac{1}{4}\log_{10}\left(\frac{1000}{2}\right) = \frac{1}{4}\left(\log_{10} 1000 - \log_{10} 2\right)$

$\qquad = \frac{1}{4}\left(\log_{10} 10^3 - \log_{10} 2\right) = \frac{1}{4}\left(3\log_{10} 10 - \log_{10} 2\right) = \frac{1}{4}(3\cdot 1 - a) = \frac{3}{4} - \frac{a}{4}$

(2) When $45^a = 27$ and $5^b = 81$, express $\dfrac{3}{a} - \dfrac{4}{b}$ with a and b.

$45^a = 27 \Rightarrow$ Taking the common logarithm of both sides to the base 10,

$$\log 45^a = \log 27 \qquad \therefore a\log 45 = \log 27 \qquad \therefore a = \frac{\log 27}{\log 45}$$

$5^b = 81 \Rightarrow$ Taking the common logarithm of both sides to the base 10,

$$\log 5^b = \log 81 \qquad \therefore b\log 5 = \log 81 \qquad \therefore b = \frac{\log 81}{\log 5}$$

$\dfrac{3}{a} - \dfrac{4}{b} = 3\dfrac{1}{a} - 4\dfrac{1}{b} = 3\dfrac{\log 45}{\log 27} - 4\dfrac{\log 5}{\log 81} = \dfrac{3\log 45}{\log 3^3} - \dfrac{4\log 5}{\log 3^4} = \dfrac{3\log 45}{3\log 3} - \dfrac{4\log 5}{4\log 3}$

$\qquad = \dfrac{\log 45}{\log 3} - \dfrac{\log 5}{\log 3} = \dfrac{\log 45 - \log 5}{\log 3} = \dfrac{\log\left(\frac{45}{5}\right)}{\log 3} = \dfrac{\log 9}{\log 3} = \log_3 9 = 2\log_3 3 = 2$

(3) When $\log_2 3 = a$ and $\log_3 5 = b$, express $\log_{30} 20$ with a and b.

$\log_{30} 20 = \dfrac{\log_3 20}{\log_3 30} = \dfrac{\log_3(2^2\cdot 5)}{\log_3(2\cdot 3\cdot 5)} = \dfrac{2\log_3 2 + \log_3 5}{\log_3 2 + \log_3 3 + \log_3 5} = \dfrac{2\frac{1}{\log_2 3} + \log_3 5}{\frac{1}{\log_2 3} + 1 + \log_3 5} = \dfrac{\frac{2}{a} + b}{\frac{1}{a} + 1 + b}$

$\qquad = \dfrac{\frac{2+ab}{a}}{\frac{1+a+ab}{a}} = \dfrac{2+ab}{1+a+ab}$

(4) When $10^a = x$, $10^b = y$, and $10^c = z$, express $\log_{xy}\sqrt{y^2 z}$ with a, b, and c.

Taking the common logarithm of both sides to the base 10,

$a = \log_{10} x$, $b = \log_{10} y$, and $c = \log_{10} z$

$\therefore \log_{xy}\sqrt{y^2 z} = \dfrac{\log_{10}\sqrt{y^2 z}}{\log_{10} xy} = \dfrac{\frac{1}{2}\log_{10} y^2 z}{\log_{10} xy} = \dfrac{\frac{1}{2}\left(\log_{10} y^2 + \log_{10} z\right)}{\log_{10} x + \log_{10} y}$

$\qquad = \dfrac{2\log_{10} y + \log_{10} z}{2(\log_{10} x + \log_{10} y)} = \dfrac{2b+c}{2(a+b)}$

(5) When $\log_2 10 = \dfrac{1}{a}$ and $\log_3 10 = \dfrac{1}{b}$, express $\log_{\sqrt{48}}\sqrt[3]{24}$ with a and b.

$\log_2 10 = \dfrac{1}{a} \Rightarrow \log_{10} 2 = a$

$\log_3 10 = \dfrac{1}{b} \Rightarrow \log_{10} 3 = b$

$\log_{\sqrt{48}}\sqrt[3]{24} = \dfrac{\log_{10}\sqrt[3]{24}}{\log_{10}\sqrt{48}} = \dfrac{\frac{1}{3}\log_{10} 24}{\frac{1}{2}\log_{10} 48} = \dfrac{2\log_{10}(2^3\cdot 3)}{3\log_{10}(2^4\cdot 3)} = \dfrac{2(3\log_{10} 2 + \log_{10} 3)}{3(4\log_{10} 2 + \log_{10} 3)} = \dfrac{2(3a+b)}{3(4a+b)}$

(6) When $x^a = y^b = 6$ for positive numbers x and y, express $\log_{xy} x^3$ with a and

b. $(xy \neq 1)$

$$x^a = y^b = 6 \;\Rightarrow\; a = \log_x 6 \,,\; b = \log_y 6 \qquad \therefore\; \log_6 x = \frac{1}{a}\,,\;\; \log_6 y = \frac{1}{b}$$

$$\log_{xy} x^3 = \frac{\log_6 x^3}{\log_6 xy} = \frac{3\log_6 x}{\log_6 x + \log_6 y} = \frac{3\cdot\frac{1}{a}}{\frac{1}{a}+\frac{1}{b}} = \frac{\frac{3}{a}}{\frac{a+b}{ab}} = \frac{3b}{a+b}$$

#21 Simplify the expression.

(1) $10^{\log 5} = 5^{\log 10} = 5^1 = 5$

(2) $\log_2 (32)^x = \log_2 (2^5)^x = \log_2 2^{5x} = 5x\log_2 2 = 5x\cdot 1 = 5x$

(3) $\log_3\left(\frac{1}{9}\right) = \log_3 9^{-1} = \log_3 (3^2)^{-1} = \log_3 3^{-2} = -2\log_3 3 = -2\cdot 1 = -2$

(4) $\log_9 3 = \log_{3^2} 3 = \frac{1}{2}\log_3 3 = \frac{1}{2}\cdot 1 = \frac{1}{2}$

(5) $\log_5\left(\frac{\sqrt[3]{5}}{25}\right) = \log_5\left(\frac{\sqrt[3]{5}}{5^2}\right) = \log_5 \sqrt[3]{5} - \log_5 5^2 = \frac{1}{3}\log_5 5 - 2\log_5 5 = \frac{1}{3}\cdot 1 - 2\cdot 1 = -\frac{5}{3}$

(6) $\log_2 3 \cdot \log_3 2 = \frac{\log_{10} 3}{\log_{10} 2}\cdot\frac{\log_{10} 2}{\log_{10} 3} = 1$

(7) $\log_2 3 \cdot \log_3 4 \cdot \log_4 2 = \frac{\log_{10} 3}{\log_{10} 2}\cdot\frac{\log_{10} 4}{\log_{10} 3}\cdot\frac{\log_{10} 2}{\log_{10} 4} = 1$

(8) $\log_2 6 - \log_4 9 = \log_2 (2\cdot 3) - \log_{2^2} 3^2 = (\log_2 2 + \log_2 3) - \frac{2}{2}\log_2 3$

$$= 1 + \log_2 3 - \log_2 3 = 1$$

(9) $2\log_2 \sqrt{3} + 4\log_2 3 = \log_2\left(\sqrt{3}\right)^2 + \log_2 3^4$

$$= \log_2 3 + \log_2 81 = \log_2 (3\times 81) = \log_2 3^5 = 5\log_2 3$$

(10) $2\log\frac{3}{2} - \log\frac{5}{3} + 3\log 2 + \frac{1}{2}\log 25 = \log\left(\frac{3}{2}\right)^2 - \log\frac{5}{3} + \log 2^3 + \log\sqrt{25}$

$$= \log\left(\frac{9}{4}\div\frac{5}{3}\times 2^3\times\sqrt{25}\right) = \log\left(\frac{9}{4}\times\frac{3}{5}\times 8\times 5\right)$$

$$= \log 54 = \log(2\times 3^3) = \log 2 + 3\log 3$$

(11) $5^{2\log_5 4 + \log_5 3 - 3\log_5 2}$

$$2\log_5 4 + \log_5 3 - 3\log_5 2 = \log_5 4^2 + \log_5 3 - \log_5 2^3 = \log_5 (4^2\times 3\div 2^3) = \log_5 6$$

$$\therefore\; 5^{2\log_5 4 + \log_5 3 - 3\log_5 2} = 5^{\log_5 6} = 6^{\log_5 5} = 6^1 = 6$$

#22 Find the value.

(1) When α and β are two different roots of an equation $x^2 - 3x + 3 = 0$,

find the value of $\log_4(\alpha + \beta^{-1}) + \log_4(\alpha^{-1} + \beta) + \log_4 \alpha\beta$.

By the relationship between roots and coefficients, $\alpha + \beta = 3$, $\alpha\beta = 3$

$$\log_4(\alpha + \beta^{-1}) + \log_4(\alpha^{-1} + \beta) + \log_4 \alpha\beta = \log_4\left(\alpha + \frac{1}{\beta}\right) + \log_4\left(\frac{1}{\alpha} + \beta\right) + \log_4 \alpha\beta$$

$$= \log_4\left\{\left(\alpha + \frac{1}{\beta}\right)\left(\frac{1}{\alpha} + \beta\right)\alpha\beta\right\}$$

$$= \log_4\left\{\left(1 + \frac{1}{\alpha\beta} + \alpha\beta + 1\right)\alpha\beta\right\}$$

$$= \log_4\{2\alpha\beta + 1 + (\alpha\beta)^2\}$$

$$= \log_4(2 \cdot 3 + 1 + 3^2) = \log_4 16 = \log_4 4^2$$

$$= 2\log_4 4 = 2$$

(2) When $\log_2 a$ and $\log_2 b$ are two different roots of an equation $x^2 - 3x - 4 = 0$

$\quad$ **($a > 0$, $a \neq 1$, $b > 0$, $b \neq 1$), find the value of $\log_{a^2} 2 + \log_b \sqrt{2}$.**

By the relationship between roots and coefficients,

$\log_2 a + \log_2 b = 3$, $\log_2 a \cdot \log_2 b = -4$

$$\log_{a^2} 2 + \log_b \sqrt{2} = \frac{1}{2}\log_a 2 + \frac{1}{2}\log_b 2 = \frac{1}{2}(\log_a 2 + \log_b 2) = \frac{1}{2}\left(\frac{1}{\log_2 a} + \frac{1}{\log_2 b}\right)$$

$$= \frac{1}{2}\left(\frac{\log_2 a + \log_2 b}{\log_2 a \cdot \log_2 b}\right) = \frac{1}{2}\left(-\frac{3}{4}\right) = -\frac{3}{8}$$

(3) For any rational numbers a and b, $a\log_{10} 20 + \dfrac{b}{\log_8 100} + 5 = 0$.

$\quad$ **Find the value of $a + b$.**

$$a\log_{10} 20 + \frac{b}{\log_8 100} + 5 = 0 \implies a(\log_{10} 2 + \log_{10} 10) + b(\log_{100} 8) + 5 = 0$$

$$\implies a(\log_{10} 2 + \log_{10} 10) + b(\log_{10^2} 2^3) + 5 = 0$$

$$\implies a(\log_{10} 2 + \log_{10} 10) + \frac{3b}{2}(\log_{10} 2) + 5 = 0$$

$$\implies \left(a + \frac{3b}{2}\right)\log_{10} 2 + a + 5 = 0$$

Since a and b are rational numbers and $\log_{10} 2$ is irrational number,

$a + \dfrac{3b}{2} = 0$ and $a + 5 = 0$. $\qquad \therefore\ a = -5$, $b = \dfrac{10}{3}$

Therefore, $a + b = -5 + \dfrac{10}{3} = -\dfrac{5}{3}$

(4) For any real numbers a, b, and c ($a > 1$, $b > 1$, $c > 1$), $\log_a c : \log_b c = 1 : 2$

$\quad$ **Find the value of $\log_a b + \log_b a$.**

$\log_a c : \log_b c = 1 : 2 \implies \log_b c = 2\log_a c \qquad \therefore\ \dfrac{1}{\log_c b} = \dfrac{2}{\log_c a}$

$\therefore\ \dfrac{\log_c a}{\log_c b} = 2 \qquad \therefore\ \log_b a = 2$

$\therefore\ \log_a b + \log_b a = \dfrac{1}{\log_b a} + \log_b a = \dfrac{1}{2} + 2 = \dfrac{5}{2}$

Alternative approach:

$$\log_a b + \log_b a = \log_a b + \frac{1}{\log_a b} \quad \text{and} \quad \log_a b = \frac{\log_c b}{\log_c a}$$

$$\log_a c : \log_b c = 1 : 2 \Rightarrow \log_b c = 2\log_a c \quad \therefore \frac{1}{\log_c b} = 2\log_a c \quad \therefore \log_c b = \frac{1}{2\log_a c}$$

$$\therefore \log_a b = \frac{\log_c b}{\log_c a} = \frac{\frac{1}{2\log_a c}}{\log_c a} = \frac{1}{2\log_a c \cdot \log_c a} = \frac{1}{2\log_a c \cdot \frac{1}{\log_a c}} = \frac{1}{2}$$

Therefore, $\log_a b + \log_b a = \frac{1}{2} + 2 = \frac{5}{2}$

(5) When $a\log_2 3 = 5$ and $\log_2 b = 1 - \log_2(\log_3 2)$ for any real numbers a and b, find the value of ab.

$$a\log_2 3 = 5 \Rightarrow a = \frac{5}{\log_2 3}$$

$$\log_2 b = 1 - \log_2(\log_3 2) = \log_2 2 - \log_2(\log_3 2) = \log_2\left(\frac{2}{\log_3 2}\right)$$

$$\therefore b = \frac{2}{\log_3 2} \qquad \therefore ab = \frac{5}{\log_2 3} \cdot \frac{2}{\log_3 2} = 10\log_3 2 \cdot \frac{1}{\log_3 2} = 10$$

#23 Compare the numbers.

(1) $\log_2 3$, $\log_3 2$

$$\log_2 3 - \log_3 2 = \frac{\log 3}{\log 2} - \frac{\log 2}{\log 3} = \frac{(\log 3)^2 - (\log 2)^2}{\log 2 \cdot \log 3} = \frac{(\log 3 + \log 2)(\log 3 - \log 2)}{\log 2 \cdot \log 3} > 0$$

$$\therefore \log_2 3 > \log_3 2$$

(2) $\log(\log 2) + \log(\log 3)$, $2\log(\log\sqrt{6})$

Since $\log 2 > 0$, $\log 3 > 0$, and $\log 2 \neq \log 3$,

$\log 2 + \log 3 > 2\sqrt{\log 2 \cdot \log 3}$ by the relationship between arithmetic and geometric means

$$\therefore \log(2 \cdot 3) > 2\sqrt{\log 2 \cdot \log 3} \qquad \therefore \log 6 > 2\sqrt{\log 2 \cdot \log 3}$$

$$\therefore \frac{1}{2}\log 6 > \sqrt{\log 2 \cdot \log 3} \qquad \therefore \log\sqrt{6} > \sqrt{\log 2 \cdot \log 3}$$

Taking common logarithm of both sides, $\log(\log\sqrt{6}) > \log(\sqrt{\log 2 \cdot \log 3})$

$$\therefore \log(\log\sqrt{6}) > \frac{1}{2}\log(\log 2 \cdot \log 3) \qquad \therefore 2\log(\log\sqrt{6}) > \log(\log 2 \cdot \log 3)$$

$\therefore 2\log(\log\sqrt{6}) > \log(\log 2) + \log(\log 3)$ Therefore, $\log(\log 2) + \log(\log 3) < 2\log(\log\sqrt{6})$

(3) $\log\frac{5}{2}$, $\frac{\log 2 + \log 3}{2}$

$$\log\frac{5}{2} - \frac{\log 2 + \log 3}{2} = \log\frac{5}{2} - \frac{1}{2}\log(2 \cdot 3) = \log\frac{5}{2} - \log\sqrt{6} = \log\frac{5}{2\sqrt{6}}$$

Since $\left(\frac{5}{2\sqrt{6}}\right)^2 = \frac{25}{24} > 1$, $\frac{5}{2\sqrt{6}} > 1$

$\therefore \ \log \frac{5}{2\sqrt{6}} > 0 \quad$ Therefore, $\ \log \frac{5}{2} > \frac{\log 2 + \log 3}{2}$

(4) $\log_a b$, $\log_b a$, $\log_a \left(\frac{a}{b}\right)$, $\log_b \left(\frac{b}{a}\right)$ **when** $1 < a < b < a^2$

$1 < a < b < a^2 \Rightarrow \log_a 1 < \log_a a < \log_a b < \log_a a^2 \qquad \therefore \ 0 < 1 < \log_a b < 2 \ \cdots\cdots$ ①

Since $\ \log_b a = \frac{1}{\log_a b}$, $\ \frac{1}{2} < \log_b a < 1 \ \cdots\cdots$ ②

$\log_a \left(\frac{a}{b}\right) = \log_a a - \log_a b = 1 - \log_a b$

Since $\ -2 < -\log_a b < -1$, $\ -1 < 1 - \log_a b < 0 \ \cdots\cdots$ ③

$\log_b \left(\frac{b}{a}\right) = \log_b b - \log_b a = 1 - \log_b a$

Since $\ -1 < -\log_b a < -\frac{1}{2}$, $\ 0 < 1 - \log_b a < \frac{1}{2} \ \cdots\cdots$ ④

By ①, ②, ③, and ④, $\ \log_a \left(\frac{a}{b}\right) < \log_b \left(\frac{b}{a}\right) < \log_b a < \log_a b$

#24 When $\log 2.34 = 0.3692$, determine each logarithm.

(1) $\log 23.4$

Since the integer part has 2 digits, the characteristic is $2 - 1 = 1$.

$\therefore \ \log 23.4 = 1 + 0.3692 = 1.3692$

(2) $\log 234000$

Since the integer part has 6 digits, the characteristic is $6 - 1 = 5$.

$\therefore \ \log 234000 = 5 + 0.3692 = 5.3692$

(3) $\log 0.00234$

Since the first non-zero number in the decimal part appears on 3rd digit,

$\log 0.00234 = \overline{3}.3692 = -3 + 0.3692 = -2.6308$

(4) $\log(0.234)^2$

$\log(0.234)^2 = 2\log 0.234 = 2(\overline{1}.3692) = 2(-1 + 0.3692) = -1.2616$

(5) $\log \sqrt[3]{2340}$

$\log \sqrt[3]{2340} = \frac{1}{3}\log 2340 = \frac{1}{3}(3.3692) = 1.1231$

#25 When $\log 34.5 = 1.5378$, find the value of x.

(1) $\log x = 4.5378$

Note that mantissa is 0.5378.

Since the characteristic is 4, the integer part has 5 digits. Therefore, $x = 34500$

(2) $\log x = \overline{4}.5378$

Since the characteristic is $\overline{4}$, the non-zero decimal digit appears on 4th.

Therefore, $x = 0.000345$

#26 When $\log 2 = 0.3010$ and $\log 3 = 0.4771$, determine how many digits are in the integer part of the number or when the first non-zero digit appears in the decimal part of the number.

(1) 6^{100}

$\log 6^{100} = 100 \log 6 = 100(\log 2 + \log 3) = 100(0.3010 + 0.4771) = 77.81$

Since the characteristic is 77, the integer part of the number 6^{100} has 78 digits.

(2) 6^{25}

Since 6^{100} has 78 digits, $\log 6^{100} = 77.\times\times\times$

$\therefore \ 77 \leq \log 6^{100} < 78 \qquad \therefore \ 77 \leq 100 \log 6 < 78 \qquad \therefore \ 0.77 \leq \log 6 < 0.78$

$\therefore \ 0.77 \times 25 \leq 25 \log 6 < 0.78 \times 25 \qquad \therefore \ 19.25 \leq 25 \log 6 < 19.50$

$\therefore \ \log 6^{25} = 19.\times\times\times$ (The characteristic is 19)

Therefore, the number 6^{25} has 20 digits in the integer part.

(3) $\left(5^5\right)^5$

$\log(5^5)^5 = \log 5^{25} = 25 \log 5 = 25(\log 10 - \log 2) = 25(1 - 0.3010) = 17.475$

Since the characteristic is 17, $(5^5)^5$ has 18 digits in the integer part.

(4) 5^{-50}

$\log 5^{-50} = -50 \log 5 = -50(\log 10 - \log 2) = -50(1 - 0.3010) = -34.95$

$\qquad\qquad = -35 + 0.05 = \overline{35}.05$

$\therefore$ The 1st non-zero number appears 35th away from the decimal point.

(5) $2^{100} \div 3^{200}$

$\log(2^{100} \div 3^{200}) = \log 2^{100} - \log 3^{200} = 100 \log 2 - 200 \log 3$

$\qquad\qquad\qquad = 100 \times 0.3010 - 200 \times 0.4771 = -65.32 = \overline{66}.68$

Since the characteristic of the number $\log(2^{100} \div 3^{200})$ is $\overline{66}$, the 1st non-zero number appears 66th away from the decimal point.

#27 (Mantissa) Find the value.

(1) When $\log x$ and $\log x^2$ have the same mantissa, find the value of x. ($10 \leq x < 100$)

Since $10 \leq x < 100$, $\log 10 \leq \log x < \log 100 \qquad \therefore \ 1 \leq \log x < 2$

$\therefore \ \log x = 1.\times\times\times \qquad \therefore$ The characteristic of $\log x$ is 1.

Let α be the mantissa of $\log x$. Then, $\log x = 1 + \alpha$, $0 \leq \alpha < 1$

$\therefore \ \log x^2 = 2 \log x = 2(1 + \alpha) = 2 + 2\alpha$

If 2 is the characteristic and 2α is the mantissa, $0 \le 2\alpha < 2$.

But, it is not true. ($\because 0 \le$ mantissa < 1)

Thus, we have to consider the following cases:

i) $0 \le \alpha < \dfrac{1}{2}$; i.e., $0 \le 2\alpha < 1$

$\therefore \log x^2 = 2 + \underline{2\alpha}$ mantissa

Since $\log x$ and $\log x^2$ have the same mantissa, $\alpha = 2\alpha$ $\therefore \alpha = 0$

$\therefore \log x = 1 + 0 = 1$ $\therefore x = 10$

ii) $\dfrac{1}{2} \le \alpha < 1$; i.e., $1 \le 2\alpha < 2$

$\therefore \log x^2 = 2 + 2\alpha = (2 + 1) + \underline{(2\alpha - 1)}$ mantissa

Since $\log x$ and $\log x^2$ have the same mantissa, $\alpha = 2\alpha - 1$ $\therefore \alpha = 1$

Since $0 \le \alpha < 1$, $\alpha = 1$ is not possible.

Therefore, by i) and ii), $x = 10$

(2) When $\log x$ and $\log x^3$ have the same mantissa,

    find the value of x. ($100 \le x < 1000$)

Let α be the mantissa of $\log x$ and $\log x^3$.

Then, $\log x = n + \alpha$ and $\log x^3 = m + \alpha$,   (n, m; integers, $0 \le \alpha < 1$)

$\therefore \log x^3 - \log x = n - m$ (integer)

$\log x^3 - \log x = 3\log x - \log x = 2\log x$ (integer)

Since $100 \le x < 1000$, $2 \le \log x < 3$ $\therefore 4 \le 2\log x < 6$

Since $2\log x$ is an integer, $2\log x = 4$ or $2\log x = 5$

$\therefore \log x = 2$ or $\log x = \dfrac{5}{2}$

Therefore, $x = 10^2$ or $x = 10^{\frac{5}{2}}$

(3) When $\log x$ and $\log \dfrac{1}{x}$ have the same mantissa, find the value of x. ($10 < x < 1000$)

Let α be the mantissa of $\log x$ and $\log \dfrac{1}{x}$.

Then, $\log x = n + \alpha$ and $\log \dfrac{1}{x} = m + \alpha$,   ($n, m$; integers, $0 \le \alpha < 1$)

$\therefore \log x - \log \dfrac{1}{x} = n - m$ (integer)

$\log x - \log \dfrac{1}{x} = \log x - \log x^{-1} = \log x - (-\log x) = 2\log x$

Since $10 < x < 1000$, $1 < \log x < 3$ $\therefore 2 < 2\log x < 6$

Since $2\log x$ is an integer, $2\log x$ is 3, 4, or 5

$\therefore \ \log x = \dfrac{3}{2}$, $\log x = \dfrac{4}{2}$, or $\log x = \dfrac{5}{2}$

Therefore, $x = 10^{\frac{3}{2}}$, $x = 10^2$, or $x = 10^{\frac{5}{2}}$

(4) When the sum of mantissas of $\log x$ and $\log \sqrt{x}$ is 1, ($100 \le x < 1000$),

find the mantissa of $\log x$.

Since $100 \le x < 1000$, $2 \le \log x < 3$

Let α be the mantissa of $\log x$. Then, $\log x = 2 + \alpha$ $(0 \le \alpha < 1)$

$\therefore \ \log \sqrt{x} = \dfrac{1}{2}\log x = \dfrac{1}{2}(2 + \alpha) = 1 + \dfrac{\alpha}{2}$

Since $0 \le \alpha < 1$, $0 \le \dfrac{\alpha}{2} < \dfrac{1}{2}$ $\qquad \therefore \dfrac{\alpha}{2}$ is the mantissa of $\log \sqrt{x}$

Since $\alpha + \dfrac{\alpha}{2} = 1$, $\dfrac{3\alpha}{2} = 1$ $\qquad \therefore \ \alpha = \dfrac{2}{3}$

Therefore, the mantissa of $\log x$ is $\dfrac{2}{3}$.

(5) When $\log x = n + \alpha$, (n is an integer, $0 \le \alpha < 1$),

find the value of $\log x$ $(x > 0)$ such that $n^2 + \alpha^2 = \dfrac{37}{9}$

Since n is an integer and $0 \le \alpha < 1$, n^2 is an integer and $0 \le \alpha^2 < 1$

Since $n^2 + \alpha^2 = \dfrac{37}{9} = 4 + \dfrac{1}{9}$, $n^2 = 4$ and $\alpha^2 = \dfrac{1}{9}$

$\therefore \ n = \pm 2$ and $\alpha = \dfrac{1}{3}$ $(\because \alpha > 0)$

$\therefore \ \log x = 2 + \dfrac{1}{3} = \dfrac{7}{3}$ or $\log x = -2 + \dfrac{1}{3} = -\dfrac{5}{3}$

Therefore, $\log x$ is $\dfrac{7}{3}$ or $-\dfrac{5}{3}$

(6) For positive real numbers x and y, the following conditions are satisfied:

    ① $\log x$ and $\log y$ have the same characteristic.

    ② $\log x$ and $\log \dfrac{1}{y}$ have the same mantissa.

    ③ $\log x^2 y^3 = 17.5$

    Find the value of $x - y$.

By ①, let $\log x = n + \alpha$, $\log y = n + \beta$ $\quad$ (n; integer, $0 \le \alpha < 1$, $0 \le \beta < 1$)

$\log \dfrac{1}{y} = \log y^{-1} = -\log y = -(n + \beta) = -n - \beta = (-n - 1) + \underbrace{(1 - \beta)}_{\text{mantissa}}$

$(\because -\beta \ (< 0)$ cannot be a mantissa of $\log \dfrac{1}{y})$

By ②, $\alpha = 1 - \beta$ $\therefore$ $\alpha + \beta = 1$

By ③, $\log x^2 y^3 = \log x^2 + \log y^3 = 2 \log x + 3 \log y = 2(n + \alpha) + 3(n + \beta)$

$$= 5n + 2(\alpha + \beta) + \beta = 5n + 2 \cdot 1 + \beta = 17.5$$

$\therefore$ $5n + 2 = 17$, $\beta = 0.5$

$\therefore$ $n = 3$, $\alpha = \dfrac{1}{2}$, $\beta = \dfrac{1}{2}$

$\therefore$ $\log x = 3 + \dfrac{1}{2}$ and $\log y = 3 + \dfrac{1}{2}$ $\therefore$ $\log x = \log y$ $\therefore x = y$

Therefore, $x - y = 0$

#28 When $10^{0.3456} = 2.2162$, determine the *anti*log of each of the following number and express it in decimal form.

(1) *anti*log $0.3456 = 10^{0.3456} = 2.2162$

(2) *anti*log $2.3456 = 10^{2.3456} = 10^{0.3456+2} = 10^{0.3456} \cdot 10^2 = 221.62$

(3) *anti*log$(0.3456 - 2) = 10^{0.3456-2} = 10^{0.3456} \cdot 10^{-2} = 0.022162$

(4) *anti*log$(-2.6544) = anti \log(0.3456 - 3) = 10^{0.3456-3} = 10^{0.3456} \cdot 10^{-3} = 0.0022162$

#29 Find the value.

(1) Let the integer part and decimal part of the number $\log_3 18$ be a and b, respectively. Find the value of $2^a + 3^b$

Since $\log_3 3^2 < \log_3 18 < \log_3 3^3$, $2 < \log_3 18 < 3$

$\therefore$ $\log_3 18 = 2.\times\times\times$ $\therefore$ $a = 2$ and $b = (\log_3 18) - 2$

$2^a + 3^b = 2^2 + 3^{(\log_3 18)-2} = 2^2 + 3^{\log_3 18} \cdot 3^{-2} = 2^2 + 18^{\log_3 3} \cdot \dfrac{1}{9} = 4 + \dfrac{18}{9} = 6$

(2) Given $\log 2 = 0.3010$, $\log 3 = 0.4771$, and $\log 7 = 0.8451$, find the sum of the two numbers in the highest place value and the ones place value of the number 6^{100}.

$\log 6^{100} = 100 \log 6 = 100 \log(2 \cdot 3) = 100 (\log 2 + \log 3) = 100(0.3010 + 0.4771)$

$$= 77.81 = 77 + 0.81 \qquad \therefore \log 6 = 0.7781$$

Since $\log 7 = 0.8451$, $\log 6 < 0.81 < \log 7$

$\therefore$ $77 + \log 6 < 77 + 0.81 < 77 + \log 7$

$\therefore$ $\log 10^{77} + \log 6 < 77 + 0.81 < \log 10^{77} + \log 7$

$\therefore$ $\log(6 \cdot 10^{77}) < 77 + 0.81 < \log(7 \cdot 10^{77})$

$\therefore$ $\log(6 \cdot 10^{77}) < \log 6^{100} < \log(7 \cdot 10^{77})$

$\therefore$ $6 \cdot 10^{77} < 6^{100} < 7 \cdot 10^{77}$ $\therefore$ $6^{100} = (\underline{6 \cdot 10^{77}}).\times\times\times$

integer part

$\therefore$ The number in the highest place value is 6

Since $6^1 = \boxed{6}$, $6^2 = 3\boxed{6}$, $6^3 = 21\boxed{6}$, $\cdots\cdots$,

the digit in the ones place value of the number 6^n (n is a positive integer) is 6.

Therefore, the sum is $6 + 6 = 12$.

(3) For any positive numbers x and y such that $\log x = 3 + \alpha$ $(0 < \alpha < \frac{1}{4})$ and

$\quad \log y = 1 + \beta$ $\left(\frac{1}{2} < \beta < 1\right)$, $\dfrac{x^2}{y}$ **has n digits in the integer part. Find the value of n.**

$\log \dfrac{x^2}{y} = \log x^2 - \log y = 2\log x - \log y = 2(3 + \alpha) - (1 + \beta) = 5 + 2\alpha - \beta$

Since $0 < \alpha < \dfrac{1}{4}$, $\quad 0 < 2\alpha < \dfrac{1}{2}$

Since $\dfrac{1}{2} < \beta < 1$, $\quad -1 < -\beta < -\dfrac{1}{2}$

$\therefore \quad -1 < 2\alpha - \beta < 0 \qquad$ That is, $2\alpha - \beta$ cannot be a mantissa.

$\therefore \quad \log \dfrac{x^2}{y} = (5 - 1) + \underline{(1 + 2\alpha - \beta)}$

$\qquad\qquad\qquad\qquad\qquad$ mantissa, $0 < 1 + 2\alpha - \beta < 1$

Since the characteristic is $5 - 1 = 4$,

$\dfrac{x^2}{y}$ has 5 digits in the integer part. Therefore, $n = 5$

#30 The characteristic and mantissa of the number $\log 2013$ are m and α, respectively.

The characteristic and mantissa of the number $\log \dfrac{1}{2013}$ are n and β, respectively.

For a triangle $\triangle OAB$ (where $O = O(0,0)$, $A = A(m, \alpha)$, $B = B(n, \beta)$ in a coordinate

plane), find the coordinate of the centroid of the triangle $\triangle OAB$.

$\log 2013 = m + \alpha$, $(m;$ integer, $0 < \alpha < 1)$

$\qquad\qquad = 3 + \alpha \quad (\because 2013$ has 4 digits$)$

$\log \dfrac{1}{2013} = \log 2013^{-1} = -\log 2013 = -3 - \alpha = (-3 - 1) + \underline{(1 - \alpha)}$

$\qquad\qquad\qquad\qquad\qquad\qquad\qquad\qquad\qquad\qquad\qquad$ mantissa

$\therefore \quad (-3 - 1) + (1 - \alpha) = n + \beta \qquad \therefore \quad n = -4, \ \beta = 1 - \alpha$

$\therefore \quad A = A(3, \alpha), \ B = B(-4, \beta) = B(-4, 1 - \alpha), \ O = O(0,0)$

$\therefore$ The coordinate of the centroid of the triangle $\triangle OAB$ is $\left(\dfrac{3 - 4 + 0}{3}, \dfrac{\alpha + 1 - \alpha + 0}{3}\right) = \left(-\dfrac{1}{3}, \dfrac{1}{3}\right)$

#31 Find the inverse of each function.

(1) $y = 2 \cdot 3^{x-2}$

$y = 2 \cdot 3^{x-2} \ \Rightarrow \ y = 2 \cdot 3^x \cdot 3^{-2} = \dfrac{2}{9} \cdot 3^x, \ y > 0$

Switching x and y, $\quad x = \dfrac{2}{9} \cdot 3^y, \ x > 0$

Solving for y, $\quad 3^y = \dfrac{9}{2}x \qquad \therefore \quad y = \log_3\left(\dfrac{9}{2}x\right), \ x > 0$

Therefore, the inverse of $y = 2 \cdot 3^{x-2}$ is $y = \log_3\left(\dfrac{9}{2}x\right), \ x > 0$

(2) $y = 1 + \log_3(x + 2)$

$y = 1 + \log_3(x + 2) \;\Rightarrow\; y - 1 = \log_3(x + 2)$ for real number x

Switching x and y, $\; x - 1 = \log_3(y + 2)\;$ for real number y

$\therefore\; y + 2 = 3^{x-1}$

Solving for y, $\; y = 3^{x-1} - 2$

Therefore, the inverse of $y = 1 + \log_3(x + 2)$ is $y = 3^{x-1} - 2$ for real number x

(3) $y = \log_{10}\left(x + \sqrt{x^2 - 1}\right),\;\; x \geq 1$

$y = \log_{10}\left(x + \sqrt{x^2 - 1}\right) \;\Rightarrow\; 10^y = 10^{\log_{10}\left(x+\sqrt{x^2-1}\right)}$

$$= \left(x + \sqrt{x^2 - 1}\right)\log_{10} 10 = x + \sqrt{x^2 - 1},\; y \geq 0$$

Switching x and y, $\; 10^x = y + \sqrt{y^2 - 1},\; x \geq 0$

$\therefore\; 10^x - y = \sqrt{y^2 - 1}$

Squaring both sides, $10^{2x} - 2y10^x + y^2 = y^2 - 1 \qquad \therefore\; 10^{2x} - 2y10^x = -1$

Solving for y, $\; y = \dfrac{10^{2x}+1}{2\cdot 10^x} = \dfrac{1}{2}\left(10^x + 10^{-x}\right),\; x \geq 0$

Therefore, the inverse of $y = \log_{10}\left(x + \sqrt{x^2 - 1}\right)$ is $y = \dfrac{1}{2}\left(10^x + 10^{-x}\right),\; x \geq 0$

(4) $y = \dfrac{1}{2^x + 1}$

Switching x and y, $\; x = \dfrac{1}{2^y + 1}$

$\therefore\; x \cdot 2^y + x = 1 \qquad \therefore\; 2^y = \dfrac{1-x}{x}$

Taking common logarithm of both sides to the base 2, $y \log_2 2 = \log_2\left(\dfrac{1-x}{x}\right)$

$\therefore\; y = \log_2\left(\dfrac{1-x}{x}\right)$

Therefore, the inverse of $y = \dfrac{1}{2^x + 1}$ is $y = \log_2\left(\dfrac{1-x}{x}\right)$

#32 Solve each exponential equation for x.

(1) $4^{2x-1} = 7^{x+2}$

$4^{2x-1} = 7^{x+2} \;\Rightarrow\; \ln 4^{2x-1} = \ln 7^{x+2} \;\Rightarrow\; (2x - 1)\ln 4 = (x + 2)\ln 7$

$\Rightarrow\; 2x \ln 4 - \ln 4 = x \ln 7 + 2 \ln 7$

$\Rightarrow\; (2\ln 4 - \ln 7)x = \ln 4 + 2\ln 7$

$\therefore\; x = \dfrac{\ln 4 + 2\ln 7}{2\ln 4 - \ln 7}$

(2) $3 \cdot 5^{2x} = 2 \cdot 4^{3x}$

$3 \cdot 5^{2x} = 2 \cdot 4^{3x} \;\Rightarrow\; 3 \cdot 25^x = 2 \cdot 64^x \;\Rightarrow\; \dfrac{25^x}{64^x} = \dfrac{2}{3}$

$\Rightarrow\; \left(\dfrac{25}{64}\right)^x = \dfrac{2}{3} \;\Rightarrow\; \ln\left(\dfrac{25}{64}\right)^x = \ln\dfrac{2}{3} \;\Rightarrow\; x\ln\left(\dfrac{25}{64}\right) = \ln\dfrac{2}{3}$

$$\therefore\ x = \frac{\ln\frac{2}{3}}{\ln\left(\frac{25}{64}\right)}$$

(3) $\dfrac{2^x + 2^{-x}}{3} = 4$

Let $2^x = X$. Then, $\dfrac{X + X^{-1}}{3} = 4$ $\quad\therefore\ X + X^{-1} = 12$

$\therefore\ X^2 - 12X + 1 = 0$ $\quad\therefore\ X = 6 \pm \sqrt{36 - 1} = 6 \pm \sqrt{35}$

Since $X = 2^x$, $2^x = 6 + \sqrt{35}$ or $2^x = 6 - \sqrt{35}$

Solving for x, $x = \dfrac{\ln(6+\sqrt{35})}{\ln 2}$ or $x = \dfrac{\ln(6-\sqrt{35})}{\ln 2}$

#33 Solve for x.

(1) $\left[\log_3(x-1)\right]^2 - 2\log_3(x-1) = 15$

Substituting $\log_3(x-1) = X$, $X^2 - 2X - 15 = 0$ $\quad\therefore\ (X-5)(X+3) = 0$

$\therefore\ X = 5$ or $X = -3$

Substituting back $X = \log_3(x-1)$, $\log_3(x-1) = 5$ or $\log_3(x-1) = -3$

By the definition, $x - 1 > 0$ $\therefore\ x > 1$ $\therefore\ \log_3(x-1) > 0$

$\therefore\ \log_3(x-1) = 5$ $\therefore\ x - 1 = 3^5$ Therefore, $x = 3^5 + 1 = 244$

(2) $2\log_3(x+1) - \log_3(x+4) = 2\log_3 2$

$2\log_3(x+1) - \log_3(x+4) = 2\log_3 2 \Rightarrow \log_3(x+1)^2 - \log_3(x+4) = \log_3 2^2$

$$\Rightarrow \log_3\frac{(x+1)^2}{x+4} = \log_3 4$$

$\therefore\ \dfrac{(x+1)^2}{x+4} = 4$ $\therefore\ x^2 + 2x + 1 = 4x + 16$ $\therefore\ x^2 - 2x - 15 = 0$ $\therefore\ (x-5)(x+3) = 0$

$\therefore\ x = 5$ or $x = -3$

If $x = -3$, then $\log_3(x+1) = \log_3(-3+1) = \log_3(-2)$ is undefined.

Therefore, $x = 5$

(3) $x^{\log x} - 1000\left(\dfrac{1}{x}\right)^2 = 0$, $x > 1$

$x^{\log x} - 1000\left(\dfrac{1}{x}\right)^2 = 0 \Rightarrow \log(x^{\log x}) = \log\left\{1000\left(\dfrac{1}{x}\right)^2\right\}$

$$\Rightarrow \log x \cdot \log x = \log 1000 + \log\left(\frac{1}{x}\right)^2 = \log 10^3 + \log x^{-2}$$

$$\Rightarrow (\log x)^2 = 3 - 2\log x$$

Let $\log x = X$. Then, $X^2 + 2X - 3 = 0$ $\quad\therefore\ (X+3)(X-1) = 0$ $\quad\therefore\ X = -3$ or $X = 1$

$\therefore\ \log x = -3$ or $\log x = 1$ $\quad\therefore\ x = 10^{-3} = \dfrac{1}{1000}$ or $x = 10^1 = 10$

Since $x > 1$, $x = 10$

(4) $\log_{\frac{1}{3}}(x-3) + \log_{\frac{1}{3}}(x-5) > -1$

Since $x - 3 > 0$ and $x - 5 > 0$, $x > 5$ $\cdots\cdots$ ①

$\log_{\frac{1}{3}}(x-3) + \log_{\frac{1}{3}}(x-5) > -1$

$\Rightarrow \log_{\frac{1}{3}}(x-3)(x-5) > \log_{\frac{1}{3}} 3$ $\qquad$ ($\because \log_{\frac{1}{3}} 3 = \log_{3^{-1}} 3 = (-1)\log_3 3 = -1$)

Since $0 < \frac{1}{3} < 1$, $(x-3)(x-5) < 3$

$\therefore x^2 - 8x + 12 < 0$ $\quad \therefore (x-2)(x-6) < 0$ $\quad \therefore 2 < x < 6$ $\cdots\cdots$ ②

By ① and ②, $5 < x < 6$

(5) $(\log_2 x)(\log_2 4x) \le 24$

$(\log_2 x)(\log_2 4x) = (\log_2 x)(\log_2 4 + \log_2 x) \le 24$

Let $\log_2 x = X$. Then, $X(2 + X) \le 24$

$\therefore X^2 + 2X - 24 = (X+6)(X-4) \le 0$ $\qquad \therefore -6 \le X \le 4$

$\therefore -6 \le \log_2 x \le 4$

Therefore, $2^{-6} \le x \le 2^4$

#34 Find the value.

(1) Find the value of a such that $\log_2(\log_3(\log_4 a)) = 0$

$\log_2(\log_3(\log_4 a)) = 0 \Rightarrow \log_3(\log_4 a) = 1$

$\therefore \log_4 a = 3$ $\qquad \therefore a = 4^3$

(2) When an equation $(\log x)(\log 2x) = \log 3x$ has two different real roots α and β, find the value of $\alpha\beta$.

$(\log x)(\log 2x) = \log 3x \Rightarrow (\log x)(\log 2 + \log x) = \log 3 + \log x$

$\therefore (\log x)^2 + \log 2 \cdot \log x - \log 3 - \log x = 0$

Let $\log x = X$. Then, $X^2 + (\log 2 - 1)X - \log 3 = 0$

Since $\log 2 - 1 = \log 2 - \log 10 = \log\frac{2}{10} = \log\frac{1}{5} = \log 5^{-1} = -\log 5$,

$X^2 - (\log 5)X - \log 3 = 0$ $\cdots\cdots$ ①

Since α and β are two different roots of the given equation,

the equation ① has roots $\log \alpha$ and $\log \beta$.

By the relationship between roots and coefficients,

$\log \alpha + \log \beta = \log 5$ and $\log \alpha \cdot \log \beta = -\log 3$

$\therefore \log \alpha\beta = \log 5$ $\qquad$ Therefore, $\alpha\beta = 5$

(3) For any real numbers x and y $(x > y)$ such that $\begin{cases} \log_2 xy = \dfrac{7}{2} \\ \log_2 x \cdot \log_2 y = \dfrac{3}{2} \end{cases}$,

find the value of $\dfrac{x}{y}$.

$\log_2 xy = \dfrac{7}{2} \Rightarrow \log_2 x + \log_2 y = \dfrac{7}{2}$

The quadratic equation of X with the two roots, $\log_2 x$ and $\log_2 y$, is

$X^2 - (\log_2 x + \log_2 y)X + (\log_2 x \cdot \log_2 y) = 0$

$\therefore \ X^2 - \dfrac{7}{2}X + \dfrac{3}{2} = 0 \quad \therefore \ 2X^2 - 7X + 3 = 0 \quad \therefore \ (2X - 1)(X - 3) = 0 \quad \therefore \ X = \dfrac{1}{2} \ \text{or} \ X = 3$

$\therefore \ \log_2 x = \dfrac{1}{2}, \ \log_2 y = 3 \ \text{ or } \ \log_2 x = 3, \ \log_2 y = \dfrac{1}{2}$

$\therefore \ x = \sqrt{2}, \ y = 2^3 \ \text{ or } \ x = 2^3, \ y = \sqrt{2}$

Since $x > y$, $x = 2^3$ and $y = \sqrt{2}$

Therefore, $\dfrac{x}{y} = \dfrac{2^3}{\sqrt{2}} = 4\sqrt{2}$

(4) For a function $f(x) = |\log_2 x - 1|$, $f(a) = f(b)$ where a, b are two different real numbers. Find the value of ab.

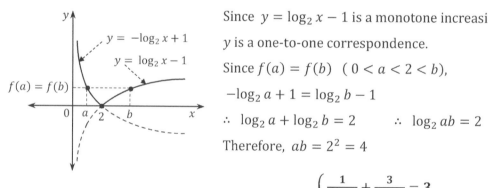

Since $y = \log_2 x - 1$ is a monotone increasing function,

y is a one-to-one correspondence.

Since $f(a) = f(b)$ $(0 < a < 2 < b)$,

$-\log_2 a + 1 = \log_2 b - 1$

$\therefore \ \log_2 a + \log_2 b = 2 \qquad \therefore \ \log_2 ab = 2$

Therefore, $ab = 2^2 = 4$

(5) When the solutions of a system of equations $\begin{cases} \dfrac{1}{\log_x 2} + \dfrac{3}{\log_y 8} = 3 \\ \log_2 3x + \log_{\sqrt{2}} y = \log_2 48 \end{cases}$ **are $x =$**

α **and $y = \beta$, find the value of $\alpha + \beta$.**

$\dfrac{1}{\log_x 2} + \dfrac{3}{\log_y 8} = 3 \ \Rightarrow \ \log_2 x + 3\log_8 y = 3 \ \Rightarrow \ \log_2 x + \log_2 y = 3 \ \cdots\cdots\ ①$

$\left(\because \log_8 y = \log_{2^3} y = \dfrac{1}{3}\log_2 y\right)$

$\log_2 3x + \log_{\sqrt{2}} y = \log_2 48$

$\Rightarrow \ \log_2 3 + \log_2 x + 2\log_2 y = \log_2(2^4 \cdot 3) = \log_2 2^4 + \log_2 3 = 4 + \log_2 3$

$\therefore \ \log_2 x + 2\log_2 y = 4 \ \cdots\cdots\ ②$ $\qquad\left(\because \log_{\sqrt{2}} y = \log_{2^{\frac{1}{2}}} y = 2\log_2 y\right)$

$② - ① : \log_2 y = 1, \ \log_2 x = 2$

$\therefore \ y = 2, \ x = 2^2 \qquad$ Therefore, $\alpha + \beta = 2^2 + 2 = 6$

(6) When an equation $\left(\log_2 \frac{x}{2}\right)^2 - 30 \log_8 x + 31 = 0$ has two roots α and β,

find the value of $\alpha\beta$.

$$\left(\log_2 \frac{x}{2}\right)^2 - 30 \log_8 x + 31 = 0 \quad \Rightarrow \quad (\log_2 x - \log_2 2)^2 - 30 \log_{2^3} x + 31 = 0$$

$$\Rightarrow \quad (\log_2 x - 1)^2 - \frac{30}{3} \log_2 x + 31 = 0$$

$$\Rightarrow \quad (\log_2 x - 1)^2 - 10 \log_2 x + 31 = 0$$

Let $\log_2 x = X$. Then, $(X - 1)^2 - 10X + 31 = 0$

$\therefore X^2 - 12X + 32 = (X - 4)(X - 8) = 0 \qquad \therefore X = 4$ or $X = 8$

When $X = 4$, $\log_2 x = 4 \qquad \therefore x = 2^4$

When $X = 8$, $\log_2 x = 8 \qquad \therefore x = 2^8$

Therefore, $\alpha\beta = 2^4 \cdot 2^8 = 2^{12}$

(7) When an equation $a^{2x} - 2a^x = 3 \ (a > 0, \ a \neq 1)$ has a solution $x = \frac{1}{5}$,

find the value of the constant a.

Let $a^x = X$. Then $X > 0$

$a^{2x} - 2a^x = 3 \quad \Rightarrow \quad X^2 - 2X - 3 = 0 \quad \Rightarrow \quad (X - 3)(X + 1) = 0 \quad \therefore X = 3$ or $X = -1$

Since $X > 0$, $X = 3$

$\therefore a^x = 3 \qquad \therefore x = \log_a 3 = \frac{1}{5}$

$\therefore a^{\frac{1}{5}} = 3 \qquad$ Therefore, $a = 3^5$

#35 Maximum and minimum values

(1) For a function $y = \log_2(x + 1)$, $0 \leq x \leq 3$, find the maximum and minimum values.

Note that the graph of the function rises

as x increases (monotone increasing).

When $x = 0$,

$y = \log_2(0 + 1) = \log_2 1 = 0$ is the minimum value.

When $x = 3$,

$y = \log_2(3 + 1) = \log_2 4 = \log_2 2^2 = 2$ is the maximum value.

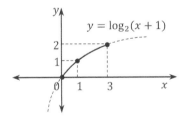

(2) Find the minimum value of $\log_4(x^2 - 4x + 8)$.

When $x^2 - 4x + 8$ is the minimum value, $\log_4(x^2 - 4x + 8)$ is the minimum value.

Since $x^2 - 4x + 8 = (x - 2)^2 - 4 + 8 = (x - 2)^2 + 4 \geq 4$,

$\log_4(x^2 - 4x + 8)$ has minimum value $\log_4 4 = 1$ when $x = 2$.

(3) For a function $y = \log_2(-x^2 + 4x + 6)$, $0 \le x \le 5$,

find the maximum and minimum values of y.

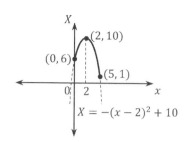

Let $X = -x^2 + 4x + 6$

Then, $X = -(x-2)^2 + 4 + 6 = -(x-2)^2 + 10$

Since $0 \le x \le 5$, $1 \le X \le 10$ (From the graph)

$\therefore \log_2 1 \le \log_2 X \le \log_2 10$ ($\because$ base($= 2$) > 1)

Since $\log_2 10 = \log_2 2 \cdot 5 = \log_2 2 + \log_2 5 = 1 + \log_2 5$, $0 \le y \le 1 + \log_2 5$

Therefore, the maximum value of y is $1 + \log_2 5$ and the minimum value is 0.

(4) For two positive numbers x **and** y **such that** $x + y = 6$,

find the maximum or minimum value of $\log_{\frac{1}{3}} x + \log_{\frac{1}{3}} y$.

$x + y = 6 \implies y = 6 - x$

$\log_{\frac{1}{3}} x + \log_{\frac{1}{3}} y = \log_{\frac{1}{3}} xy$

Since the base is less than 1, $\log_{\frac{1}{3}} xy$ is the minimum value when xy is the maximum value.

$xy = x(6-x) = -x^2 + 6x = -(x^2 - 6x) = -(x-3)^2 + 9 \le 9$

$\therefore$ The minimum value of $\log_{\frac{1}{3}} xy$ is $\log_{\frac{1}{3}} 9 = \log_{3^{-1}} 3^2 = \dfrac{2}{-1} = -2$

(5) When the minimum value of $y = 2^{a+x} + 2^{a-x}$ **is 4, find the value of** a.

Since $2^{a+x} > 0$ and $2^{a-x} > 0$ for any real number x,

$2^{a+x} + 2^{a-x} \ge 2\sqrt{2^{a+x} \cdot 2^{a-x}}$ (by the relationship between arithmetic and geometric means)

$$= 2\sqrt{2^{a+x+a-x}} = 2\sqrt{2^{2a}} = 2\sqrt{(2^a)^2} = 2 \cdot 2^a = 2^{a+1}$$

$\therefore y \ge 2^{a+1}$ $\therefore 2^{a+1} = 4 = 2^2$ $\therefore a = 1$

(6) Find the minimum value of $4\log_a b + 9\log_b a$ ($a > 1$, $b > 1$).

Since $a > 1$ and $b > 1$, $\log_a b > 0$ and $\log_b a > 0$

$4\log_a b + 9\log_b a \ge 2\sqrt{4\log_a b \cdot 9\log_b a}$

$$= (2 \cdot 6)\sqrt{\log_a b \cdot \dfrac{1}{\log_a b}} = 12$$

$\therefore$ The minimum value of $4\log_a b + 9\log_b a$ is 12.

(7) Find the maximum value of $y = \log 5x \cdot \log \dfrac{20}{x}$ $\left(\dfrac{1}{5} < x < 20\right)$.

$\dfrac{1}{5} < x \implies 1 < 5x \quad \therefore \log_{10} 1 < \log_{10} 5x \quad \therefore \log_{10} 5x > 0$

$x < 20 \implies 1 < \dfrac{20}{x} \quad \therefore \log_{10} 1 < \log_{10} \dfrac{20}{x} \quad \therefore \log_{10} \dfrac{20}{x} > 0$

Since $\log_{10} 5x > 0$ and $\log_{10} \dfrac{20}{x} > 0$,

$$\log_{10} 5x + \log_{10} \dfrac{20}{x} \geq 2\sqrt{\log_{10} 5x \cdot \log_{10} \dfrac{20}{x}}$$

Since $\log_{10} 5x + \log_{10} \dfrac{20}{x} = \log_{10}(5x \cdot \dfrac{20}{x}) = \log_{10} 100 = 2, \quad 2 \geq 2\sqrt{\log_{10} 5x \cdot \log_{10} \dfrac{20}{x}}$

$\therefore \ \log_{10} 5x \cdot \log_{10} \dfrac{20}{x} \leq 1$

$\therefore$ The maximum value of y is 1.

(8) Find the minimum value of $y = 2(\log_2 2x)^2 + \log_2(2x)^2 + 6\log_2 x + 12$.

Let $\log_2 x = X$.

$y = 2(\log_2 2x)^2 + \log_2(2x)^2 + 6\log_2 x + 12$

$\quad = 2(\log_2 2 + \log_2 x)^2 + 2(\log_2 2 + \log_2 x) + 6\log_2 x + 12$

$\quad = 2(1 + X)^2 + 2(1 + X) + 6X + 12$

$\quad = 2X^2 + 12X + 16 = 2(X^2 + 6X) + 16 = 2(X + 3)^2 - 18 + 16$

$\quad = 2(X + 3)^2 - 2$

When $X = -3$, y has the minimum value -2.

(9) Find the maximum and minimum values of $y = (\log_2 x)^2 - \log_2 x^4 + 7, \ 4 \leq x \leq 8$.

Let $\log_2 x = X$.

Since $4 \leq x \leq 8, \ \log_2 4 \leq \log_2 x \leq \log_2 8 \quad \therefore \ 2 \leq X \leq 3$

$y = (\log_2 x)^2 - \log_2 x^4 + 7 = X^2 - 4X + 7$

$\quad = (X - 2)^2 - 4 + 7 = (X - 2)^2 + 3$

From the graph, y has minimum value 3 when $X = 2$.

When $X = 3$, y has maximum value 4.

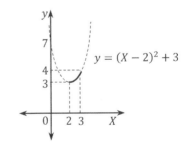

(10) Find the maximum value of $y = 100x^4 \div x^{\log x}$.

$\log y = \log(100x^4 \div x^{\log x}) = \log 100x^4 - \log x^{\log x}$

$\quad = (\log 100 + \log x^4) - (\log x \cdot \log x) = 2 + 4\log x - (\log x)^2$

Let $\log x = X$.

Then, $\log y = -X^2 + 4X + 2 = -(X - 2)^2 + 6$

$\therefore$ When $X = 2$, $\log y$ has the maximum value 6. Therefore, maximum of y is 10^6.

(11) Find the maximum and minimum values of $y = 100x^{6-\log x}$, $\dfrac{1}{10} \le x \le 1000$.

$\log y = \log\left(100x^{6-\log x}\right) = \log 100 + \log\left(x^{6-\log x}\right) = 2 + (6 - \log x)\log x$

$\qquad = -(\log x)^2 + 6\log x + 2$

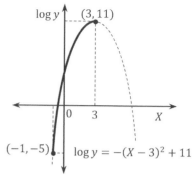

Let $\log x = X$.

Then, $\log y = -X^2 + 6X + 2 = -(X-3)^2 + 9 + 2$

$\qquad\qquad = -(X-3)^2 + 11$

Since $\dfrac{1}{10} \le x \le 1000$, $\log 10^{-1} \le \log x \le \log 10^3$

$\therefore -1 \le X \le 3$

From the graph, the minimum value is $\log y = -5$ when $X = -1$

and the maximum value is $\log y = 11$ when $X = 3$.

Therefore, the minimum value of y is 10^{-5} and maximum value is 10^{11}.

(12) When the minimum value of $y = (\log_3 x)^2 - 3\log_3 x^2 + a$, $1 \le x \le 9$ is 0, find the maximum value of y.

Let $\log_3 x = X$.

Since $1 \le x \le 9$, $\log_3 1 \le \log_3 x \le \log_3 3^2$ $\quad\therefore 0 \le X \le 2$

$y = (\log_3 x)^2 - 3\log_3 x^2 + a = X^2 - 6X + a = (X-3)^2 - 9 + a$

When $X = 2$, y has the minimum value $-8 + a$.

$\therefore -8 + a = 0$ $\quad\therefore a = 8$

$\therefore y$ has the maximum value 8 when $X = 0$.

(13) When $x > 1$ and $y > 1$, find the minimum value of $\dfrac{\log_x 2 + \log_y 2}{\log_{xy} 2}$

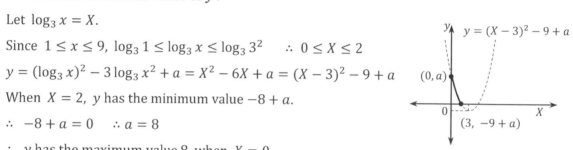

$\dfrac{\log_x 2 + \log_y 2}{\log_{xy} 2} = \dfrac{\frac{1}{\log_2 x} + \frac{1}{\log_2 y}}{\frac{1}{\log_2 xy}} = \dfrac{\frac{\log_2 x + \log_2 y}{\log_2 x \cdot \log_2 y}}{\frac{1}{\log_2 xy}} = \dfrac{\frac{\log_2 xy}{\log_2 x \cdot \log_2 y}}{\frac{1}{\log_2 xy}} = \dfrac{(\log_2 xy)^2}{\log_2 x \cdot \log_2 y}$

Since $x > 1$ and $y > 1$, $\log_2 x > 0$ and $\log_2 y > 0$

$\therefore \log_2 xy = \log_2 x + \log_2 y \ge 2\sqrt{\log_2 x \cdot \log_2 y}$

$\qquad\qquad\qquad$ by the relationship between arithmetic and geometric means

$\therefore (\log_2 xy)^2 \ge 4(\log_2 x \cdot \log_2 y)$

$\therefore \dfrac{\log_x 2 + \log_y 2}{\log_{xy} 2} \ge \dfrac{4(\log_2 x \cdot \log_2 y)}{\log_2 x \cdot \log_2 y} = 4$

$\therefore$ The minimum value is 4.

(14) When the difference of the maximum and minimum values of x

such that $|a - \log_2 x| \leq 1$ is 15, find the value of a.

$|a - \log_2 x| \leq 1 \Rightarrow |\log_2 x - a| \leq 1$

$\therefore -1 \leq \log_2 x - a \leq 1 \qquad \therefore a - 1 \leq \log_2 x \leq a + 1 \qquad \therefore 2^{a-1} \leq x \leq 2^{a+1}$

$\therefore$ The minimum value of x is 2^{a-1} and the maximum value of x is 2^{a+1}.

Since $2^{a+1} - 2^{a-1} = 15, \quad 2 \cdot 2^a - \dfrac{2^a}{2} = 15$

$\therefore \dfrac{3 \cdot 2^a}{2} = 15 \qquad \therefore 2^a = 10$

$\therefore a = \log_2 10 = \log_2 2 \cdot 5 = \log_2 2 + \log_2 5 = 1 + \log_2 5$

(15) When $x \geq 0$, $y \geq 0$, and $xy = 1000$,

find the maximum and minimum values of $\log x \cdot \log y$.

$\log x \cdot \log y = -\left(X - \dfrac{3}{2}\right)^2 + \dfrac{9}{4}$

Since $xy = 1000$, $\log xy = \log 1000$

$\therefore \log x + \log y = \log 10^3 = 3$

Since $x \geq 0$ and $y \geq 0$, $\log x \geq 1$ and $\log y \geq 1$

Let $\log x = X$, $\log y = Y$.

Then $X \geq 1$, $Y \geq 1$, $X + Y = 3 \qquad \therefore 1 \leq X \leq 2$

$\log x \cdot \log y = XY = X(3 - X) = -X^2 + 3X = -\left(X - \dfrac{3}{2}\right)^2 + \dfrac{9}{4}$

From the graph, $\log x \cdot \log y$ has maximum value $\dfrac{9}{4}$ when $X = \dfrac{3}{2}$ and

minimum value 2 when $X = 1$ or $X = 2$

#36 Find the range.

(1) For any real number x, an inequality $(1 - \log_2 a)x^2 + 2(1 - \log_2 a)x + \log_2 a > 0$ is

always true. Find the range of a positive number a.

i) $1 - \log_2 a = 0 \Rightarrow \log_2 a = 1 \qquad \therefore a = 2$

Then, given inequality is $0 \cdot x^2 + 2 \cdot 0x + \log_2 2 > 0$; i.e., $1 > 0$ (True)

$\therefore a = 2$

ii) $1 - \log_2 a \neq 0 \Rightarrow \log_2 a \neq 1 \qquad \therefore a \neq 2$

Let D be the discriminant of the equation $(1 - \log_2 a)x^2 + 2(1 - \log_2 a)x + \log_2 a = 0$

Then $D < 0$ and $1 - \log_2 a > 0$

$\therefore (1 - \log_2 a)^2 - (1 - \log_2 a)\log_2 a = 1 - 2\log_2 a + (\log_2 a)^2 - \log_2 a + (\log_2 a)^2$

$$= 2(\log_2 a)^2 - 3\log_2 a + 1 < 0$$

Let $\log_2 a = X$.

Then, $2X^2 - 3X + 1 < 0$ $\quad \therefore (2X - 1)(X - 1) < 0$ $\quad \therefore \dfrac{1}{2} < X < 1$

$\therefore \dfrac{1}{2} < \log_2 a < 1$ $\quad \therefore \sqrt{2} < a < 2$ $\cdots\cdots$ ①

Since $1 - \log_2 a > 0$, $\log_2 a < 1$ $\quad \therefore \log_2 a < \log_2 2$ $\quad \therefore 0 < a < 2$ $\cdots\cdots$ ②

By ① and ②, $\sqrt{2} < a < 2$

Therefore, by i) and ii), $\sqrt{2} < a \leq 2$

(2) When the roots of an equation $(\log_2 x)^2 - \log_2 x^3 + a = 0$ are in-between $\dfrac{1}{2}$ and 8, find the range of real number a.

Let $\log_2 x = X$. Then, $X^2 - 3X + a = 0$

Since $\dfrac{1}{2} < x < 8$, $\log_2 2^{-1} < \log_2 x < \log_2 2^3$ $\quad \therefore -1 < X < 3$

To have the roots of $X^2 - 3X + a = 0$; i.e., $\left(X - \dfrac{3}{2}\right)^2 - \dfrac{9}{4} + a = 0$, in-between -1 and 3,

  i)   The discriminant D of the equation is $D \geq 0$.

       $\therefore 9 - 4a \geq 0$ $\quad \therefore a \leq \dfrac{9}{4}$

  ii) Let $f(x) = X^2 - 3X + a$ $\quad$ Then, $f(-1) > 0$ and $f(3) > 0$

       $\therefore 4 + a > 0$ and $a > 0$ $\quad \therefore a > 0$

By i) and ii), $0 < a \leq \dfrac{9}{4}$

(3) Find the range of x such that $\begin{cases} \log_3 |x - 1| < 3 \\ \log_2 x + \log_2 (x + 2) \geq 3 \end{cases}$

$\log_3 |x - 1| < 3$ $\Rightarrow$ $x \neq 1$ and $|x - 1| < 3^3$

$\qquad\qquad\qquad\quad \Rightarrow$ $x \neq 1$ and $-27 < x - 1 < 27$

$\qquad\qquad\qquad\quad \Rightarrow$ $x \neq 1$ and $-26 < x < 28$ $\cdots\cdots$ ①

$\log_2 x + \log_2 (x + 2) \geq 3$ $\Rightarrow$ $\log_2 x(x + 2) \geq \log_2 2^3$

$\qquad\qquad\qquad\qquad\quad \Rightarrow$ $x(x + 2) \geq 2^3$ $\;$ ($\because$ The base $2 > 1$)

$\qquad\qquad\qquad\qquad\quad \Rightarrow$ $x^2 + 2x - 8 = (x + 4)(x - 2) \geq 0$

$\qquad\qquad\qquad\qquad\quad \Rightarrow$ $x \geq 2$ or $x \leq -4$ $\cdots\cdots$ ②

$\qquad\qquad\qquad\qquad$ Since $x > 0$ and $x + 2 > 0$, $x > 0$ $\cdots\cdots$ ③

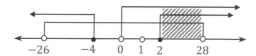

By ①, ②, and ③, $2 \leq x < 28$